Lecture Notes in Networks and Systems 1219

The series "Lecture Notes in Networks and Systems" publishes the latest developments in Networks and Systems—quickly, informally and with high quality. Original research reported in proceedings and post-proceedings represents the core of LNNS.

Volumes published in LNNS embrace all aspects and subfields of, as well as new challenges in, Networks and Systems.

The series contains proceedings and edited volumes in systems and networks, spanning the areas of Cyber-Physical Systems, Autonomous Systems, Sensor Networks, Control Systems, Energy Systems, Automotive Systems, Biological Systems, Vehicular Networking and Connected Vehicles, Aerospace Systems, Automation, Manufacturing, Smart Grids, Nonlinear Systems, Power Systems, Robotics, Social Systems, Economic Systems and other. Of particular value to both the contributors and the readership are the short publication timeframe and the world-wide distribution and exposure which enable both a wide and rapid dissemination of research output.

The series covers the theory, applications, and perspectives on the state of the art and future developments relevant to systems and networks, decision making, control, complex processes and related areas, as embedded in the fields of interdisciplinary and applied sciences, engineering, computer science, physics, economics, social, and life sciences, as well as the paradigms and methodologies behind them.

Indexed by SCOPUS, EI Compendex, INSPEC, WTI Frankfurt eG, zbMATH, SCImago.

All books published in the series are submitted for consideration in Web of Science.

For proposals from Asia please contact Aninda Bose (aninda.bose@springer.com).

Roman Szewczyk · Cezary Zieliński ·
Małgorzata Kaliczyńska · Vytautas Bučinskas
Editors

Automation 2024: Advances in Automation, Robotics and Measurement Techniques

Springer

Editors
Roman Szewczyk
Łukasiewicz Research Network
Industrial Research Institute for Automation and Measurements
Warsaw, Poland

Cezary Zieliński
Łukasiewicz Research Network
Industrial Research Institute for Automation and Measurements
Warsaw, Poland

Małgorzata Kaliczyńska
Łukasiewicz Research Network
Industrial Research Institute for Automation and Measurements
Warsaw, Poland

Vytautas Bučinskas
Department of Mechatronics, Robotics and Digital Manufacturing
Faculty of Mechanics
Vilnius, Lithuania

ISSN 2367-3370 ISSN 2367-3389 (electronic)
Lecture Notes in Networks and Systems
ISBN 978-3-031-78265-7 ISBN 978-3-031-78266-4 (eBook)
https://doi.org/10.1007/978-3-031-78266-4

This Springer imprint is published by the registered company Springer Nature Switzerland AG
The registered company address is: Gewerbestrasse 11, 6330 Cham, Switzerland

Foreword

Since the Industrial Revolution, we have witnessed the reduction of human labour in all aspects of our activities, initially in industry, agriculture, and services. At the onset, this was due to mechanization, but gradually, the cause shifted mainly to automatisation. Currently, the reason for the introduction of automatisation is not only the reduction of human labour, often tedious, and dangerous, but also due to the necessity of increasing production output, its repeatability, energy saving, and diversity of products produced, thus the requirement of rapid changeover, etc.

The rapid development of automatisation results from using computers and control sciences. Computer control, regardless of the object of control, besides the computer hardware, requires receptors (sensors) and effectors (e.g. actuators, heaters, valves). Hence, the fundament of control engineering is based on measurements and means of influencing the object or process of control. The link between receptors and effectors is in decision-making in hardware and software. Control decisions are based on sensor feedback and compared to the desired values, frequently using prediction utilising the process model.

This volume focuses on recent developments in measurements and control applied to diverse processes and devices. One prominent group of those devices are robots. The papers deal with applying artificial neural networks and other machine learning methods in perception, modelling and control, utilisation of fractional order systems, and novel sensors and measurement techniques. The economic viability of a system depends on the quality of exerted control. Thus, optimisation techniques are in high demand. Recent developments in robotics are also prominent in this volume. Specific aspects of the design of diverse robots, as well as their modelling and control, are described in depth.

We strongly believe that the solutions and guidelines presented in this volume will be useful for both researchers of technical sciences and engineers solving practical problems both in the industry and the service sector.

March 2024

Roman Szewczyk
Cezary Zieliński
Małgorzata Kaliczyńska
Vytautas Bučinskas

Contents

About the Editors

Prof. Roman Szewczyk received both his PhD and DSc in the field of mechatronics. He specializes in modelling of properties of magnetic materials as well as in sensors and sensor interfacing, in particular magnetic sensors for security applications. He leads the development of: a sensing unit for a mobile robot developed for the Polish Police Central Forensic Laboratory, and methods of non-destructive testing based on magnetoelastic effect. Professor Szewczyk has been involved in over 10 European Union funded research projects within the FP6 and FP7 as well as projects financed by the European Defence Organization. Moreover, he has lead two regional and national scale technological foresight projects and was active in the organization and implementation of technological transfer between companies and research institutes. Roman Szewczyk is Secretary for Scientific Affairs in the Industrial Research Institute for Automation and Measurements PIAP. He is also Associate Professor at the Faculty of Mechatronics, Warsaw University of Technology, and Vice-chairman of the Academy of Young Researchers of the Polish Academy of Sciences.

Prof. Cezary Zieliński received his MSc/Eng., PhD and habilitation degrees in control and robotics from the Faculty of Electronics and Information Technology, Warsaw University of Technology (WUT), Warsaw, Poland. He is: a full professor of WUT, head of the Robotics Research Group, the director of the Institute of Control and Computation Engineering, and professor of Industrial Research Institute for Automation and Measurement PIAP. His research concentrates on robot control and programming methods. His research interests focus on robotics, in general, and in particular include: robot programming methods, formal approach to the specification of architectures of multi-effector and multi-receptor systems, robot kinematics, robot position-force control, visual servo control, and design of digital circuits. He is the author/co-author of over 200 conference and journal papers as well as books concerned with the above-mentioned research subjects.

Dr. Małgorzata Kaliczyńska received her MSc/Eng. degree in cybernetics from the Faculty of Electronics, Wrocław University of Technology, and her PhD degree in the field of fluid mechanics from the Faculty of Mechanical and Power Engineering in the same university. Now she is Assistant Professor in the Industrial Research Institute for Automation and Measurement PIAP and Editor of the scientific and technological magazine "Measurements Automation Robotics". Her areas of research interest include distributed control systems, Internet of Things, Industry 4.0, information retrieval and webometrics.

Prof. Vytautas Bučinskas received engineer's degree in automotive engineering from the Vilnius Civil Engineering Institute in 1985. The same year he started as an engineer of work safety in VISI, Lithuania. In 1987, he became involved in research at the same institute. In 2000, he was employed as a researcher in the Faculty of Mechanics, Dept. of

Machine Engineering. In 2002, he obtained a Doctor's Degree in the field of Theory of Machines and became an associate professor in the Department of Machine Engineering, Vilnius Technical University. In 2012, he became a full professor of the same Department. In 2013, the Department of Mechatronics and Robotics was established where he became the chairman. After merging of two Departments, he heads Department of Mechatronics, Robotics, and Digital Manufacturing. His current research interests include design of mechatronic systems, dynamical properties of mechatronic systems, energy harvesting from vibrations. He is Fellow of the Lithuanian Association of Engineering Industry (LINPRA), Lithuanian Association of Robotics. He is an author of Lithuanian, European, US, and Japanese patents. Currently, he is an investigator in four H2020 ECSEL JU projects 3Ccar, Autodrive, A4DI and AI4CSM, where research focuses on autonomous vehicles and electric mobility.

Control and Automation

Research Towards an Optimal Method of Modeling and Regulating a Cement Mill Using AI Algorithms

Dawid Pawuś(✉) and Szczepan Paszkiel

Faculty of Electrical Engineering, Automatic Control and Informatics, Opole University of Technology, Prószkowska Street 76, 45–758 Opole, Poland
dawid.pawus@student.po.edu.pl

Abstract. This research explores the optimization of modeling and control strategies for a cement mill using advanced Artificial Intelligence (AI) techniques. Neural Nonlinear AutoRegressive with eXogenous input (NNARX) models are employed to capture the nonlinear, multivariable dynamics of the cement grinding process. The Levenberg-Marquardt algorithm demonstrated superior training performance, achieving high accuracy in predicting key process parameters like inlet chamber fill and main drive current. The study also presents a neural control approach based on inverse modeling, enabling precise input predictions to maintain desired output conditions, stabilizing product quality and minimizing energy usage. Comparisons of neural network configurations suggest that simpler models are effective, with only slight benefits from more complex networks.

Keywords: non-linear models · artificial neural networks · expert system · NARX · comparative study · cement mill · process control · practical approach

1 Introduction

The object of the research is the problem of searching for an optimal way of modeling and regulating the control parameters of an industrial facility, such as a cement mill. For this purpose, artificial intelligence tools, such as nonlinear neural networks (NNNs), working in an appropriately selected structure, will be examined [12,13].

In all branches of industry, constant development is observed, both in terms of economy, ecology and technology. The galloping industrial revolution forces companies to search for optimal solutions for development and increase in production, in order to maintain competitiveness on the market. Especially in recent years, modern numerical algorithms, including those based on artificial intelligence (AI) and machine learning (ML), have enjoyed increasing interest.

Plants engaged in cement production, despite the presence of this industry on the market for several hundred years, are constantly looking for modern methods

R. Szewczyk et al. (Eds.): AUTOMATION 2024, LNNS 1219, pp. 3–16, 2024.
https://doi.org/10.1007/978-3-031-78266-4_1

that can optimize both production processes and those related to automatic regulation and mathematical modeling of processes occurring in various facilities. In connection with this, the work addresses the problem of optimal modeling and regulation of the parameters of a cement mill in a closed loop. For years, the cement grinding process in many plants in Poland has been carried out under the watchful eye of operators, without the use of automatic regulation. Since this object has a low coefficient of energy input to the system in relation to the efficiency of the grinding process, it is worth looking at ways of optimizing using alternative control methods, such as NNNs algorithms [3,6,9].

The authors of [8] investigate issues related to raw material mixing in the cement industry. They analyze the ways of identification and approaches to process control, although the facility was observed to be disrupted by delays. Article [2] explored the use of NARX models and wavelet networks in modeling cement mill processes, utilizing MATLAB for system identification and model validation. Frequent breakdowns in raw mill machinery were analyzed by [15], who applied TPM, Fishbone, and FMEA methods to identify and mitigate failure causes. Work [22] developed a adaptive model predictive controller to address the challenges in controlling roller mills, demonstrating its superiority over traditional controllers.

The work by [10] introduced a predictive control strategy for cement grinding, focusing on improving cement strength using a fuzzy control algorithm validated with real data. Authors of [7] presented a modular hardware control system for cement mills, emphasizing local processing and efficient data transmission through FFT-based algorithms. A review by [14] highlighted the effectiveness of Model Predictive Control (MPC) in optimizing cement kiln operations, particularly in maintaining stable Burning Zone Temperature. [20] implemented a feedforward-feedback control system to improve sulfate regulation in cement production, achieving better consistency and target adherence.

Article [19], in turn, discussed the digital transformation of Chinese cement plants, showing how AI and APC technologies can enhance efficiency and reduce operational costs. Finally, [4] used a Multi-Layer Perceptron (MLP) neural network to model the rotary cement kiln, finding it more accurate than traditional approaches.

In [1], a rough neural network (R-NN) was used for the first time to identify the mill without using MISO (Multi Input Single Output) structures. The simulation results show the effectiveness of the methodology and the validity of the solutions presented by the authors.

Artificial neural networks also have many other important applications. Articles [16–18,21], present the use of artificial intelligence methods based on neural networks in the tasks of electroencephalographic signal classification, control and musical notation classification using an FPGA (Field Programmable Gate Array) system.

Artificial neural networks also have many other important applications in classification and control, as well as many different fields [16,18,21].

2 Closed-Loop Cement Grinding: a Multivariable Mill System Presentation

Identification of dynamic, multidimensional, non-stationary and discrete MIMO (Multi Input Multi Output) objects is never a trivial issue. Numerous automation approaches have been considering dozens of approaches to this problem for decades. The matter is often further complicated by the presence of stochastic factors influencing the system.

This is precisely what characterizes a cement mill. Until now, control of this type of object has been possible thanks to well-trained and experienced engineers who manually select the appropriate control settings. Such operations are inextricably linked to the possibility of a factor such as human error. Add to this the fact that the mill is burdened with many disturbing factors, and the problem becomes much more complicated. Interferences include, among others, changing parameters of input signals resulting from non-stationary interactions of raw material feeds, which change, for example, depending on the season or humidity. The system features also depend on many factors, including the degree of wear of grinding media, bearings, vibrations, machine wear. Finally, measurement errors that accompany output and process measurements also constitute a significant problem.

Fig. 1. Input-Output representation of the system (based on [13])

When considering the automatic MIMO system, it was decided to isolate 6 inputs and 6 outputs of the systems that would reliably reflect the real object without an unnecessary multitude of additional parameters. Details are included in Fig. 1.

In the analysis of the cement production process, several key components and output variables can be distinguished. The input raw materials include the supply of cement clinker, slag (granulated blast furnace slag), additives (including calcium sulphate, which acts as a setting time regulator and other components), fly ash, dust (waste from gases, which are losses from the ongoing process) and the IMOFLOV grinding activator.

The results of the process are measured by several variables: the filling level of the first and second mill chambers, the value of the current for the main mill drive, the value of the current for the bucket conveyor drive transporting the product to the air separator, the pressure at the mill inlet and the pressure at the mill outlet.

In the next chapter, an effective approach to the problem of identifying the cement grinding process using a neural network model will be described.

3 An Approach to the Problem of Grinding Process Identification Using a Neural Network Model

Nonlinear models play a key role in the identification tasks of systems with a significant degree of mathematical complexity. One of them may be the model described by the relationship 1. In our case, it will be the NNARX model (Neural Nonlinear AutoRegressive with eXogenous input). The application of this type of solution is broad and multidisciplinary, from business and economic fields, through time series forecasting, to automatic control and modeling. The strength of this application is the nonlinear activation function and the provision of measurements from the model output to its input.

$$\tilde{y}(k+1) = g(y(k), y(k-1), \ldots, y(k-n+1), u(k-d), u(k-1-d), \ldots, u(k-m-d+1)), \quad (1)$$

in this case $y(k)$ is the delayed output signal of the object, while $u(k)$ is the delayed input signal. Continuing, $g(\cdot)$ is a function that can be approximated, for example, by a nonlinear function, while $\tilde{y}(k+1)$ is the value of the output signal that is predicted for the future step.

Figure 2 effectively shows the dependencies presented in Eq. 1 and is its reflection. It represents the way the discussed nonlinear structure works.

4 Description of Quality Indicators

Simulation tests were conducted to assess the control of the analyzed facility, followed by practical verification. The evaluation was based on the following quality indicators [11] and [5]:

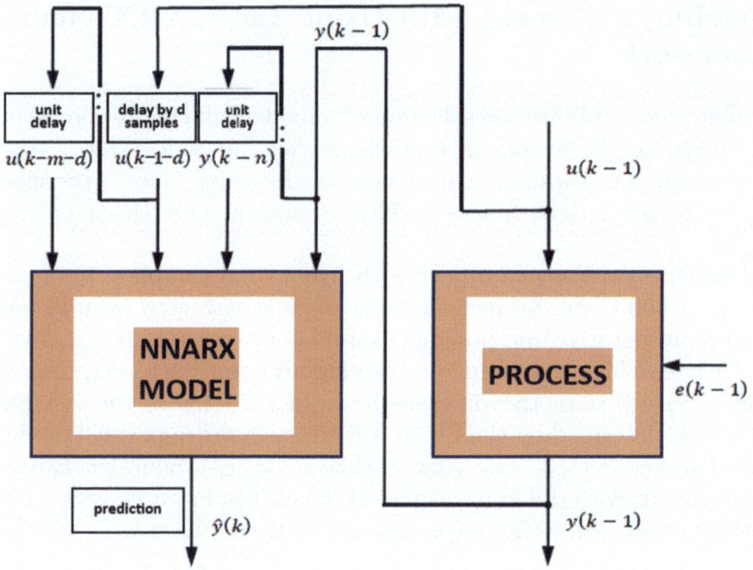

Fig. 2. General scheme the NNARX system

- Q_1 – the first quality indicator is the square of the error $e(t)$ that was integrated

$$Q_1 = \int_{t_0}^{t} e^2(t)dt. \tag{2}$$

- Q_2 – the second quality indicator is the absolute value of the error that was integrated

$$Q_2 = \int_{t_0}^{t} |e(t)|dt. \tag{3}$$

- Q_3 – the third quality indicator is the square of the control value $c(t)$ that has been integrated

$$Q_3 = \int_{t_0}^{t} c^2(t)dt. \tag{4}$$

- Q_4 the fourth quality indicator is a popular correlation indicator, the Pearson correlation coefficient.

The quality indicators used in the evaluation include the integrated square of the error (Q_1) and the integrated absolute value of the error (Q_2), both of which assess the deviation of the system's response from the desired behavior. Additionally, the integrated square of the control value (Q_3) measures the energy or effort used in control, while Q_4 is the Pearson correlation coefficient, which evaluates the linear relationship between variables.

5 Modeling a Cement Mill Using the NARX Neural Framework

The research process of optimal selection of a model sufficiently representing the physical conditions of the cement mill was carried out as a result of conducting many tests using various nonlinear structures. However, due to the clarity and compactness of the article, it was decided to present only the most important achievements.

Measurements on the real object with five-second sampling intervals lasted about 576 h. A 400 thousand measurement set was extracted from it, which was considered representative and possible to use in our research. It was divided into two equal parts, which made up the training and test data sets. The designed structure is described in the diagram from Fig. 3. This is the NARX neural model, which is described by the Eq. 1. It works as a series-parallel model, which is described in the Sect. 3. The core of this model is a neural network, whose input is fed signals with different degrees of delay. The input layer is a tangential activation function, while the output layer is a linear function.

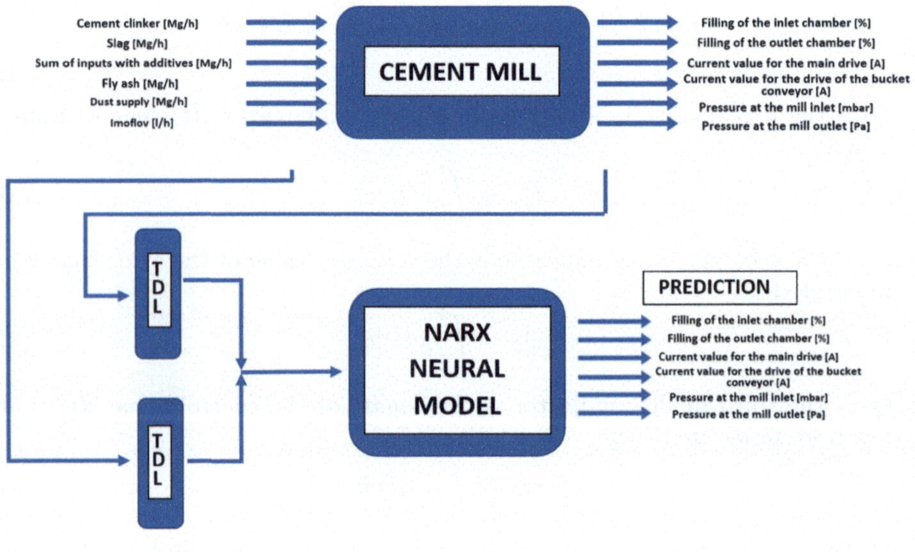

Fig. 3. Schematic diagram of the model operation in the studied structure (based on [13])

During the research, the possibility of different methods of training the neural model was checked, which resulted in different research results included in Table 1. The research tests were performed on Lenovo Legion 5-16 i7-14650HX hardware with 32 GB of RAM and the Win11 operating system. Since in the case of the NARX neural model, three different options could be manipulated, i.e. the number of signal delays, the number of neurons and the number of first layer neurons, it was decided to check different possibilities. So, the system was tested using sample delays of 5, 15, and 35, and with three learning algorithms. The Levenberg-Marquardt algorithm (LMa) performed best, followed by backpropagation with Bayesian regularization. The last method was the gradient method with "inertia". The number of hidden layer neurons was set to 30, and due to the good results, other attempts were omitted in the research for this article.

Table 1. Research results regarding the model

TDL	Q_1	Q_2
Inlet Chamber Fill [%]		
5	3.35×10^3	1.29×10^4
15	3.52×10^3	1.31×10^4
35	3.46×10^3	1.33×10^4
Outlet Chamber Fill [%]		
5	6.67×10^4	1.57×10^3
15	1.42×10^5	1.51×10^4
35	9.18×10^4	1.86×10^3
Main Drive Current [A]		
5	3.68×10^3	1.36×10^4
15	3.58×10^3	1.34×10^4
35	3.94×10^3	1.28×10^4
Elevator Drive Current [A]		
5	3.63×10^3	1.50×10^4
15	3.22×10^3	1.35×10^4
35	3.14×10^3	1.31×10^4
Pressure At The Mill Inlet [mbar]		
5	3.63×10^3	1.50×10^4
15	3.33×10^3	1.40×10^4
35	2.88×10^3	1.66×10^4
Pressure At The Mill Outlet [Pa]		
5	1.01×10^2	2.03×10^3
15	1.29×10^2	2.86×10^3
35	1.25×10^2	2.76×10^3

The set of tests presented in Table 1 presents the results for the model based on the LM learning method, which makes up the architecture of the nonlinear ARX neural model. The number of sample delays fed to the input of the neural structure was modified. The results table presents the results for several trials conducted as part of the research.

The results show that a larger number of TDL delays does not always give better results. It is important to note the difference between the quality of the model and its sophistication and composition. Moreover, with the increase of the number of delay samples, the demand for computational resources of the processor increases. Each additional set of samples is then multiplied by the number of 6 inputs and 6 outputs of the MIMO object.

The analysis of the research results for modeling the cement mill as a multivariable system, considering different numbers of delayed samples in the NNARX model, reveals the key points. Increasing TDL generally leads to a slight rise in both ISE and IAE across most parameters, indicating a decrease in model accuracy as the number of delayed samples increases. However, certain parameters, such as the Elevator Drive Current and Pressure at the Mill Inlet, show a decrease in ISE or more stable behavior with higher delay, suggesting that in some cases, a higher number of delays may improve system performance. Optimal delay selection should thus balance control precision with error minimization, depending on the specific system parameter. It is also worth noting that the correlation coefficient for each test was 0.999 or 1.

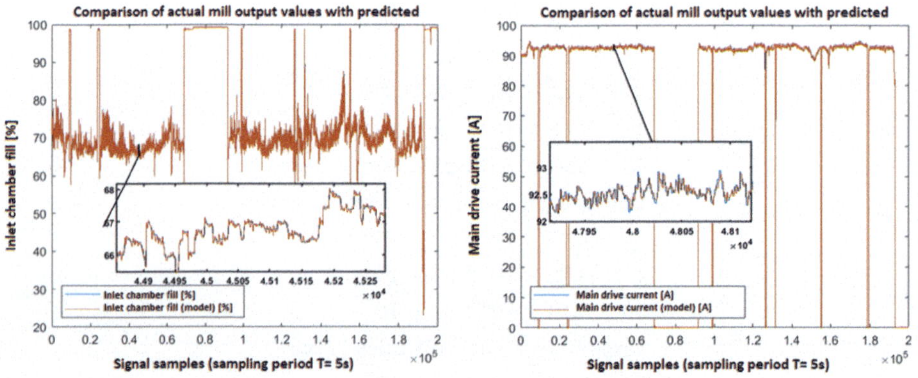

Fig. 4. Representation of data from the mill along with model predictions (based on [13])

The graphs in Fig. 4 show a sample comparison of actual mill output values and those predicted by the model, with particular emphasis on percentage of inlet chamber fill and main drive current in amps.

The left graph shows the actual and predicted inlet chamber fill levels over time. The two lines, representing the actual and predicted values, are closely aligned, indicating that the model accurately captures the behavior of the inlet chamber fill. The inset zooms into a smaller time window, where the model's prediction closely follows the actual measurements.

The right graph compares the actual and predicted main drive current over time. Again, the predicted values (model) closely track the actual current, with both lines showing similar trends and magnitudes. The inset highlights a specific period where the model's accuracy is especially evident, with very little deviation between the predicted and actual values. This suggests that the model effectively captures the dynamics of the main drive current, including minor fluctuations.

The close matching of both the inlet chamber filling and the main drive current means that the model can be reliably used to predict these critical operating parameters, which is essential for effective control and optimization of mill performance.

The methodology and results of the neural control research are described later in this paper.

6 Proposed Neural Control of the Mill System

The chapter will explore the concept of inverse modeling, where the proposed neurocontroller learns to predict the necessary inputs to maintain desired output conditions. By doing so, the system aims to stabilize the degree of cement grinding, ensure consistent product quality, and minimize energy consumption. The research presented here builds upon previous work (see Sect. 5), integrating expert knowledge and advanced neural modeling techniques to develop a robust and efficient control system for the cement mill. A graphical representation of the controller architecture using the inverse model is shown in Fig. 5.

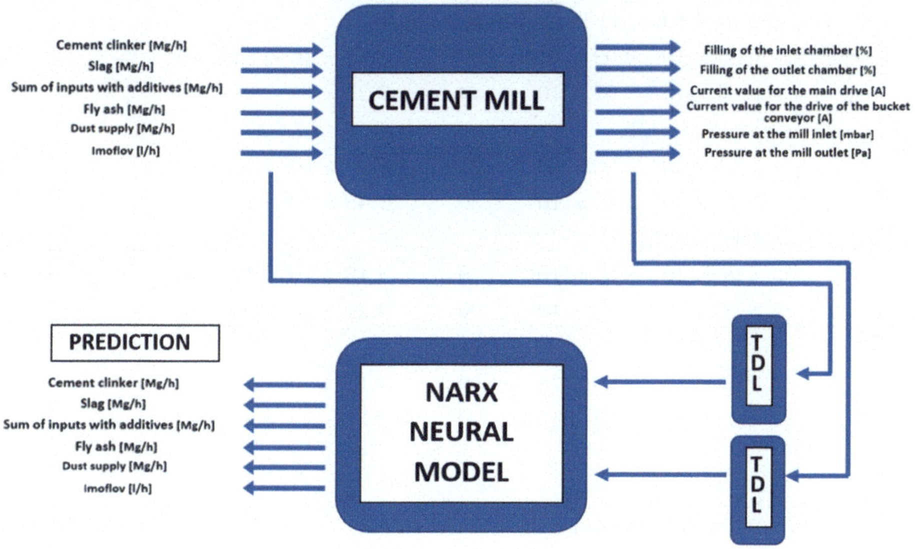

Fig. 5. Schematic diagram of the control system in the studied structure (based on [13])

The cement milling system is challenging for automatic control due to its nonlinear and multivariable nature, where multiple inputs interact in complex, unpredictable ways. Significant time delays between input adjustments and observable output changes make real-time control difficult. Additionally, the system is subject to disturbances and variability in raw materials and environmental conditions, complicating consistent process management. The lack of direct real-time quality measurements forces reliance on indirect indicators, further increasing control complexity. Finally, the need to balance energy efficiency with stable operation adds another layer of difficulty, making traditional control methods often inadequate.

The methodology of the current research work, the amount of test and training data, and the methods of training the neural algorithms are analogous to those discussed in the previous section.

Table 2. Research results regarding the neural control

Q_3 Process	Q_3 Controller	Neurons	Q_1	Q_2
Cement Clinker [t/h]				
2.72×10^7	2.71×10^7	20	6.03×10^3	2.21×10^4
2.72×10^7	2.71×10^7	30	5.66×10^3	2.19×10^4
Slag [t/h]				
8.58×10^6	8.56×10^6	20	1.71×10^3	9.16×10^3
8.58×10^6	8.57×10^6	30	1.76×10^3	9.32×10^3
Sum Of Inputs With Additives [t/h]				
9.03×10^7	9.02×10^7	20	8.37×10^1	2.54×10^3
9.03×10^7	9.02×10^7	30	7.22×10^1	2.37×10^3
Fly Ash [t/h]				
1.31×10^6	1.30×10^6	20	7.80×10^1	1.56×10^3
1.31×10^6	1.30×10^6	30	7.45×10^1	1.50×10^3
Dusts [t/h]				
8.96×10^5	8.95×10^5	20	4.16×10^1	9.17×10^2
8.96×10^5	8.95×10^5	30	4.20×10^1	9.24×10^2
IMOFLOV [l/h]				
1.24×10^7	1.23×10^7	20	2.55×10^3	8.42×10^3
1.24×10^7	1.23×10^7	30	2.38×10^3	8.36×10^3

Table 2 contains a summary of tests resulting from research on the NARX model, described graphically in Fig. 5. In this case, two sample studies were performed involving changes in the number of neurons in the hidden layer, but with the same delays TDL = 15.

The research findings highlight the effectiveness of the neural controller across a range of critical cement mill process parameters, including cement clinker throughput, slag handling, and the management of various material inputs like fly ash and dust. The controller's outputs show a strong alignment with the actual process values, indicating that the neural network model is well-calibrated and capable of accurately predicting and controlling these complex industrial processes.

In terms of neural network architecture, the study compares configurations with 20 and 30 neurons to assess their impact on control performance. Across all examined parameters, the controller with 30 neurons generally exhibits slightly improved metrics, particularly in terms of the Q_1 and Q_2 values, which represent measures of control accuracy and variability. Notably, the Q_3 metric, which reflects energy consumption, remains nearly identical between the process and the controller for both neuron configurations (or even slightly better). This suggests that the neural controller is highly efficient in its energy use, regardless of the network size, while still maintaining precise control over the process.

However, the gains from increasing the neuron count are relatively modest. For instance, in the case of cement clinker and slag throughput, the difference between the Q_3 values for the process and the controller is minimal, regardless of whether 20 or 30 neurons are used. This indicates that the 20-neuron configuration is already highly effective at capturing the essential characteristics of the system, and the additional neurons provide only incremental benefits.

The results also show a consistent trend across different process parameters, where the neural controller maintains close alignment with the actual process outputs. This robustness across various control scenarios highlights the reliability of the neural network-based controller in managing the complex, multi-variable environment of a cement mill.

Moreover, the slight improvements observed with the 30-neuron model, though not substantial, could be crucial in scenarios where maximum precision is required, such as in optimizing energy consumption or ensuring product quality consistency. However, considering the minimal difference in performance and the potential increase in computational complexity with a larger network, the 20-neuron model might offer a more balanced solution, combining effective control with computational efficiency.

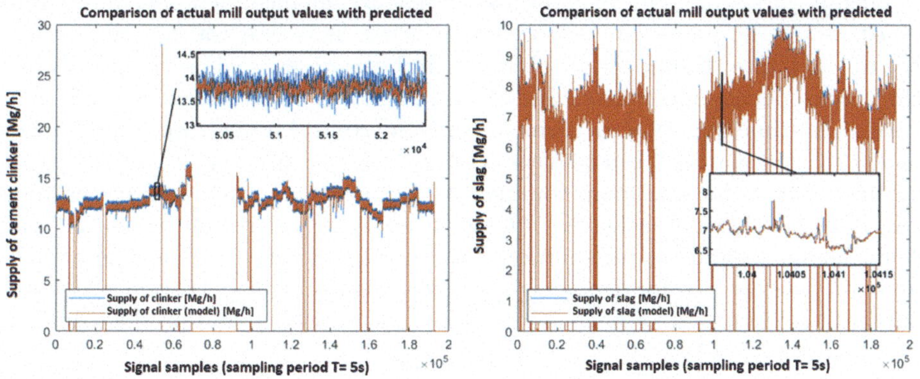

Fig. 6. Representation of data from the mill along with controller predictions (based on [13])

In summary, the neural controller demonstrates strong performance and reliability across different configurations, with the simpler 20-neuron model providing a highly effective and computationally efficient solution for cement mill control. The modest improvements seen with the 30-neuron model suggest that it may be useful in specific, precision-critical applications, but overall, the 20-neuron configuration appears to be sufficient for most practical purposes, especially when considering energy consumption. This indicates the possibility of developing this research into practice in the future. All the issues discussed are reflected in Fig. 6, which shows a comparison of the actual feed measurement values with those proposed by the algorithm.

7 Conclusions and Open Problems

The study presented in this paper explores the application of artificial intelligence algorithms, specifically neural NARX models, for modeling and controlling the complex dynamics of a cement mill. The results demonstrate that the NNARX-based model, particularly when trained with the Levenberg-Marquardt algorithm, effectively captures the nonlinear and multivariable nature of the cement grinding process. The developed neural network models show high accuracy in predicting critical process parameters such as the inlet chamber fill percentage, main drive current, and other key indicators, with correlation coefficients close to 1.

Furthermore, the research highlights the successful implementation of a neural controller designed using inverse modeling techniques. The neural controller effectively predicts the necessary input parameters to achieve desired output conditions, maintaining consistent product quality while minimizing energy consumption. Comparative analysis between different neural network configurations (20 vs. 30 neurons) reveals that while a larger network can yield slight improvements in precision, a smaller network is generally sufficient for effective control with reduced computational complexity.

Future work will focus on several key areas to enhance the applicability and robustness of the proposed control system. First, it is essential to conduct further research into the impact of environmental variations, such as changes in raw material properties and climatic conditions, on model performance. Second, integrating real-time quality measurements into the control framework could further refine the accuracy and responsiveness of the neural controller. Finally, expanding the study to include more complex AI architectures, such as deep learning models or hybrid approaches combining fuzzy logic and neural networks, may offer even more sophisticated solutions for cement mill optimization. These directions will pave the way for deploying such intelligent control systems in real-world industrial environments, enhancing both process efficiency and product quality.

References

1. Ahmadi, G., Teshnelab, M.: Identification of multiple input-multiple output non-linear system cement rotary kiln using stochastic gradient-based rough-neural network. J. AI Data Min. **8**(3), 417–425 (2020)
2. Anand, P., Godwin, R., Priya, K., Subbaraj, P.: Application of ICT in multivariable system identification of cement mill process. Int. J. Res. Innov. Sci. Technol. **2** (2015)
3. Bursy, G.: Neural control of the cement grinding process (in polish). Ph.D. thesis, Opole University of Technology (2008)
4. Doghmane, M., Kidouch, M., Eladj, S., Ouali, A.: Identification and modeling of a rotary kiln in cement plant based on ANN (MLP). In: Artificial Intelligence and Heuristics for Smart Energy Efficiency in Smart Cities: Case Study: Tipasa, Algeria, pp. 825–836. Springer (2022)
5. Elbeltagi, A., Pande, C.B., Kouadri, S., Islam, A.R.M.T.: Applications of various data-driven models for the prediction of groundwater quality index in the Akot basin, Maharashtra, India. Environ. Sci. Pollut. Res. 1–15 (2022)
6. Ghalandari, V., Iranmanesh, A.: Energy and exergy analyses for a cement ball mill of a new generation cement plant and optimizing grinding process: a case study. Adv. Powder Technol. **31**(5), 1796–1810 (2020)
7. Gouveia, G., Alves, J., Sousa, P., Araújo, R., Mendes, J.: Edge computing-based modular control system for industrial environments. Processes **12**(6), 1165 (2024)
8. Kural, A., Özsoy, C.: Identification and control of the raw material blending process in cement industry. Int. J. Adapt. Control Signal Process. **18**(5), 427–442 (2004)
9. Liu, Z., Wang, X., Zhang, Q., Huang, C.: Empirical mode decomposition based hybrid ensemble model for electrical energy consumption forecasting of the cement grinding process. Measurement **138**, 314–324 (2019)
10. Ma, T., Li, Z., Liu, J., Alkhateeb, A.F., Jahanshahi, H.: A novel self-learning fuzzy predictive control method for the cement mill: simulation and experimental validation. Eng. Appl. Artif. Intell. **120**, 105,868 (2023)
11. Majewski, P., Pawuś, D., Szurpicki, K., Hunek, W.P.: Toward optimal control of a multivariable magnetic levitation system. Appl. Sci. **12**(2), 674 (2022)
12. Pawuś, D.: Identification and control of the cement milling process using artificial neural networks and other non-linear models - preliminary research (thesis - in polish) (2023)
13. Pawuś, D., Paszkiel, S.: Identification and expert approach to controlling the cement grinding process using artificial neural networks and other non-linear models. IEEE Access **12**, 26364–26383 (2024). https://doi.org/10.1109/ACCESS.2024.3366703
14. Ramasamy, V., et al.: A comprehensive review on advanced process control of cement kiln process with the focus on MPC tuning strategies. J. Process Control **121**, 85–102 (2023)
15. Sari, R., Syahputri, K., Rizkya, I., Nadhirah, N.: Analysis of raw mill machines maintenance in cement industry. In: IOP Conference Series: Materials Science and Engineering, vol. 851, p. 012051. IOP Publishing (2020)
16. Shanthi, R., Kalyani, S., Devie, P.: Design and performance analysis of adaptive neuro-fuzzy controller for speed control of permanent magnet synchronous motor drive. Soft. Comput. **25**, 1519–1533 (2021)
17. Shi, Q., Lam, H.K., Xuan, C., Chen, M.: Adaptive neuro-fuzzy PID controller based on twin delayed deep deterministic policy gradient algorithm. Neurocomputing **402**, 183–194 (2020)

18. Sokół, S., Pawuś, D., Majewski, P., Krok, M.: The study of the effectiveness of advanced algorithms for learning neural networks based on FPGA in the musical notation classification task. Appl. Sci. **12**(19), 9829 (2022)
19. Tong, R., Sui, T., Feng, L., Lin, L.: The digitization work of cement plant in China. Cem. Concr. Res. **173**, 107,266 (2023)
20. Tsamatsoulis, D.: An industrial control system for cement sulfates content using a feedforward and feedback mechanism. ChemEngineering **8**(2), 33 (2024)
21. Vignesh, R., Ashok, B., Kumar, M.S., Szpica, D., Harikrishnan, A., Josh, H.: Adaptive neuro fuzzy inference system-based energy management controller for optimal battery charge sustaining in biofuel powered non-plugin hybrid electric vehicle. Sustain. Energy Technol. Assess. **59**, 103,379 (2023)
22. Vijayabhaskar, B., Jayalalitha, S.: Dual adaptive model predictive controller application to vertical roller mill process used in the cement industry. IEEE Access **8**, 226,705–226,723 (2020)

New Sliding Hyperplane for Achieving Bounded Output Performance in DSMC

Paweł Latosiński(✉)

Institute of Automatic Control, Lodz University of Technology, B. Stefanowskiego 18, 90-537 Lodz, Poland
pawel.latosinski@p.lodz.pl

Abstract. Sliding mode control strategies for discrete-time systems offer many advantages, such as high computational efficiency and a high degree of robustness with respect to disturbance. However, such strategies are typically only concerned with guiding the evolution of a virtual sliding variable, rather than the real output of the controlled plant. As a result, imposing practical bounds on the motion of the system may be challenging in sliding mode control. In response to this problem, we have introduced a new sliding hyperplane for discrete-time systems, with the aim of establishing a direct proportional relationship between the virtual sliding variable and the actual system output. As a result, when the sliding variable becomes strictly bounded during the control process, system output will become explicitly limited as well.

Keywords: discrete-time systems · robust control · sliding mode control · output constraints

1 Introduction

Sliding mode control strategies have traditionally been employed in cases, where robustness of the system with respect to disturbance and model uncertainties is crucial [12,16]. Such strategies aim to confine the system representative point to an appropriate area defined in the state space, typically referred to as a sliding hyperplane. Once the representative point is driven onto this hyperplane, the effect of disturbance and model uncertainties on system dynamics can be effectively rejected [7,8]. Although sliding mode control strategies have been conventionally designed for continuous-time systems, they have later been adapted to discrete-time ones as well [9,17]. In the case of discrete-time systems, the system representative point is driven to a vicinity of the sliding hyperplane, often referred to as a quasi-sliding mode band. As a result, rather than complete insensitivity to disturbance and model uncertainties, only a degree of robustness to these factors can be obtained. However, discrete-time sliding modes are still very significant because of the prevalence of digitally controlled systems, in which no continuous states exist [5,18].

R. Szewczyk et al. (Eds.): AUTOMATION 2024, LNNS 1219, pp. 17–23, 2024.
https://doi.org/10.1007/978-3-031-78266-4_2

An important aspect of sliding mode controller design for both continuous and discrete-time systems is the selection of an appropriate sliding hyperplane in the state space. The choice of this hyperplane has a significant effect on closed-loop performance of the system, the degree of robustness with respect to uncertainties, as well as presence of some undesirable phenomena such as chattering [3]. Various authors have proposed the use of non-linear sliding surfaces in order to improve the transient response of the system [1,15]. Others have employed time-varying hyperplanes to eliminate the initial reaching phase of sliding motion during which the plant is vulnerable to disturbance [4,10]. Another significant development occurred in the so-called higher-order sliding mode control, in which specific hyperplanes are designed to ensure a smooth motion of the system, devoid of harmful oscillations [2,13].

In this paper we focus on the issue of output boundedness in discrete-time sliding mode control. Such strategies aim to confine the system representative point on an appropriately defined hyperplane in the state space or in its immediate vicinity. However, they typically do not provide any insight into the evolution of individual system states and do not allow one to directly influence the values of the system output. In response to this shortcoming, we have designed a novel sliding hyperplane for discrete-time systems, which allows one to easily limit the magnitude of the system output when the system is operating in sliding motion. This is achieved by analyzing the properties of inverse transfer function between the sliding variable and the output of the system, and ensuring that this function is monotonic and bounded.

2 Considered Class of Systems

In this paper, a new control approach will be proposed for discrete-time systems with the aim of ensuring uniform ultimate boundedness of their output. Dynamics of the considered systems can be expressed in the state space as

$$\begin{aligned} \boldsymbol{x}(k+1) &= \boldsymbol{A}\boldsymbol{x}(k) + \boldsymbol{b}u(k) + \boldsymbol{b}d(k), \\ y(k) &= \boldsymbol{c}^{\mathrm{T}}\boldsymbol{x}. \end{aligned} \tag{1}$$

In this relation $\boldsymbol{x} \in \mathbb{R}^n$ represents the state vector, $u(k) \in \mathbb{R}$ is the scalar control signal, $y(k) \in \mathbb{R}$ is the system output, $d(k) \in \mathbb{R}$ represents matched uncertainties affecting the plant and $\boldsymbol{A}, \boldsymbol{b}, \boldsymbol{c}$ are constant elements of appropriate dimensions. The objective is to drive the system representative point from its initial position $\boldsymbol{x}(0)$ to $\mathbf{0}$ and maintain this value for the remainder of the control process. In order to make any robust control strategy feasible for such plants, it is reasonably assumed that $d(k)$ is finite, i.e. there exists a constant $d_{\max}$ such that for all k

$$|d(k)| \leq d_{\max}. \tag{2}$$

This disturbance is assumed to satisfy matching conditions, which means it affects the system through the same input channels as the control signal.

Although this assumption is entirely possible to omit in discrete-time sliding mode control, it has been used in this paper in order to simplify notation in the upcoming analysis.

The objective of this paper is to design a novel sliding mode control strategy that ensures uniform ultimate boundedness of the system for any initial conditions of the system. In order to enable this property, system dynamics in the so-called controllable canonical form will be considered. Since state dynamics of any controllable plant can be expressed in the controllable canonical form, it is assumed that (1) already reflects this form. Consequently, state matrix $\boldsymbol{A}$ and vectors $\boldsymbol{b}, \boldsymbol{c}$ are expressed as

$$\boldsymbol{A} = \begin{bmatrix} 0 & 1 & 0 & \cdots & 0 \\ 0 & 0 & 1 & \cdots & 0 \\ \vdots & \vdots & \vdots & \ddots & \vdots \\ 0 & 0 & 0 & \cdots & 1 \\ -a_n & -a_{n-1} & -a_{n-2} & \cdots & -a_1 \end{bmatrix}, \quad \boldsymbol{b} = \begin{bmatrix} 0 \\ 0 \\ \vdots \\ 0 \\ 1 \end{bmatrix}, \quad \boldsymbol{c} = \begin{bmatrix} 1 \\ 0 \\ \vdots \\ 0 \\ 0 \end{bmatrix}. \tag{3}$$

3 Proposed Sliding Mode Control Strategy

In this section, a sliding mode control strategy for discrete-time systems (1) will be introduced. In particular, the main novelty of this work involves designing a new sliding hyperplane for systems in the controllable canonical form (3). This hyperplane, expressed using mixed notation [14], is defined as

$$s(k) = (z - \alpha)^{n-1} y(k) = 0 \tag{4}$$

for $k \geq 0$. In this relation $s(k)$ is referred to as the sliding variable, z^i represents time advance by i discrete steps and $\alpha \in (0, 1)$ is a constant selected by the designer. Considering state matrix (3) of the system in controllable canonical form, one observes that for any $i = 1, ..., n-1$ and for all k

$$z^i y(k) = y(k+i) = x_{i+1}(k). \tag{5}$$

With this in mind, one can express sliding variable $s(k)$ as

$$s(k) = \sum\nolimits_{i=1}^{n} x_i(k) \alpha^{n-i} \binom{n-1}{i-1} = \boldsymbol{\sigma}^{\mathrm{T}} \boldsymbol{x}(k), \tag{6}$$

where

$$\boldsymbol{\sigma}^{\mathrm{T}} = \begin{bmatrix} \alpha^{n-1}\binom{n-1}{0} & \alpha^{n-2}\binom{n-1}{1} & \cdots & \alpha^{1}\binom{n-1}{n-2} & \alpha^{0}\binom{n-1}{n-1} \end{bmatrix}. \tag{7}$$

The appropriate control signal which aims to drive the system representative point onto sliding hyperplane (4) can now be stated. In particular, in this work we use a relatively simple control strategy [11] with the following form

$$u(k) = (\boldsymbol{\sigma}^{\mathrm{T}} \boldsymbol{b})^{-1} \left\{ s(k) - s_0 \mathrm{sgn}[s(k)] \min\left\{1, |s_0^{-1} s(k)|\right\} - \boldsymbol{\sigma}^{\mathrm{T}} \boldsymbol{A} \boldsymbol{x}(k) \right\}, \tag{8}$$

where $s_0 \in \mathbb{R}_+$ is a constant selected by the designer. It can be easily demonstrated that this control signal ensures a favorable sliding mode performance of the controlled plant, which will be summarized in the following theorem.

Theorem 1. *If the control signal for plant* (1) *is defined by* (8)*, then the system representative point will monotonically approach sliding hyperplane* (4)*, enter the quasi-sliding mode band*

$$\{\boldsymbol{x} : |\boldsymbol{\sigma}^T \boldsymbol{x}(k)| \leq |\boldsymbol{\sigma}^T \boldsymbol{b}| d_{\max}\} \tag{9}$$

in a finite number of steps not greater than $k_0 = \lceil s(0)/s_0 \rceil$ *and remain inside that band for all* $k \geq k_0$.

Proof. The proof can be found in [11] and it will not be repeated here.

One further notices that when the proposed control signal (8) is applied to system (1), the following closed-loop system state matrix can be obtained:

$$\boldsymbol{A}_{cl} = \left[\boldsymbol{I}_{n \times n} - \boldsymbol{b}(\boldsymbol{\sigma}^{\mathrm{T}} \boldsymbol{b})^{-1} \boldsymbol{\sigma}^{\mathrm{T}}\right] \boldsymbol{A}. \tag{10}$$

Using the proposed sliding hyperplane (4), one ensures that all poles of this matrix are real and equal to α, which places them inside the unit circle and thus ensures closed-loop stability of the system. In the next section it will be demonstrated that the newly proposed hyperplane (4) allows one to establish a direct relationship between sliding variable $s(k)$ and system output $y(k)$.

4 Achieving Output Boundedness in DSMC

In this section it will be demonstrated that uniform ultimate boundedness of sliding variable $s(k)$ as per relation (9) is directly translated to boundedness of system output $y(k)$. This will be achieved by analyzing the discrete-time transfer function between these two values. Before this analysis can be conducted, two well known properties of Z-transforms will be cited as lemmas.

Lemma 1 [6]**.** *If* $F(z)$ *is the Z-transform of function* $f(k)$ *and all of its poles are placed inside the unit circle around the origin of the Z-plane, then*

$$\lim_{z \to \infty} F(z) = \lim_{k \to 0} f(k). \tag{11}$$

Lemma 2 [6]**.** *If* $F(z)$ *is the Z-transform of function* $f(k)$ *and all of its poles are placed inside the unit circle around the origin of the Z-plane, then*

$$\lim_{z \to 1} (z-1) F(z) = \lim_{k \to \infty} f(k). \tag{12}$$

These properties, respectively referred to as *initial value theorem* and *final value theorem*, will now allow us to establish a proportional relationship between the sliding variable and the system output. This property will be summarized in the following theorem.

Theorem 2. *If the control signal for system* (1) *is defined by* (8) *and the sliding hyperplane is selected according to* (4), *then for all* $k \geq k_0$ *the output of the system is bounded in the following way*

$$|y(k)| \leq |\boldsymbol{\sigma}^T \boldsymbol{b}| d_{\max} (1-\alpha)^{1-n}. \tag{13}$$

Proof. The goal of this proof is to demonstrate that the boundedness of $s(k)$ expressed by relation (9) directly leads to output constraints denoted by (13). This will be done by obtaining bounds of the inverse transfer function $f(k)$ between these two values. One can first calculate the Z-transform of $y(k)$ using relation (4) to get

$$Y(z) = Z\{y(k)\} = Z\left\{\frac{1}{(z-\alpha)^{n-1}} s(k)\right\} = (z-\alpha)^{1-n} S(z), \tag{14}$$

where $S(z)$ is the Z-transform of sliding variable $s(k)$. With this in mind, one can calculate the transfer function $F(z)$ between $s(k)$ and $y(k)$ as

$$F(z) = \frac{Z\{\mathbb{1}(k) y(k)\}}{Z\{s(k)\}} = \frac{(z-\alpha)^{1-n} S(z)}{S(z)} Z(\mathbb{1}(k)) = (z-\alpha)^{1-n} \frac{z}{z-1}. \tag{15}$$

In this relation function $\mathbb{1}(k)$, which equals 1 for $k \geq 0$ and 0 for $k < 0$, is included due to the fact that sliding variable $s(k)$ is only defined for $k \geq 0$. As such, no relation is possible to establish between $y(k)$ and $s(k)$ for any $k < 0$.

Considering relation (15) and Lemma 1, the inverse transfer function $f(k)$ satisfies the following equality

$$f(0) = \lim_{k \to 0} f(k) = \lim_{k \to \infty} (z-\alpha)^{1-n} \frac{z}{z-1} = 0. \tag{16}$$

Furthermore, from Lemma 2 one obtains the final value of $f(k)$ as

$$\lim_{k \to \infty} f(k) = \lim_{z \to 1} (z-1)(z-\alpha)^{1-n} \frac{z}{z-1} = \lim_{z \to 1} z(z-\alpha)^{1-n} = (1-\alpha)^{1-n}. \tag{17}$$

One further notices that all poles of transfer function (15) are equal to α, which makes them positive real values. Therefore, since this function is devoid of an oscillatory component, $f(k)$ will be a strictly monotonic function in its entire domain. Further considering (16) and (17), one concludes that for all $k \geq 0$

$$0 \leq f(k) \leq (1-\alpha)^{1-n}. \tag{18}$$

Finally, since Theorem 1 implies that $|s(k)| = |\boldsymbol{\sigma}^{\mathrm{T}} \boldsymbol{x}(k)| \leq |\boldsymbol{\sigma}^{\mathrm{T}} \boldsymbol{b}| d_{\max}$ for all $k \geq k_0$, then for the same k the relationship between $y(k)$ and $s(k)$ expressed via the inverse transfer function $f(k)$ yields

$$|y(k) = |f(k) s(k)| \leq (1-\alpha)^{1-n} |s(k)| \leq (1-\alpha)^{1-n} |\boldsymbol{\sigma}^{\mathrm{T}} \boldsymbol{b}| d_{\max}, \tag{19}$$

which is consistent with (13). □

In conclusion, Theorem 2 has demonstrated that for the proposed novel sliding hyperplane (4), it is possible to develop a strict formula for uniform ultimate boundedness of system output $y(k)$. The obtained bound is relatively small and linearly dependent on the maximum value of the disturbance affecting the plant.

5 Discussion and Conclusions

In this paper we have proposed a new way to design a sliding hyperplane for sliding motion of discrete-time systems. The goal of the proposed approach is to make sure that the output of the controlled systems is always strictly bounded in the sliding phase. Indeed, Theorem 2 has shown that when the sliding variable is confined to a narrow quasi-sliding mode band, the output of the system is limited by a proportionally small constant.

The proposed hyperplane is specified to make sure that all poles of the closed-loop system are positive real values in order to prevent unwanted oscillations during the control process. This is done via the selection of a singular constant parameter α in the formula defining the hyperplane. The choice of this parameter offers an indirect compromise between the robustness of the obtained result and the magnitude of the control signal needed to achieve it. In particular, values of α closer to zero reduce the magnitude of bound (1), but may result in unfeasibly large values of the control signal in the initial stages of the control process.

It is worth remarking that the proposed approach has been designed with systems in controllable canonical form in mind. This may appear to be a restrictive assumption that could limit the applicability of the new approach. However, systems with dynamics expressed by (1) and (3) can in fact represent any discrete-time controllable plant. Indeed, any controllable plant with dynamics

$$\begin{aligned} \hat{\boldsymbol{x}}(k+1) &= \hat{\boldsymbol{A}}\hat{\boldsymbol{x}}(k) + \hat{\boldsymbol{b}}u(k) + \hat{\boldsymbol{b}}d(k), \\ y(k) &= \hat{\boldsymbol{c}}^{\mathrm{T}}\hat{\boldsymbol{x}}. \end{aligned} \tag{20}$$

can be transformed to (1) with $\boldsymbol{A}$, $\boldsymbol{b}$, $\boldsymbol{c}$ defined as (3) by using a linear state transformation $\boldsymbol{x} = \boldsymbol{T}\hat{\boldsymbol{x}}$, where $\boldsymbol{T} = \boldsymbol{\Sigma}^{\mathrm{T}}\hat{\boldsymbol{\Sigma}}$ and controllability matrices

$$\begin{aligned} \boldsymbol{\Sigma} &= [\boldsymbol{b} \quad \boldsymbol{A}\boldsymbol{b} \quad \dots \quad \boldsymbol{A}^{n-1}\boldsymbol{b}]^{\mathrm{T}}, \\ \hat{\boldsymbol{\Sigma}} &= [\hat{\boldsymbol{b}} \quad \hat{\boldsymbol{A}}\hat{\boldsymbol{b}} \quad \dots \quad \hat{\boldsymbol{A}}^{n-1}\hat{\boldsymbol{b}}]^{\mathrm{T}}. \end{aligned} \tag{21}$$

In other words, as long as a simple linear state transformation is considered, the result proposed in this paper is applicable to any controllable discrete-time plant (20).

References

1. Bandyopadhyay, B., Fulwani, D.: High-performance tracking controller for discrete plant using nonlinear sliding surface. IEEE Trans. Industr. Electron. **56**(9), 3628–3637 (2009)
2. Bartoszewicz, A.: Discrete time quasi-sliding mode control strategies. IEEE Trans. Industr. Electron. **45**(4), 633–637 (1998)
3. Bartoszewicz, A.: Chattering attenuation in sliding mode control systems. Control. Cybern. **29**(2), 585–584 (2000)

4. Bartoszewicz, A., Nowacka-Leverton, A.: ITAE optimal sliding modes for third order systems with input signal and state constraints. IEEE Trans. Autom. Control **55**(8), 1928–1932 (2010)
5. Corradini, M., Orlando, G.: Variable structure control of discretized continuous-time systems. IEEE Trans. Autom. Control **43**(9), 1329–1334 (1998)
6. Dorsey, J.: Continuous and Discrete Control Systems. Publishing House of Electronics Industry (2002)
7. Draženović, B.: The invariance conditions in variable structure systems. Automatica **5**(3), 287–295 (1969)
8. Gao, W., Hung, J.: Variable structure control of nonlinear systems: a new approach. IEEE Trans. Industr. Electron. **40**(1), 45–55 (1993)
9. Gao, W., Wang, Y., Homaifa, A.: Discrete-time variable structure control systems. IEEE Trans. Industr. Electron. **42**(2), 117–122 (1995)
10. Ha, Q., Rye, D., Durrant-Whyte, H.: Fuzzy moving sliding mode control with application to robotic manipulators. Automatica **35**(4), 607–616 (1999)
11. Latosinski, P., Bartoszewicz, A., Lesniewski, P.: A new reaching law based DSMC for inventory management systems. In: 21st International Conference on System Theory, Control and Computing (2017)
12. Sabanovic, A.: Variable structure systems with sliding modes in motion control - a survey. IEEE Trans. Industr. Inf. **7**(2), 212–223 (2011)
13. Sharma, N.K., Janardhanan, S.: Discrete higher order sliding mode: concept to validation. IET Control Theory Appl. **11**(8), 1098–1103 (2017)
14. Slotine, J.J.E., Li, W.: Applied Nonlinear Control, vol. 199. Prentice Hall, Englewood Cliffs (1991)
15. Tokat, S., Eksin, I., Guzelkaya, M., Soylemez, M.T.: Design of a sliding mode controller with a nonlinear time-varying sliding surface. Trans. Inst. Meas. Control. **25**(2), 145–162 (2003)
16. Utkin, V.: Variable structure systems with sliding modes. IEEE Trans. Autom. Control **22**(2), 212–222 (1977)
17. Utkin, V., Drakunov, S.V.: On discrete-time sliding mode control. In: IFAC Conference on Nonlinear Control, pp. 484–489 (1989)
18. Yu, X., Wang, B., Li, X.: Computer-controlled variable structure systems: the state of the art. IEEE Trans. Industr. Inf. **8**(2), 197–205 (2012)

Optimizing Ambulance Routing for Timely Delivery and Pick-Up of Nursing Teams Visiting Patients Undergoing Home Therapy

Eryk Szwarc[1], Robert Wójcik[2], Grzegorz Bocewicz[1](✉), and Zbigniew Banaszak[1]

[1] Faculty of Electronics and Computer Science, Koszalin University of Technology, Koszalin, Poland
{eryk.szwarc,grzegorz.bocewicz,zbigniew.banaszak}@tu.koszalin.pl
[2] Department of Computer Engineering, Faculty of Information and Communication Technology, Wrocław University of Science and Technology, Wrocław, Poland
robert.wojcik@pwr.edu.pl

Abstract. This paper proposes a comprehensive approach to optimize ambulance scheduling, considering various factors such as patient needs, nursing team availability, route optimization, and real-time tracking. The goal is to enhance efficiency, reduce delays, and prioritize urgent cases, ultimately improving the overall quality of home healthcare services by routing a fleet of ambulances responsible for delivering and receiving nursing teams without the constraint of requiring the same vehicle for both operations. Focusing on proactive/reactive routing strategies, the proposed declarative modeling-based framework aims to optimize resource utilization while contributing to the overall success of home therapy programs. Through a case study and results of conducted computer experiments, we demonstrate the effectiveness of the developed reference model in ensuring efficiency and adaptability in scenarios where vehicle consistency is not a strict requirement.

Keywords: OR in health services · Vehicle routing · Pick-ups and deliveries · Time windows · Home Healthcare

1 Introduction

Increased life expectancy, especially within the last three decades, has stimulated a need for home healthcare services (HHS) [5] or home healthcare (HHC) [10], perceived as a dynamically growing service industry in the medical services business, mainly in Europe and North America [8]. A multidisciplinary team provides HHC services at patients' homes using a distribution network facing NP-hard logistics problems such as districting, health care routing and scheduling, inventory management, and the lack of integrated approaches to address them [5]. The districting problem partitioning the HHC service territory into clusters of patients and aggregately assigning each cluster to a set of resources, e.g., care worker, and resource dimensioning identifying the level of resources required in HHC, e.g., the number of care workers to deliver care services or

R. Szewczyk et al. (Eds.): AUTOMATION 2024, LNNS 1219, pp. 24–38, 2024.
https://doi.org/10.1007/978-3-031-78266-4_3

the number of technical devices required, and assigning these resources to the districts are the subject of interest at a strategic and a tactical level, respectively. In turn, the operator assignment problem decides which care workers deliver care services to which patients and the home health care routing and scheduling problem (HHCRSP) is a subject of interest at an operational level of HHC [14].

The escalating, increasingly prevalent demand for home therapies necessitates the construction of a schedule and the daily routes for each medical or nursing/caregivers team (hereinafter referred to as NCT) to maximize service effectiveness and cost/time efficiency. This demand necessitates a robust routing and scheduling of NCTs to ensure timely visits aimed at carrying out pre-arranged therapies and treatments. Satisfying these needs is part of a vehicle routing problem with pick-up and delivery time windows, where a pick-up and delivery time window refers to a specific period during which NCTs visit patients undergoing home therapies.

In this context, the problem considered presents a variant of the vehicle routing problem with pick-up and delivery time windows (VRPPDTW) where ambulance travel times and routes are coordinated according to the requirements imposed by the distances separating patients dispersed in a network and adopted strategy of service delivery [6, 15]. The literature on the subject is dominated by cases in which medical/rehabilitation assistance is provided by the NCT, which has a vehicle that allows access to the visited patient. Due to traffic difficulties and the lack of parking spaces, this mode of providing medical care does not guarantee timely completion of scheduled visits and effective use of the assigned NCT time, i.e., quality/effectiveness and cost/time efficiency of provided medical services. In this situation, a more rational solution seems to be using specially equipped and appropriately marked technical ambulances housing several specialized NCTs delivered to the scheduled patients. The privileged status of such a vehicle can guarantee the timely completion of door-to-door visits. Assuming that the medical service provider has an appropriate fleet of such vehicles, it is interesting to assess the effectiveness of the above solution.

Assuming that each vehicle in the ambulance fleet accommodates four one- or two-person teams of NCTs, the following two strategies correspond to the situation where NCTs are delivered and picked up by the same ambulance and corresponding to routing of the fleet delivering and receiving transported NCTs without requiring the delivered team to be picked up by the same ambulance, are considered. The last strategy, optimizing the movement of fleets involved in the delivery and pick-up of NCTs, specifically addressing scenarios where the vehicle delivering an NCT does not need to be the same as the one picking it up, has received scant coverage in the literature. VRPPDTW-I stands for VRPPDTW, allowing interspersed delivery and collection of service teams. Of course, since different strategies for the distribution of NCTs have different impacts on shortening the time and reducing costs, their implementation should be determined each time by assessing objective criteria such as the total cost (i.e., the sum of fixed cost and traveling cost).

In this context, the paper proposes a comprehensive approach to optimize ambulance scheduling, considering various factors such as patient needs, nursing team availability, fleet size, and route optimization, as well as for proactive and reactive routing planning of the fleet of technical ambulances used in HHC missions in time horizons covering several days. The goal is to enhance efficiency, reduce delays, and prioritize urgent cases, ultimately improving the overall quality of HHS. The long-term goal is to check the usefulness of the proposed approach and the possibility of its implementation in interactive DSS systems.

The remainder of the paper is organized as follows. Section 2 gives related literature on emphasizing proactive mission planning of home healthcare services while focusing on the scheduling and routing of NCTs. Section 3 provides an illustrative example of the problem of proactive mission planning of a fleet of medical ambulances. Section 4 describes the constraints of programming-driven model assumptions and presents problem formulation. Section 5 presents the computational experiments conducted to evaluate the performance of different strategies both in quantitative and qualitative aspects. Finally, Sect. 6 offers some conclusions and insights drawn from this work.

2 Literature Review

The aging populations of European and North American countries have underpinned the rapid growth of the HHC medical services industry [8]. This dynamic development, observed for over 30 years [10, 11], is widely reflected in the newly established home care agencies and research on the search for efficient scheduling and routing of their staff. The importance of this research stems from the nature of the home care sector, which is never wholly stable; the pool of patients constantly evolves, and the staff is subject to a high turnover. The range of services that can be requested is vast: it can be medical care, housecleaning, meal preparation, etc. [14].

Numerous scientific publications in this area include review papers attempting to build a taxonomy of problems and methods used to solve them [6, 9, 10] and work focusing on selected problems of planning and distribution of medical care services [15–17, 19]. Among the most frequently addressed problems, it is necessary to distinguish those related to constructing a schedule and the daily routes for each nurse to maximize both service effectiveness and cost/time efficiency, subject to the daily total working hours of a nurse and the time window requests of the patients [3, 18]. It is worth noting that the research does not highlight the methods related to the implementation of visits, including how to move NCTs and the specificity of public transport. They tacitly assume that individual NCTs decide the choice and use of means of transport.

The presented research gap has inspired the research undertaken in this work. The starting point is the assumption that home care agencies have a fleet of technical ambulances that carry out the delivery and pick-up of NCTs. The proposed approach departs from the scheme in which the vehicle delivering a given NCT must also collect it. As an alternative solution, a situation is being considered in which NCTs distributed by some vehicles can be received by others belonging to the fleet carrying out a daily (shift) home care mission.

Most research papers focus on single-period optimization problems, where a single working day is assumed as the planning horizon. Increasingly, multi-period problems are also being addressed, assuming that not all service requests arise in the same planning period. Consequently, nurses may work multiple days, and clients request multiple services spread over different days [5, 18].

Among the most frequently addressed issues, shift scheduling, staff dimensioning, staff assignment, and staff routing should be distinguished [5, 6, 10, 16, 18, 19]. These problems are often addressed through requirements such as matching nurses' skills and clients' requirements, re-assigning care workers, re-scheduling the tours, etc. It is worth noting that their complexity makes it impossible to solve them for real-size instances.

These problems are often formulated in deterministic models that allow them to be solved by operations research (OR) methods, often exact, such as Mixed Integer Programming [4, 11, 12], and most often by approximate methods, including heuristic and metaheuristic methods such as Genetic Algorithm, Adaptive Large Neighborhood Search [5, 15]. Recent literature reviews regarding OR models applied to HHC can be found in [16, 17, 19]. Stochastic representations of scheduling and routing models are rare [22]. A similar situation occurs when trying to model problems described by parameters with uncertain, vague values using the formalism of fuzzy sets [12, 13, 21] and those implementing the declarative modeling paradigm enabling constraint programming environments [1, 2, 7, 20]. The declarative model allows us to formulate and solve the VRPPDTW-I problem as an instance of the constraint satisfaction problem.

3 Illustrative Example

To illustrate the nature of the problem, consider the following situation. There are 10 patients (hereinafter referred to as $C_1, \ldots, C_{10}$) requiring cyclic care. Due to their different health conditions, each of them requires a different type of care. Care is provided by NCTs offering 4 types of medical care and/or therapy (hereinafter referred to as MCT), i.e. $o_1, \ldots, o_4$.. Every day, patients are visited by one of these teams. NCTs are delivered by ambulance (two technical ambulances are available: $NA = 2$). Patients are located in the urban space of 15 km × 15 km (city of Warsaw). Their location in relation to the ambulance base is shown in Fig. 1.

The individual needs of patients are known, which are summarized in Table 1 in the form of a binary OR matrix. A value of $r_{l,u} = 1$ of the OR matrix means that the patient C_u requires o_l MCT, a value of 0 means the opposite.

NCTs $k_1, \ldots, k_4$ with different qualifications, collected in Table 2, in the form of a binary matrix OK, are used to care for patients. The value of $ok_{i,l} = 1$ of the OK matrix means that the NCT k_i has MCT o_l, the value of 0 means the opposite.

NCTs visit patients (duration of a single visit: $nt_u = 120$ min) on specific days of the 7-day time horizon. This schedule (hereinafter referred to as WH) is repeated periodically in the following weeks. The adopted schedule is illustrated in Table 3. It assumes that on the first day, patients to whom NCTs k_1 and k_4 have been assigned are visited. In addition, it is assumed that the visit to each patient must be completed within 8 h (working hours 7:00–15:00), $TH_u = [0, 480]$ min.

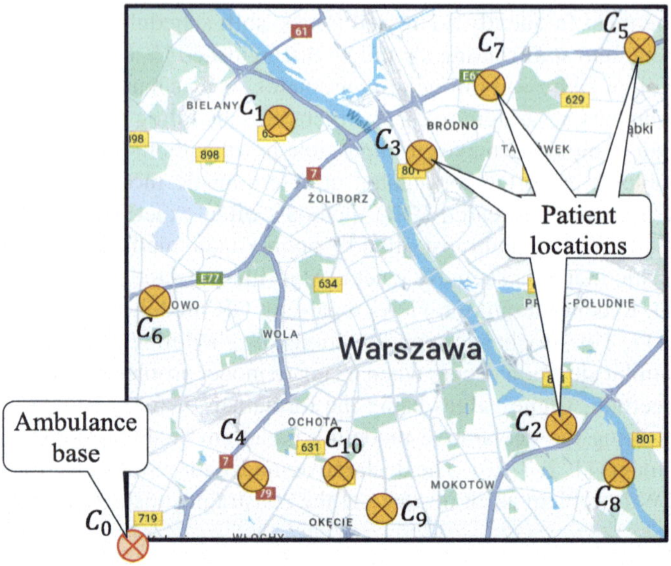

Fig. 1. Locations of patients of the group under consideration – the city of Warsaw

Table 1. *OR* matrix of the considered group of patients

Functionality	Patients									
	C_1	C_2	C_3	C_4	C_5	C_6	C_7	C_8	C_9	C_{10}
o_1	1	1	1	1	0	0	1	0	1	1
o_2	1	1	0	0	1	1	0	1	0	0
o_3	0	0	0	1	1	1	1	1	0	0
o_4	0	1	1	1	0	0	1	1	1	1

Table 2. *OK* matrix of the considered group of patients

Functionality	Nursing team			
	k_1	k_2	k_3	k_4
o_1	0	0	1	1
o_2	1	0	1	0
o_3	1	1	0	0
o_4	1	1	0	1

For the above assumptions, the considered problem boils down to answering the following questions:

Table 3. NCTs Work Schedule

Team	Day						
	1	2	3	4	5	6	7
k_1	1	0	0	0	0	0	1
k_2	0	0	1	0	1	0	0
k_3	0	1	0	1	0	1	0
k_4	1	0	1	0	1	0	1

- Is there an assignment of NCTs to patients to schedule their appointments (schedule and travel plan) each day within a given time window?
- If so, what is the fastest way to complete the mission?

4 Problem Formulation

Searching for the mission of home healthcare services that guarantee visits to all patients comes down to solving an optimization problem, which could be described using the declarative modeling paradigm.

Parameters:

K: a set of types of NCTs that are available, $K = \{k_1, \ldots, k_i, \ldots, k_{LK}\}$ (where LK is a number of types of NCTs),

O: a set of all functionalities (MCTs) of the set of NCTs, $O = \{o_1, \ldots, o_l, \ldots, o_{LO}\}$,

$\mathbb{C}$: a set of all visited patients, $\mathbb{C} = \{C_1, \ldots, C_u, \ldots, C_{LC}\}$ (where LC is the number of patients),

OK: MCTs provided by individual NCT of the set K are represented by matrix $OK = \left[ok_{l,i}\right]_{l=1\ldots LO;i=1\ldots LK}$, in which value $ok_{i,l} = 1$ means that team k_i has o_l MCT and value 0 means the opposite situation,

OR: patients' requirements represented by a matrix $OR = \left[r_{l,u}\right]_{l=1\ldots LO;u=1\ldots LC}$, in which the values of the elements $r_{l,u}$ denote the number of resources with o_l MCT that customer C_u expect,

TH: a sequence of time windows TH_u defining the period in which the patient C_u expect the visit, $TH = (TH_1, \ldots, TH_u, \ldots, TH_{LC})$,

WH: a work schedule of nursing teams: $WH = (WH_1, \ldots, WH_h, \ldots, WH_{NH})$ where WH_h contains the NCTs working in day h,

NA: a number of ambulances,

$vt_{u,v}$: ambulance travel time between patients C_u and C_v,

vw_u: ambulance waiting time at the patient C_u,

nt_u: nominal visit time of the patient C_u.

Decision Variables:

X: a set of NCT assignments for patients constituting the set $\mathbb{C}$, $X = \{X_1, \ldots, X_u, \ldots, X_{LC}\}$, where X_u is assignment determining whether the team k_i ($x_{u,i} = 1$) visit the patient C_u, $X_u = \left(x_{u,1}, \ldots, x_{u,i}, \ldots, x_{LC,LK}\right)$,

$CP(h, X)$: the set of patients visited on the h-th day, $CP \subset \mathbb{C}$,

$dx_{u,v,h,a}$: binary variable used to indicate if the ambulance a travels on day h to C_u from C_v to deliver a nursing team,

$px_{u,v,h,a}$: binary variable used to indicate if the ambulance a travels on day h to C_u from C_v to pick up a nursing team,

$\overline{dy_{u,h,a}}$: time at which ambulance a arrives on day h at the patient C_u to deliver a nursing team,

$\underline{dy_{u,h,a}}$: time at which ambulance a starts to travel on day h to the patient C_u to deliver a nursing team,

$\overline{py_{u,h,a}}$: time at which ambulance a arrives on day h at the patient C_u to pick up a nursing team,

$\underline{py_{u,h,a}}$: time at which ambulance a starts on day h travel to patient C_u to pick up the NCT,

$\overline{sy_{u,h,a}}$: time at which the NCT ends the occupation of the patient C_u,

$\underline{sy_{u,h,a}}$: time at which the NCT starts occupation of the patient C_u,

$st_{u,h,a}$: service time of patient C_u.

Constraints:

- functionalities $ok_{l,i}$ must be consistent with the requirements $r_{l,u}$ of patient C_u:

$$\sum\nolimits_{i=1}^{LK} x_{u,i} \times ok_{l,i}\mathrm{S} \geq r_{l,u}, \text{ for } u = 1 \ldots LC; l = 1 \ldots LO, \tag{1}$$

- constraints determining the permissible routes of the ambulances.

$$\sum\nolimits_{v \in CP(h,X)} dx_{1,v,h,a} = 1, h = 1 \ldots NH, a = 1 \ldots NA, \tag{2}$$

$$\sum\nolimits_{u \in CP(h,X)} dx_{u,1,h,a} = 0, h = 1 \ldots NH, a = 1 \ldots NA, \tag{3}$$

$$dx_{u,u,h,a} = 0, \text{ for } u \in CP(h, X), h = 1 \ldots NH, a = 1 \ldots NA, \tag{4}$$

$$\sum\nolimits_{u \in CP(h,X)} dx_{u,j,h,a} = 1, \text{ for } v \in CP(h, X), h = 1 \ldots NH, \tag{5}$$

$$\sum\nolimits_{v \in CP(h,X)} px_{1,v,h,a} = 0, h = 1 \ldots NH, a = 1 \ldots NA, \tag{6}$$

$$\sum\nolimits_{u \in CP(h,X)} px_{u,1,h,a} = 1, h = 1 \ldots NH, a = 1 \ldots NA, \tag{7}$$

$$\left(\sum\nolimits_{u \in CP(h,X)} dx_{u,v,h,a} = 1\right) \Rightarrow \left(\sum\nolimits_{u \in CP(h,X)} px_{u,v,h,a} = 1\right), \text{ for } h = 1 \ldots NH, \tag{8}$$

$$\sum\nolimits_{v \in CP(h,X)} \left(dx_{u,v,h,a} + px_{u,v,h,a}\right) = \sum\nolimits_{v \in CP(h,X)} \left(dx_{v,u,h,a} + px_{v,u,h,a}\right), \tag{9}$$

- constraints, the fulfillment of which guarantees the timely delivery and pick-up of NCTs.

$$\left(dx_{u,v,a,h}=1\right)\Rightarrow\left(\overline{dy_{v,a,h}}\geq\overline{dy_{u,a,h}}+vw_u+vt_{u,v}\right),\ \text{for } u,v\in CP(h,X),u\neq v, \tag{10}$$

$$\left(dx_{u,v,a,h}=1\right)\Rightarrow\left(\underline{dy_{v,a,h}}=\overline{dy_{v,a,h}}-vt_{u,v}-vw_u\right),\ \text{for } u,v\in CP(h,X),u\neq v, \tag{11}$$

$$\overline{dy_{u,a,h}}\in TH_u,\ \text{for } u\in CP(h,X),h=1\ldots NH, \tag{12}$$

$$\left(px_{u,v,a,h}=1\right)\Rightarrow\left(\overline{py_{v,a,h}}\geq\overline{dy_{u,a,h}}+vw_u+vt_{u,v}\right),\ \text{for } u,v\in CP(h,X),u\neq v, \tag{13}$$

$$\left(px_{u,v,a,h}=1\right)\Rightarrow\left(\underline{py_{v,a,h}}=\overline{py_{v,a,h}}-vt_{u,v}-vw_u\right),\ \text{for } u,v\in CP(h,X),u\neq v, \tag{14}$$

$$\left(px_{u,u,a,h}=1\right)\Rightarrow\left(\underline{py_{u,a,h}}=\overline{dy_{u,a,h}}\right),\ \text{for } u\in CP(h,X),h=1\ldots NH, \tag{15}$$

$$\overline{py_{u,a,h}}\in TH_u,\ \text{for } u\in CP(h,X),h=1\ldots NH, \tag{16}$$

$$\overline{py_{u,a,h}}\geq\overline{dy_{u,a,h}}+st_{u,a,h},\ \text{for } u\in CP(h,X),h=1\ldots NH, \tag{17}$$

$$\left(\overline{dy_{u,a,h}}\leq\underline{dy_{v,a,h}}\right)\vee\left(\overline{dy_{v,a,h}}\leq\underline{dy_{u,a,h}}\right),\ \text{for } u,v\in CP(h,X),u\neq v, \tag{18}$$

$$\left(\overline{py_{u,a,h}}\leq\underline{py_{v,a,h}}\right)\vee\left(\overline{py_{v,a,h}}\leq\underline{py_{u,a,h}}\right),\ \text{for } u,v\in CP(h,X),u\neq v, \tag{19}$$

$$\left(\overline{py_{u,a,h}}\leq\underline{py_{v,a,h}}\right)\vee\left(\overline{py_{v,a,h}}\leq\underline{py_{u,a,h}}\right),\ \text{for } u,v\in CP(h,X),u\neq v, \tag{20}$$

$$\left(\overline{py_{u,a,h}}\leq\underline{dy_{v,a,h}}\right)\vee\left(\overline{dy_{v,a,h}}\leq\underline{py_{u,a,h}}\right),\ \text{for } u,v\in CP(h,X),u\neq v. \tag{21}$$

Objective Function: a function that minimizes the time mission on each day h of the horizon WH.

$$\min\left(\max_{h=1\ldots NH}\left\{T_h=\max_{a=1\ldots NA}\left\{\overline{py_{0,a,h}}\right\}\right\}\right) \tag{22}$$

Among the constraints introduced, relation (1) describes the condition of an acceptable allocation, i.e., one that provides the functionalities expected by the patient. Constraints (2)–(9) concern the relationship between the proposed assignment and ambulance routes. Constraints (10)–(21) guarantee timely visits to patients. This defined model answers the following question: *What allocation of NCTs to patients (if any) guarantees the shortest appointment schedule for each day during the time horizon under consideration?*

Related problems can be formulated in terms of the following Constraint Optimization Problem (COP):

$$COP = ((\mathcal{V}, \mathcal{D}), \mathcal{C}, \mathcal{C}_{OPT}), \tag{23}$$

where: $\mathcal{V}$ a set of decision variables representing assignment X, routes of ambulances ($dx_{u,v,h,a}$, $px_{u,v,h,a}$, $\overline{dy_{u,h,a}}$, $dy_{u,h,a}$, $\overline{py_{u,h,a}}$, $py_{u,h,a}$) and NCT schedule ($\overline{sy_{u,h,a}}$, $sy_{u,h,a}$, $st_{u,h,a}$); $\mathcal{D}$ is a set of domains of decision variables: $x_{u,i}$, $dx_{u,v,h,a}$, $px_{u,v,h,a} \in \{0,1\}$, $\overline{dy_{u,h,a}}, \ldots, st_{u,h,a} \in \mathbb{N}$; $\mathcal{C}$ is a set of constraints specified in inequalities (1)–(21); $\mathcal{C}_{OPT}$ is an objective function (22).

To solve problem *COP* (23), one should determine such values of decision variables $\mathcal{V}$, for which all the constraints given in the set $\mathcal{C}$ are satisfied and objective function $\mathcal{C}_{OPT}$ is minimal. Solving *COP* means determining the visit plan, which guarantees the patient's expectation.

5 Computational Experiments

The proposed model *COP* (23) was used to solve the problem presented in Sect. 3. The considered problem comes down to an analysis problem in which the answer to the following question is sought: *What allocation of NCTs to patients X (if any) guarantees the shortest T_h appointment schedule for each day of the time horizon under consideration (see* Table 3*)?* In the experiments, assuming that two vehicles make up the considered fleet of technical ambulances, two strategies for the delivery of NCTs were considered. The first strategy, marked with the symbol $\mathcal{A}$, assumes that NCTs can be delivered and collected by different ambulances. The marked, with the symbol $\mathcal{B}$, assumes that NCTs are delivered and picked up by the same ambulance. For each of the analyzed strategies, an attempt was made to determine a mission schedule for visiting patients in the given time windows, minimizing its duration (appropriate $T_h^{(A)}$ and $T_h^{(B)}$). Patient service is provided from 7:00 a.m. to 3:00 p.m.: $TH_u = [0, 480] min$). Alternative missions are determined by the total distance traveled by ambulances $D_h^{(A)}$ and $D_h^{(B)}$) respectively) and the number of NCTs involved in visits.

The answer, i.e., a solution to the *CS* (23) problem, has been obtained in the IBM ILOG CPLEX environment (CPU: Intel i7-10510U, 16 GB RAM). The results were obtained in $t = 120$ s. The obtained allocation of NCTs to patients of set $\mathbb{C}$ is presented in Table 4. According to this assignment, the patient C_1 was assigned to NCT k_3 and the patient C_2 was assigned to NCTs k_3 and k_4, etc.

This assignment determines the following NCT delivery routes:

- For days 1 and 7 of the adopted time horizon in which NCTs k_1 and k_4, are delivered, and the plans of visits carried out according to strategies $\mathcal{A}$ and $\mathcal{B}$ are illustrated in Fig. 2 and Fig. 3, respectively.

The obtained schedules (Fig. 2) guarantee the implementation of visits within the same time-windows $TH_u = [0, 480]$ min (each patient is visited before 15:00). It is easy to see that strategy $\mathcal{A}$ provides a shorter mission duration ($T_{h=1,7}^{(A)} = 284$ min) than strategy $\mathcal{B}$ ($T_{h=1,7}^{(B)} = 372$). However, it requires a longer distance covered by ambulances,

Table 4. Allocation of X NCTs to patients set $\mathbb{C}$

teams	Patients									
	C_1	C_2	C_3	C_4	C_5	C_6	C_7	C_8	C_9	C_{10}
k_1	0	0	1	0	1	1	0	0	0	0
k_2	0	0	0	1	0	0	1	1	0	1
k_3	1	1	1	1	0	0	1	1	0	1
k_4	0	1	0	0	0	0	0	0	1	0

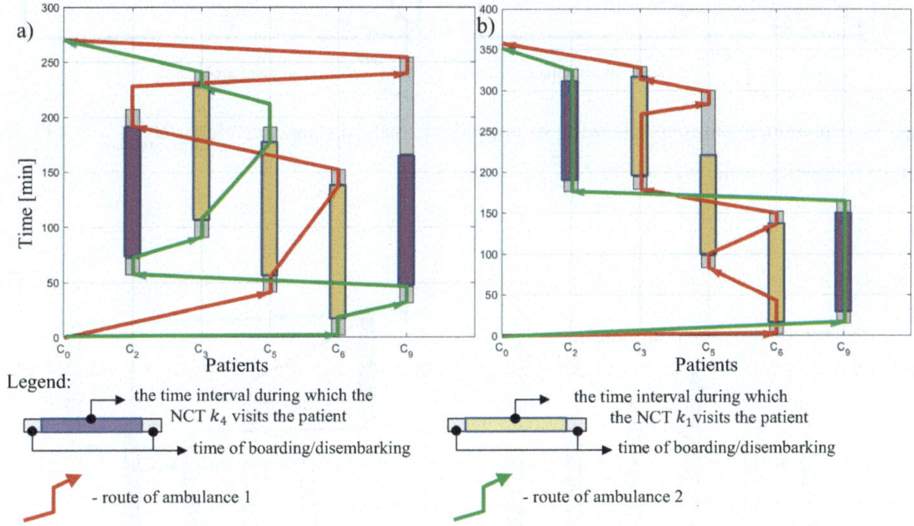

Fig. 2. Schedule for NCTs k_1 and k_4 visited patients: a) distribution strategy NCTs $\mathcal{A}$ where $T^{(A)}_{h=1,7} = 284$ min, b) distribution strategy NCTs $\mathcal{B}$ where $T^{(B)}_{h=1,7} = 372$ min

i.e. $D^{(A)}_{h=1,7} = 111.36$ km, compared to strategy $\mathcal{B}$ $D^{(B)}_{h=1,7} = 100.3$ km). Strategy $\mathcal{A}$ also requires the use of a more significant number of teams: 3 k_1 NCTs and 2 k_4 NCTs, while strategy $\mathcal{B}$ requires only: 2 k_1 NCTs and 1 k_4 NCT.

Strategy $\mathcal{A}$ (implemented on the 1st and 7th day of the time horizon) allows for visits to be completed in almost 1.5 h (88 min) shorter than in the case of using strategy $\mathcal{B}$. However, the use of this strategy is associated with the need to cover more than 11 km distance and use two additional NCTs (1 k_1 and 1 k_4 NCT).

- For days 3 and 5 of the adopted time horizon in which NCTs k_2 and k_4, are delivered, and the plans of visits carried out according to strategies $\mathcal{A}$ and $\mathcal{B}$ are illustrated in Fig. 4 and Fig. 5, respectively.

As before, the received schedules, see Fig. 4, guarantee the completion of the ordered visits within the given time window$TH_u = [0, 480] min$. Strategy $\mathcal{A}$ provides a shorter

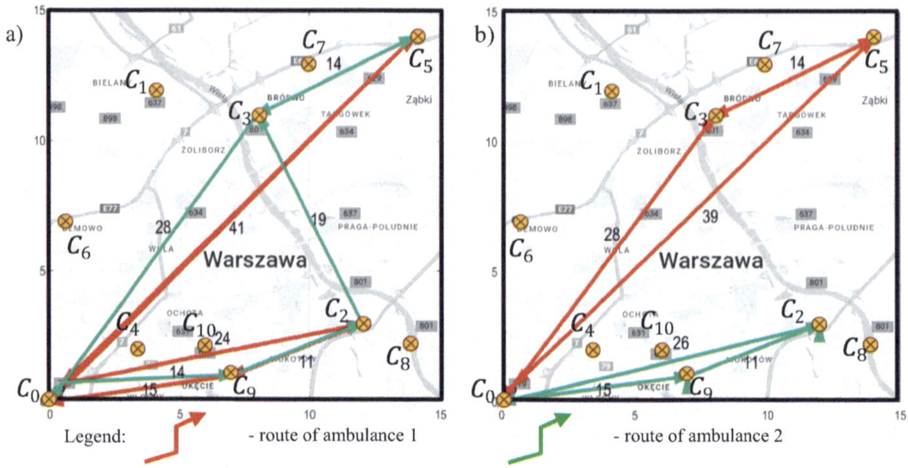

Fig. 3. Ambulance routes: a) following strategy $\mathcal{A}$ where the distance traveled is $D^{(A)}_{h=1,7} = 111.36$ km, b) following strategy $\mathcal{B}$ where $D^{(B)}_{h=1,7} = 100.3$ km

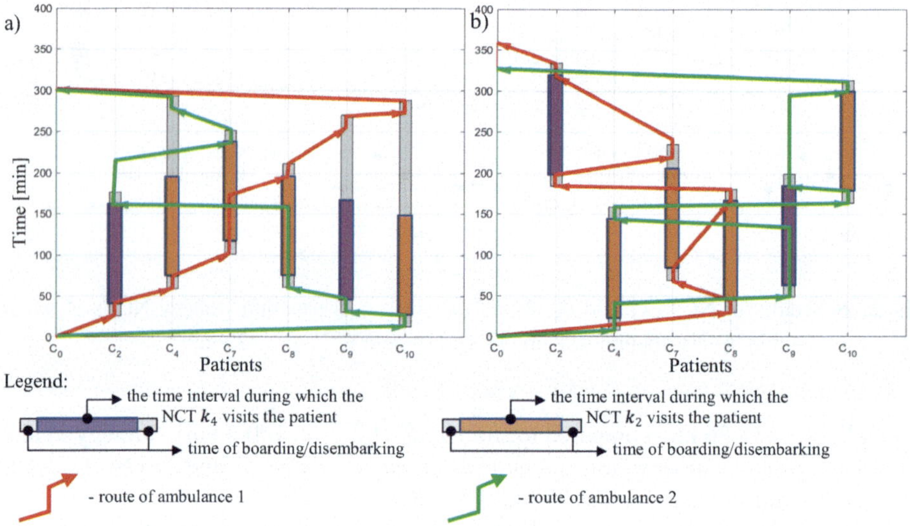

Fig. 4. Schedule of patient visits provided by NCTs k_2 and k_4: a) following strategy $\mathcal{A}$ where $T^{(A)}_{h=3,5} = 316$ min b) following strategy $\mathcal{B}$ where $T^{(B)}_{h=3,5} = 375$ min

mission duration ($T^{(A)}_{h=3,5} = 316$ min) strategy $\mathcal{B}$ ($T^{(B)}_{h=3,5} = 375$). However, it requires a longer distance: $D^{(A)}_{h=3,5} = 104.16$ km compared to the strategy $B : D^{(B)}_{h=3,5} = 95$ km. Strategy $\mathcal{A}$ also requires the use of more teams: 4 k_2 NCTs and 2 k_4 NCTs, compared to strategy $\mathcal{B}$ using 3 k_2 NCTs and $1k_4$) NCTs.

Strategy $\mathcal{A}$ (implemented on days 3 and 5) allows visits to be completed in almost 1 h (59 min) shorter than when using strategy $\mathcal{B}$. However, the use of this strategy is associated with the need to cover more than 9 km more distance and use two additional teams (1 k_2 NCT and 1 k_4 NCT).

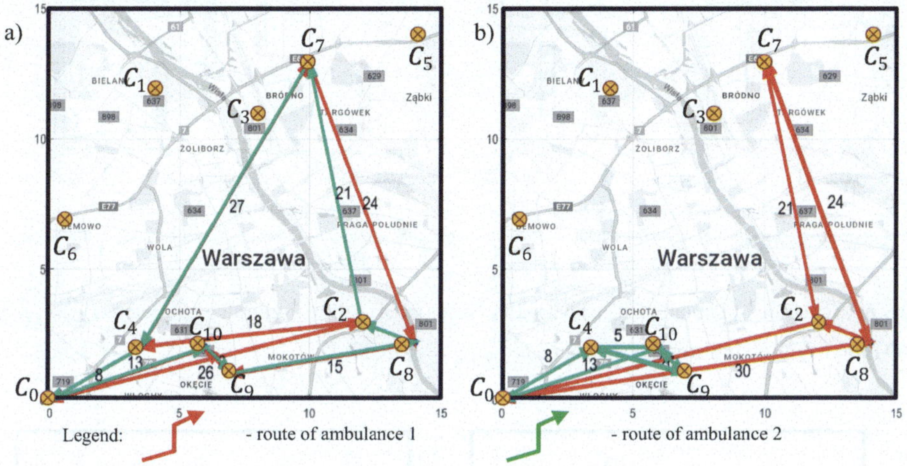

Fig. 5. Ambulance routes: a) following strategy $\mathcal{A}$ where the distance traveled is $D^{(A)}_{h=3,5} = 104.16$ km, b) following strategy $\mathcal{B}$ where $D^{(B)}_{h=3,5} = 95$ km

- For days 2, 4, and 6 of the adopted time horizon in which NCTs k_3 are delivered, the plans of visits carried out according to strategies $\mathcal{A}$ and $\mathcal{B}$ are illustrated in Fig. 6 and Fig. 7, respectively.

In contrast to the previous cases, only strategy $\mathcal{A}$ ensures that visits are completed within a given time window $TH_u = [0, 480]$ min (Fig. 6). The mission duration for Strategy $\mathcal{A}$ is $T^{(A)}_{h=2,4,6} = 460$ min and for strategy $\mathcal{B}$: $T^{(B)}_{h=2,4,6} = 502$ min. Strategy $\mathcal{A}$, however, requires covering more than 27 km of greater distance ($D^{(A)}_{h=2,4,6} = 122{,}4$ km, $D^{(B)}_{h=2,4,6} = 95{,}4$ km). Strategy $\mathcal{A}$ also requires the use of a larger number of teams (5 k_3 NCTs), then strategy $\mathcal{B}$ (4 k_3 NCTs).

The experiments show that strategy $\mathcal{A}$ (assuming the possibility of delivery of the NCT and its collection by different ambulances) allows set plans with a shorter duration of the T_h mission. In the cases considered, the missions assigned to strategy $\mathcal{A}$ have an average of 16% shorter duration. However, this comes at the cost of a longer travel distance (8% higher on average) and the need to use more NCTs. It is worth noting that it is not always possible to set a mission to guarantee timely visits to patients (Fig. 6b). The use of strategy $\mathcal{A}$, which allows for the delivery and reception of NCTs by different ambulances, increases the flexibility of the planned solutions, thus reducing the number of such situations (Fig. 6a).

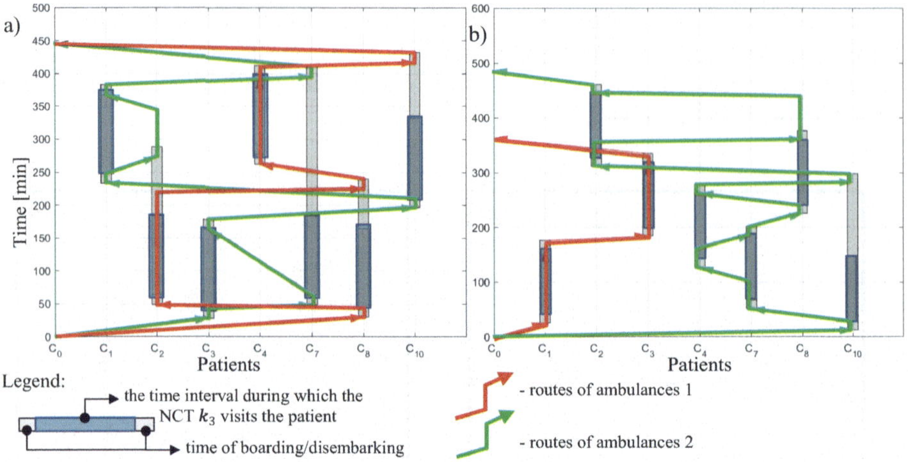

Fig. 6. Schedule of patient visits provided by NCTs k_3: a) following strategy $\mathcal{A}$ where $T^{(A)}_{h=2,4,6} =$ 460 min, b) following strategy $\mathcal{B}$ where $T^{(B)}_{h=2,4,6} = 502$ min

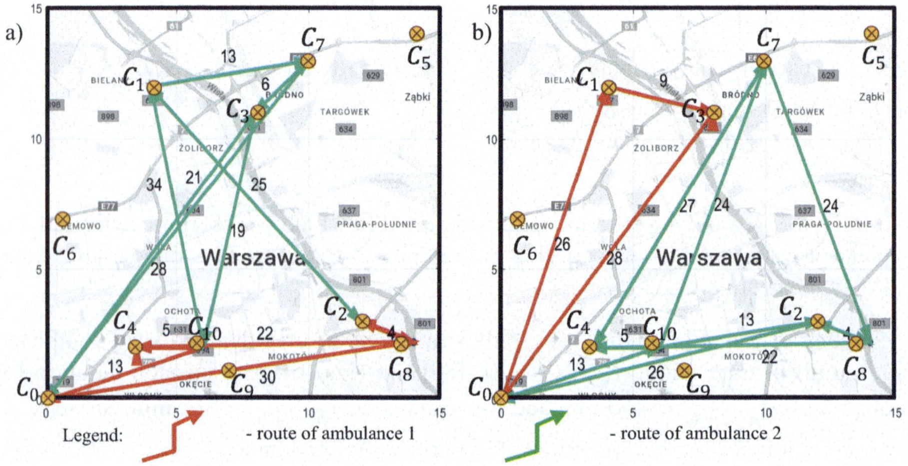

Fig. 7. Ambulance routes: a) following strategy $\mathcal{A}$ where the distance traveled is $D^{(A)}_{h=2,4,6} =$ 104.16 km, b) following strategy $\mathcal{B}$ where $D^{(B)}_{h=2,4,6} = 95$ km

6 Conclusions

This paper proposed the constraints programming model to deal with the VRPPDTW-I problem, considering maximizing the number of patients visited in their ordered time windows. This study's main contribution is developing a framework capable of integrating routing and scheduling of the fleet of vehicles delivering and receiving NCTs while decoupling delivery and pick-up assignments. Further work should extend the model to cases related to the uncertainty of operation times, including service team operation and constraints assumed by soft time windows.

References

1. Akjiratikarl, C., Yenradee, P., Drake, P.R.: PSO-based algorithm for home care worker scheduling in the UK. Comput. Ind. Eng. **53**(4), 559–583 (2007). https://doi.org/10.1016/j.cie.2007.06.002
2. Alade, O.M., Amusat, A.O.: Solving nurse scheduling problem using constraint programming technique. Artif. Intell. (2019). https://doi.org/10.48550/arXiv.1902.01193
3. Braekers, K., Hartl, R.F., Parragh, S.N., Tricoire, F.: A bi-objective home care scheduling problem: analyzing the trade-off between costs and client inconvenience. Eur. J. Oper. Res. **248**(2), 428–443 (2016). https://doi.org/10.1016/j.ejor.2015.07.028
4. Brailsford, S.C., Vissers, J.M.H.: OR in healthcare: a European perspective. Eur. J. Oper. Res. **212**, 223–234 (2011). https://doi.org/10.1016/j.ejor.2010.10.026
5. Cinar, A., Salman, S., Bozkaya, B.: Prioritized single nurse routing and scheduling for home healthcare services. Eur. J. Oper. Res. **289**(3), 867–878 (2021). https://doi.org/10.1016/j.ejor.2019.07.009
6. Cissé, M., Yalçındağ, S., Kergosien, Y., Şahin, E., Lenté, C., Matta, A.: OR problems related to home health care: a review of relevant routing and scheduling problems. Oper. Res. Health Care **13–14**, 1–22 (2017). https://doi.org/10.1016/j.orhc.2017.06.001
7. Djeffal, L., Goncalves, G., Babes, M.: Constraint satisfaction problem based modelling and value ordering based heuristic for nurse scheduling. J. Appl. Comput. Sci. **14**(2), 7–28 (2006). https://doi.org/10.34658/jacs.2006.14.2.7-28
8. Emiliano, W., Telhada, J., do Sameiro Carvalho, M.: Home health care logistics planning: a review and framework. Procedia Manuf. **13**, 948–955 (2017). https://doi.org/10.1016/j.promfg.2017.09.165
9. Erdem, M., Bulkan, S.: A literature review on home healthcare routing and scheduling problem. Eur. J. Health Technol. Assess. **2**(1), 19–32 (2017)
10. Fikar, C., Hirsch, P.: Home health care routing and scheduling: a review. Comput. Oper. Res. **77**, 86–95 (2017). https://doi.org/10.1016/j.cor.2016.07.019
11. Goodarzian, F., Garjan, H.S., Ghasemi, P.: A state-of-the-art review of operation research models and applications in home healthcare. Healthc. Anal. **4** (2023). https://doi.org/10.1016/j.health.2023.100228
12. Hosseinpour-Sarkarizi, A., Davari-Ardakani, H., Izadbakhsh, H.: Home health care routing and scheduling problem under uncertainty considering patient preferences and service desirability. Sci. Iranica E **29**(5), 2647–2669 (2022). https://doi.org/10.24200/sci.2020.54313.3694
13. Lin, C.C., Kang, J.R., Liu, W.Y., Deng, D.J.: Modelling a nurse shift schedule with multiple preference ranks for shifts and days-off. Math. Probl. Eng. (2014). https://doi.org/10.1155/2014/937842
14. Martinez, C., Espinouse, M.-L., Di Mascolo, M.: An exact two-phase approach to re-optimize tours in home care planning. Comput. Oper. Res. **161** (2024). https://doi.org/10.1016/j.cor.2023.106408
15. Méndez-Fernández, I., Lorenzo-Freire, S., González-Rueda, Á.M.: An adaptive large neighbourhood search algorithm for a real-world home care scheduling problem with time windows and dynamic breaks. Comput. Oper. Res. **159** (2023). https://doi.org/10.1016/j.cor.2023.106351

16. Moussavi, S.E., Mahdjoub, M., Grunder, O.: A metaheuristic approach to the integration of worker assignment and vehicle routing problems: application to home healthcare scheduling. Expert Syst. Appl. **125**, 317–332 (2019). https://doi.org/10.1016/j.eswa.2019.02.009
17. Nikzad, E., Bashiri, M., Abbasi, B.: Home healthcare staff dimensioning problem for temporary caregivers: a metaheuristic solution approach. Comput. Oper. Res. **152** (2023). https://doi.org/10.1016/j.cor.2022.106126
18. Santa González, R., Cherkesly, M., Crainic, T.G., Rancourt, M.-E.: Multi-period location routing: an application to the planning of mobile clinic operations in Iraq. Comput. Oper. Res. **159** (2023). https://doi.org/10.1016/j.cor.2023.106288
19. Somar, S., Urazel, B., Sahin, Y.B.: A modified metaheuristic algorithm for a home health care routing problem with health team skill levels. Appl. Soft Comput. **148** (2023). https://doi.org/10.1016/j.asoc.2023.110912
20. Szwarc E., Bocewicz G., Bach-Dąbrowska I., Banaszak Z.: Declarative model of competences assessment robust to personnel absence. In: Damaševičius R., Vasiljevienė G. (eds) ICIST 2019. CCIS, vol. 1078, pp. 12–23. Springer, Cham (2019)
21. Topaloglu, S., Selim, H.: Nurse scheduling using fuzzy modeling approach. Fuzzy Sets Syst. **161**(11), 1543–1563 (2010). https://doi.org/10.1016/j.fss.2009.10.003
22. Wang, R., Shehadeh, K.S., Xie, H., Li, L.: Data-driven integrated home service staffing and capacity planning: Stochastic optimization approaches. Comput. Oper. Res. **159** (2023). https://doi.org/10.1016/j.cor.2023.106348

Applicability of Fractional-Order PID Controllers for Twin Rotor Aerodynamic System Objects

Faisal Saleem(✉) and Józef Wiora

Department of Measurements and Control Systems, Silesian University of Technology, ul. Akademicka 16, 44-100 Gliwice, Poland
faisal.saleem@polsl.pl

Abstract. The availability of two additional tunable parameters compared with Proportional Integral Derivative (PID) controllers makes it possible to incorporate fractional-order (FO) integral/derivative actions. The selection of the optimization algorithm and objective function is important for tuning the fractional-order PID controller. This paper presents the formulation of the FOPID controller design as an optimization problem and its solution using four built-in algorithms in MATLAB. Each algorithm solves the optimization by minimizing two cost functions resulting eight sets of controller parameters. To overcome the sensitivity of algorithms to the initial and boundary values of the parameters, we considered nine different initial and boundary conditions. The proposed method considers the minimum of the optimal values obtained in each case. This provides globally-optimal FOPID controllers. To validate its efficiency, the proposed method uses a Twin Rotor Aerodynamic System (TRAS) as an example. Comparison with a nonlinear approach shows that the proposed approach ensures tracking with minimal and smooth control efforts.

Keywords: Fractional order PID · optimization · twin-rotor · and parameters tuning

1 Introduction

Proportional Integral Derivative (PID) controllers are categorized into Integer Order (IO) and Fractional Order (FO), based on the order of integral and derivative actions. It is commonly acknowledged that a fraction of loops can be tuned using IOPID [1]. The emergence of fractional calculus in the recent era has made use of non-integer modeling and control of practical systems; thus, the use of FOPID has become popular [2, 3].

Different algorithms based on fuzzy neural networks and artificial intelligence have been applied to tune the parameters of FOPID [4–6]. Obtaining the globally-optimal values of tuning parameters is difficult because it depends on the boundary and initial conditions [6]. However, solving the optimization problem using heuristic algorithms is computationally complex. This requires simple and easy-to-implement algorithms to tune FOPID parameters for desired closed-loop performance.

R. Szewczyk et al. (Eds.): AUTOMATION 2024, LNNS 1219, pp. 39–48, 2024.
https://doi.org/10.1007/978-3-031-78266-4_4

This study focuses on the impact of the selection of the cost function and optimization algorithm on tuning the FOPID parameters. Therefore, we applied standard numerical tools for optimization to find solutions that result in better closed-loop performance. Simplicity in use and less computational complexity have convinced us to use built-in algorithms in the optimization toolbox of MATLAB. The built-*in fmincon* and *fminunc* functions allow the solution of constrained and unconstrained optimization problems. These functions can solve optimization problems using any of five different algorithms: Sequential Quadratic Programming (SQP), Interior Point (IP), Active Set (AS), SQP-Legacy (SQPL), and Trust Region Reflective (TRR). We applied the SQP, IP, AS, and SQPL algorithms. Because TRR requires gradients of the objective function [7], we skip this algorithm in our research. A fast nonlinear system, Twin Rotor Aerodynamic Systems (TRAS), has been used as an example for the practicality and demonstration of the proposed tuning approach. For space limitations, the reader is referred to [8] for a detailed study of the algorithms.

The contributions of this study can be summarized as:

- Formulation of the simplest optimization problem for parameter tuning of FOPID controller.
- Validation of built-in, easy-to-use methods for effectiveness by designing FOPID controller for TRAS.
- Performance assessment of FOPID controllers in terms of reference tracking and control efforts when applied to nonlinear system.
- Comparison of the proposed method with a nonlinear controller.

2 Optimization for Parameter Tuning

Tracking error subjected to constraints based on a Linear Time Invariant (LTI) system is considered an objective of this research. The error between a unit step reference signal and the output of the closed-loop system is minimized to obtain the tracking subjected to the constraints of percent overshoot, undershoot, and settling time. The optimization problem can be given by:

$$\min_{k_P, k_I, k_D, \lambda, \mu} \quad e$$

subject to:

$$M_p \leq 0.01 \tag{1}$$

$$M_u \leq 0.01$$

$$t_s \leq 1$$

where M_p, M_u, and t_s denote the percent overshoot, percent undershoot, and settling time of the closed loop system, respectively. Closed loop system, $C(s)$, is:

$$C(s) = \frac{G(s)G_c(s)}{1 + G(s)G_c(s)}, \tag{2}$$

where $G_c(s)$ and $G(s)$ denote controller and system transfer functions, respectively. In addition, e is the error between the reference r and the output y of the closed-loop system. In this work, Integral Absolute Error (IAE) and Root Mean Squared Error (RMSE) are considered as cost functions [9]. Mathematical representations of IAE and RMSE are:

$$E_{IAE} = \frac{1}{N}\sum_{i=1}^{N}|e_i|, \tag{3}$$

$$E_{RMSE} = \frac{1}{N}\sqrt{\sum_{i=1}^{N}(e_i)^2}, \tag{4}$$

here i, and N denote the i-th sample and total number of samples, respectively.

3 Mathematical Modeling of TRAS

TRAS is a laboratory setup widely used in research and education to study control strategies for helicopters, such as trajectory tracking, stabilization, and hovering. Nonlinear equations of motion for the TRAS are [10]:

$$J_\phi \ddot{\psi} = u_\phi + J\dot{\theta}\dot{\psi}\sin(2\theta) - K_\psi \phi - C_\psi \dot{\psi}, \tag{5}$$

$$J_\theta \ddot{\theta} = u_\theta - J\frac{\dot{\psi}^2}{2}\sin(2\theta) - C_\theta \dot{\theta} - K_g g,$$

where

$$J_\psi = \left(m_m l_m^2 + m_t l_t^2\right)\cos^2(\theta) + 2m_{c\omega} l_{c\omega}\sin^2(\theta) + J_z,$$

$$J_\theta = m_m l_m^2 + m_t l_t^2 + 2m_{c\omega} l_{c\omega} + J_x,$$

$$J = m_m l_m^2 + m_t l_t^2 - 2m_{c\omega} l_{c\omega},$$

$$K_g = (m_m l_m - m_t l_t)\cos(\theta) + 2m_{c\omega} l_{c\omega}\sin(\theta).$$

The reader can find the detailed derivation of the mathematical model in [11]. Since this research adopts the nonlinear model of TRAS presented in [10, 11], we use the same model parameters.

We linearized the nonlinear model for the operating points $\theta = \theta_o = 0^o$ and $\psi = \psi_o = 0^o$. After putting the values of TRAS parameters, the transfer function is:

$$G(s) = \begin{bmatrix} \frac{1.246}{s^3+0.922s^2+4.97s+3.917} & 0 \\ \frac{1.481s+0.4233}{s^4+6.34s^3+7.06s^2+2.09s} & \frac{3.6}{s^3+6s+5.01s} \end{bmatrix}, \tag{6}$$

where $G_{11}(s) = \frac{\theta(s)}{u_\theta(s)}$, $G_{12}(s) = \frac{\theta(s)}{u_\psi(s)}$, $G_{21}(s) = \frac{\psi(s)}{u_\theta(s)}$, and $G_{22}(s) = \frac{\psi(s)}{u_\psi(s)}$. $\theta(s)$ and $\psi(s)$ denote pitch and yaw angles in s-domain while $u_\theta(s)$ and $u_\psi(s)$ denote control inputs to the motors operating main and tail rotors, respectively. Transfer function matrix in Eq. (6) advocates the coupling in the linearized model of the TRAS. Therefore, decoupling is required to design Single-Input Single-Output (SISO) controllers for yaw and pitch angles separately. The decoupled transfer function, $G_D(s)$, is obtained as $G_D(s) = D(s) \times G(s)$ where $G(s)$ is the current coupled transfer function while $G_D(s)$ and $D(s)$ are the desired decoupled and decoupler transfer functions, respectively. In simple terms, $D(s)$ is a transformation matrix which can convert $G(s)$ in to a decoupled model.

To design SISO FOPID controllers for pitch and yaw control, the TRAS model should be decoupled as:

$$G_D(s) = \begin{bmatrix} \frac{1.246}{s^3+0.922s^2+4.97s+3.917} & 0 \\ 0 & \frac{3.6}{s^3+6s+5.01s} \end{bmatrix} \tag{7}$$

The decoupled model can be obtained by the transformation matrix $D(s) = G_D(s) \times G(s)^{-1}$. The decoupler transfer function is:

$$D(s) = \begin{bmatrix} 1 & \frac{-1.189s^7-8.567s^6-20.54s^5-49.29s^4-68.88s^3-39.41s^2-6.667s}{s^7+12.34s^6+50.11s^5+76.21s^4+47.91s^3+10.47s^2} \\ 0 & 1 \end{bmatrix} \tag{8}$$

4 Controller Design

In this work, we use the Caputo definition for the fractional derivative and the Riemann-Liouville definition for the fractional integral operator as [6]:

$$I^{\lambda}f(t) = \frac{1}{\Gamma(\lambda)}\int_0^t (t-\tau)^{\lambda-1}f(\tau)d\tau, t > 0, \lambda \in \Re^+ \tag{9}$$

$$D^{\mu}f(t) = \frac{1}{\Gamma(m-\mu)}\int_0^t \frac{f^{(m)}(\tau)}{(t-\tau)^{\mu-m+1}}d\tau, \mu \in \Re^+ \tag{10}$$

where, $m-1 < \mu < m$, $m \in N$ and $t > 0$. I^{λ}, and D^{μ} denote fractional integral and differential operators, respectively. Fractional variables λ and μ denote the order of fractional integration and fractional derivative, respectively. Γ denotes Euler's gamma function [6].

In case of unity feedback control, the mathematical representation of the FOPID controller is:

$$u_k = k_P e(t) + k_I I^{\lambda} e(t) + D^{\mu} e(t) \tag{11}$$

where $e(t) = r(t) - y(t)$. Laplace transform of Eq. (11) considering zero initial conditions results in controller transfer function as [12]:

$$\frac{U_k(s)}{F(s)} = G_c(s) = k_P + k_I \frac{1}{s^{\lambda}} + k_D s^{\mu} \tag{12}$$

where k_P, k_I, λ, k_D and μ denote proportional gain, integral gain, order of the integral action, derivative gain, and order of the derivative action, respectively.

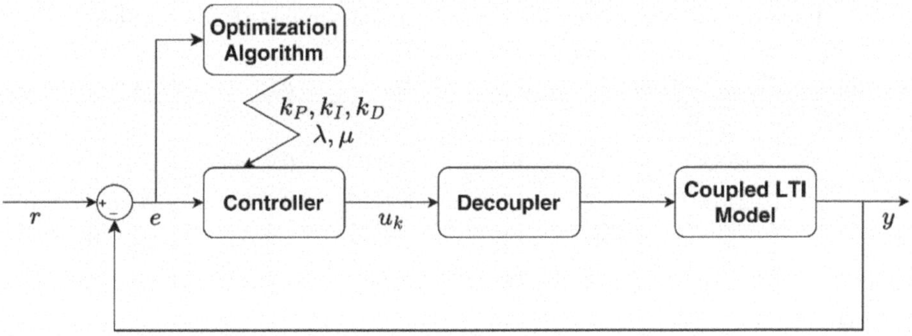

Fig. 1. Online configuration for FOPID controller design

We solved Eq. (1) to obtain two SISO FOPID controllers for the decoupled transfer functions of the TRAS given in Eq. (7). The online configuration for designing FOPID controller is:

Coupled LTI model is the MIMO model of TRAS. Decoupler converts the MIMO model to two SISO transfer functions for pitch and yaw control of the TRAS. Online configuration uses both SISO models turn by turn to design SISO FOPID controllers. Optimization algorithm takes error of the closed-loop system (e) as an input and generates controller gains which are provided to the controller. Controller takes these parameters and generates FOPID model. Since Fig. 1 represents online configuration, we use this diagram only to design controller with unit step as a reference input (r).

With four optimization algorithms (AS, SQP, SQP-L, IP) and two cost functions, we obtained eight different sets of controller parameters. We considered nine different random initial conditions; $0 \leq X_i \leq 1, X_i \in \mathfrak{R}^{5\times 1}$ (where X_i contains values for controller parameters) for nine different boundary values of k_P, k_I, and k_D. Lower boundary values for each case were 0 while the upper boundary B_u is from:

$$B_u = \{600, 800, 1000, 1200, 1800, 2000, 2400, 3000, 3500\}.$$

Boundary values of λ and μ were fixed to $0 \leq \lambda, \mu \leq 1$ for each case. Numerical values of controller gains and optimization results using IAE for pitch and yaw angles are given in Table 1.

It shows that the values of controller parameters for pitch and yaw controllers with different algorithms are comparable. The minimal difference is due to different results obtained using individual algorithms by fixing number of iterations. Closeness of values between the parameters obtained with SQP and SQPL is due to similarity in the way of implementation of these algorithms [8].

Numerical values of controller gains and optimization results using RMSE for pitch and yaw angles are given in Table 2. Although optimization of RMSE results in similar values, the controller parameters for notably different for each algorithm. This change in controller parameters is due to the non-convexity of RMSE [8]. Therefore, while the optimal value is the same, we obtained different parameters which satisfy the design criteria of Eq. (1). Since the objective function is non-convex, parameter values are sensitive to both initial and boundary conditions.

Table 1. Controller parameters and optimal values using IAE cost function

Algorithm	k_P	k_I	λ	k_D	μ	Optimal Value
Pitch angle						
AS	1.6253	8.0527	0.7686	8.6808	0.9548	0.90×10^{-4}
IP	2.2709	7.2444	0.7744	8.1717	0.9380	1.00×10^{-4}
SQP	3.7517	6.3580	0.7923	7.8453	1.0000	0.97×10^{-4}
SQPL	3.7337	6.2350	0.8241	7.8973	1.0000	0.92×10^{-4}
Yaw angle						
AS	2.3206	2.4440	0.3463	6.8664	0.9960	1.00×10^{-4}
IP	2.6101	0.6675	0.2190	4.8659	0.9488	1.00×10^{-4}
SQP	2.7235	1.6224	0.1590	6.4496	1.0000	0.50×10^{-4}
SQPL	1.5727	0.8257	0.2468	4.0187	0.8812	0.56×10^{-4}

Table 2. Controller parameters and optimal values using RMSE cost function

Algorithm	k_P	k_I	λ	k_D	μ	Optimal Value
Pitch angle						
AS	0.0038	1.5761	0.9086	1.5636	0.9950	1.00×10^{-4}
IP	0.1988	12.3581	0.8334	9.4795	0.9691	1.00×10^{-4}
SQP	0.2770	1.5800	0.9021	2.1465	1.0000	1.00×10^{-4}
SQPL	0.1409	1.9089	0.8844	2.7920	0.8900	1.00×10^{-4}
Yaw angle						
AS	0.0062	3.2317	0.5312	22.8939	1.0000	0.99×10^{-4}
IP	0.1972	39.8228	0.9250	26.9522	1.0000	0.99×10^{-4}
SQP	0.6314	3.8441	0.1828	7.6087	0.9964	0.99×10^{-4}
SQPL	2.6435	1.5124	0.1612	6.4915	1.0000	0.99×10^{-4}

5 Controller Performance

We formulated the closed-loop shown in Fig. 1 by replacing the LTI model with the nonlinear TRAS model in Eq. (5). The decoupler is given in Eq. (8) while the controller transfer function $K(s)$ is:

$$K(s) = \begin{bmatrix} G_{cp}(s) & 0 \\ 0 & G_{cy} \end{bmatrix}. \tag{13}$$

$G_{cp}(s)$ and G_{cy} are the controller transfer functions for pitch and yaw angles obtained by Eq. (12). To evaluate the closed-loop performance in different scenarios, we applied the following reference trajectory:

- Yaw and pitch angles are kept to 0° for first 6 s.
- From 6 to 10 s, yaw angle is 0° while pitch angle linearly increases and becomes constant to 13.8° at time $t = 10$ s.
- Pitch angle is fixed from $t = 10$ s to $t = 20$ s while yaw angle linearly increases and gains a constant value of 57° at $t = 20$ s.
- Yaw angle is then fixed for next 10 s while pitch angle linearly increases and gains a value of 31° at $t = 30$ s.
- For next 10 s, pitch angle is fixed and yaw angle in linearly increased to 86°. It is then kept to this value until $t = 80$ s.
- At $t = 40$ s, the value of pitch angle is linearly decreased and it reaches to 19.5° at $t = 50$ s where it is kept fixed until $t = 70$ s. Then it is further decreased until it reaches to 14° at $t = 80$ s.
- After $t = 80$ s, yaw angle is decreased until it reaches the value 57.5° at $t = 90$ s and then it constant for the rest of the simulation time.

Figure 2 shows that FOPID controller obtained after solving IAE using IP for pitch angle control shows worst tracking. The remaining controllers present a comparable performance. The zoomed version of the performance shows that pitch angle controllers results in oscillations in the main rotor of TRAS. The magnitude of oscillations with FOPID controller obtained through AS algorithm is minimum. Therefore to avoid wear-and-tear over time in the real-time system, this control should be selected. On the other hand, the zoomed version of the yaw response shows tracking error for each algorithm which is minimum for the case of AS.

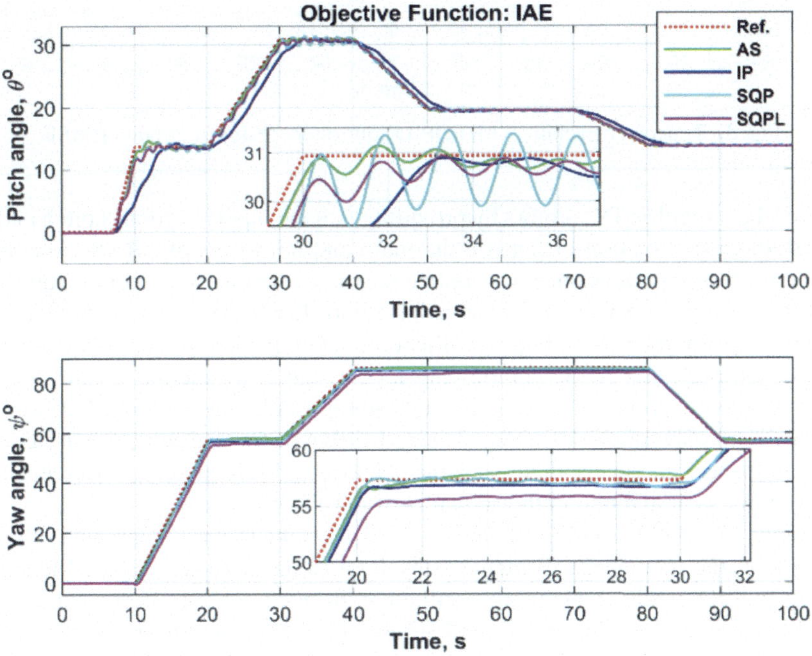

Fig. 2. Trajectory tracking with FOPID controllers designed through IAE

Figure 3 shows that AS-based FOPID controller for the pitch angle response outperforms the others. Although IP-based controller has minimal oscillation in pitch control, it has a considerable amount of time delay in reference tracking. Yaw angle tracking is comparable for IP, SQP, and SQPL based controllers. The zoomed view of yaw response in Fig. 3 shows that AS outperforms the others in terms of tracking error and time delay.

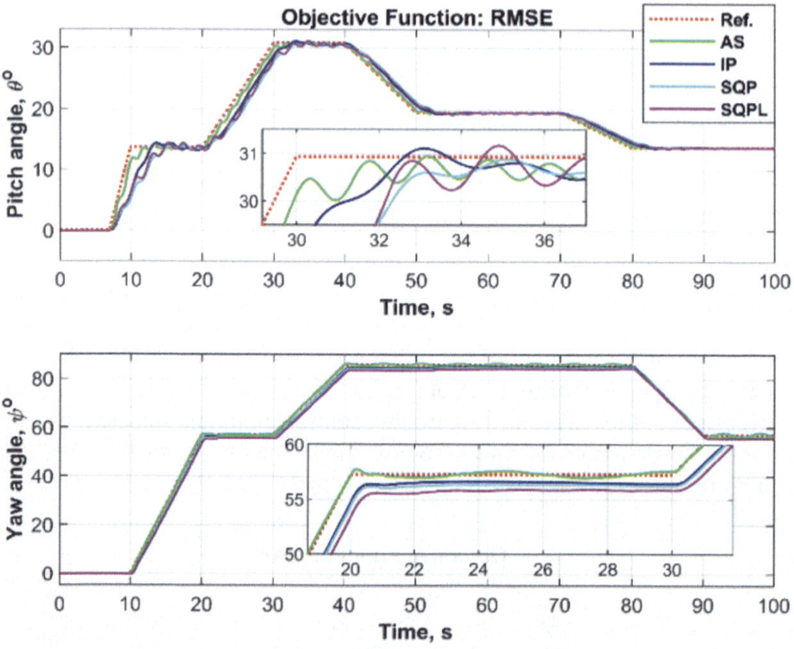

Fig. 3. Trajectory tracking with FOPID controllers designed through RMSE

Controllers regulate DC motors by providing voltage inputs to control pitch and yaw angles through the rotation of fans on the main and tail rotors of TRAS. To compare control effort based performance, we applied the same reference trajectory to the closed-loop system using the Adaptive Backstepping Controller (ABSC) [11] and recorded the corresponding control efforts. Figure 4 illustrates a comparison of control efforts of the proposed approach and ABSC, employing IAE as the cost function for the proposed method. While the proposed method exhibits some oscillations in control efforts, their frequency is notably smaller than that observed with ABSC. This reduction in frequency directly influences energy oscillations in the plant input. Minimal oscillations are crucial for preventing wear and tear in the actuators of the TRAS rotors.

Notably, the SQP-based FOPID in Fig. 4 displays the highest magnitude of control effort among the methods, accompanied by high-frequency oscillations. Meanwhile, IP-based FOPID, while showing no oscillations in pitch control efforts, exhibits sub-optimal tracking performance (Fig. 2). AS-based FOPID proves most suitable for pitch control using IAE (Fig. 2), demonstrating minimal oscillation magnitudes in control effort (Fig. 4). For yaw control, IP-based FOPID is suitable based on its lowest magnitude and frequency of oscillation in control effort (Fig. 4) and comparable trajectory tracking in (Fig. 3).

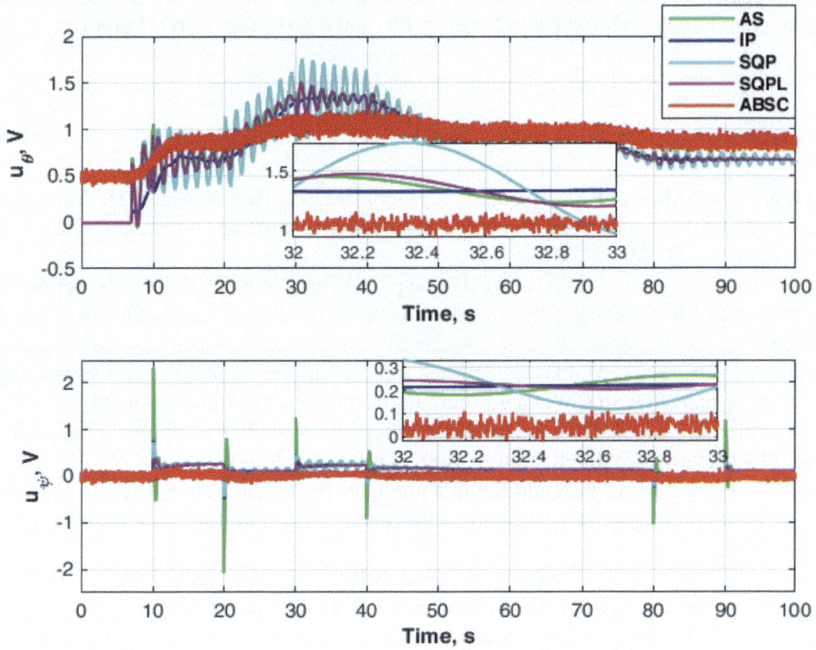

Fig. 4. Comparison of proposed approach with ABSC

6 Conclusion

In this work, we addressed the FOPID controller design problem by formulating it as an optimization task, considering maximum-peak-overshoot, maximum-peak-undershoot, and settling time constraints. Employing four optimization algorithms, we optimized two objective functions. The resulting controller parameters from IAE yielded comparable values, while those from RMSE varied significantly because of its sensitivity to initial and boundary values. We applied the proposed approach on TRAS. The investigation reveals that an AS-based FOPID controller, optimized by IAE, excels in reference trajectory tracking with minimal high-frequency oscillations in control effort. On the other hand, controller gains obtained with RMSE are far from each other. Notably, the IP-based FOPID controller demonstrated favorable control effort attributes.

Comparison of control efforts with a nonlinear control design method shows that the proposed approach has smaller magnitude and frequency. This reduction in frequency directly influences energy oscillations in the input of TRAS which minimizes wear and tear of the actuators.

Practical implementation of the proposed method requires integer-order (IO) approximation of the FOPID controllers, which introduces modeling errors. In future, we will apply the IO approximation approaches which have minimal modeling errors.

Acknowledgement. This research was funded by the Polish Ministry of Science and Higher Education, grant number 02/050/BKM23/0034 (FS) and 02/050/BK-23/0032 (JW).

References

1. Tepljakov, A., Alagoz, B.B., Yeroglu, C., Gonzalez, E., HosseinNia, S.H., Petlenkov, E.: FOPID controllers and their industrial applications: a survey of recent results. IFAC-PapersOnLine. **51**, 25–30 (2018)
2. Żegleń-Włodarczyk, J.: FOPID and PID-comparison of control quality and execution time on the example of two rotor aerodynamical system. In: Conference on Automation, pp. 29–38 (2023)
3. Żegleń-Włodarczyk, J., Wajda, K.: PID and fopid controllers combinations during control of 3D crane optimized with GWO algorithm. In: Conference on Automation, pp. 54–63 (2022)
4. Kanagaraj, N.: An adaptive neuro-fuzzy inference system to improve fractional order controller performance. Intell. Autom. Soft Comput. **35** (2023)
5. Mok, R., Ahmad, M.A.: Fast and optimal tuning of fractional order PID controller for AVR system based on memorizable-smoothed functional algorithm. Eng. Sci. Technol. Int. J. **35**, 101264 (2022)
6. Pradhan, R., Majhi, S.K., Pradhan, J.K., Pati, B.B.: Optimal fractional order PID controller design using ant lion optimizer. Ain Shams Eng. J. **11**, 281–291 (2020)
7. Moubarak, P.M.: Trust-region reflective adaptive controller for time varying systems. IET Control Theory Appl. **9**, 240–247 (2015)
8. Nocedal, J., Wright, S.J.: Numerical Optimization. Springer, Heidelberg (1999)
9. Bingül, Z., Karahan, O.: Fractional PID controllers tuned by evolutionary algorithms for robot trajectory control. Turk. J. Electr. Eng. Comput. Sci. **20**, 1123–1136 (2012)
10. Saleem, F., Ali, A., Shaikh, I.U.H., Wasim, M.: A hybrid H infinity control based ILC design approach for trajectory tracking of a twin rotor aerodynamic system. Mehran. Univ. Res. J. Eng. Technol. **40**, 169–179 (2021)
11. Butt, S.S., Sun, H., Aschemann, H.: Comparison of backstepping-based sliding mode and adaptive backstepping for a robust control of a twin rotor helicopter. In: Variable-Structure Approaches: Analysis, Simulation, Robust Control and Estimation of Uncertain Dynamic Processes, pp. 3–30 (2016)
12. Wiora, J., Wiora, A.: Influence of methods approximating fractional-order differentiation on the output signal illustrated by three variants of Oustaloup filter. Symmetry (Basel) **12**, 1898 (2020)

Employing Generative Artificial Intelligence in Replacement of Traditional Backend Systems

Artur Moroz(✉), Illia Solohubov, Andrii Oliinyk, Sergey Subbotin, and Stepan Skrupsky

National University Zaporizhzhya Politechnic, Zaporizhzhya, Ukraine
arthur.official.moroz@gmail.com

Abstract. This paper investigates the application and benefits of using Generative Artificial Intelligence (GenAI) techniques, with an emphasis on large language models (LLMs) as an innovative approach in the development of information systems. The focus lies primarily on systems employing open datasets where data confidentiality is not a focal concern. By replacing the traditional programmed logic with these advanced AI models, this paper presents the measurable improvements in the efficiency and adaptability of system design, development, and maintenance. The key advantages discussed include an enhanced rate and accuracy in query processing and response provision. Despite accentuating these substantial benefits, the paper acknowledges potential limitations with the alignment of these models to certain user-specific needs.

A noteworthy component of this study includes an in-depth exploration of Large Language Models usage and its implications in system interactivity and personalisation realms. The results suggest that the implementation of LLMs not only ensures precise and contextually relevant responses to user queries, but it also contributes to a significant reduction in time and resources expended during system design and development phases. Thus, replacing standard programmed logic with LLMs in systems utilizing open datasets accelerates the development timeline, simplifies maintenance procedures and boost the overall dynamism of the system.

Keywords: Generative Artificial Intelligence · Large Language Models · Information System Development · Open Datasets Systems · Query Processing · Response Provision · System Efficiency and Dynamism

1 Introduction

The development of web systems, a task of considerable complexity, is streamlined by integrating Large language Models trained on extensive datasets. These networks excel in data analysis, classification, and content generation, offering a more flexible and efficient alternative to traditional manual methods. They can adapt to diverse project requirements, though they may not always align perfectly with specific user needs [1].

A key innovation in this field is the application of Generative Artificial Intelligence (GenAI) techniques, with an emphasis on Large Language Models (LLMs) models like

R. Szewczyk et al. (Eds.): AUTOMATION 2024, LNNS 1219, pp. 49–58, 2024.
https://doi.org/10.1007/978-3-031-78266-4_5

GPT [2, 3], which stands out for its exceptional content creation and processing abilities. It supports a range of functions, from algorithm optimization to literary composition, and is invaluable across various sectors. Its public sector availability marks a significant advancement in the practical application of artificial intelligence.

The release of OpenAI's language model specifications and API marks a significant stride, facilitating the incorporation of this model into diverse systems. This bypasses the need for creating bespoke LLM's, a process often fraught with time and complexity. Such integration bolsters the practicality of artificial intelligence, paving the way for new solutions to complex problems traditionally requiring extensive research and human resources [4, 5].

2 Literature Review

The integration of Large Language Models across various sectors holds significant relevance due to its capacity to transform information processing, enhance process optimization, and foster innovative solutions. In the context of current artificial intelligence and automation trends, Large Language Models marks a notable advancement in natural language processing, potentially revolutionizing data handling [4].

Utilizing Large Language Models enables efficient use of neural network capabilities for extensive text analysis, pattern identification, response generation to queries, and creating new content. This method greatly accelerates information processing, offering a marked reduction in time and resources compared to conventional data processing techniques [5, 6].

Moreover, Large Language Models' role in augmenting databases with neural network-generated responses offers an alternative to expensive and time-intensive expert consultations. This facilitates gathering, structuring, and analysing domain-specific knowledge without direct expert input, potentially lowering costs and easing access to quality information.

Direct access to trained neural networks via Large Language Models also presents opportunities for refining existing algorithms and developing new, optimized solutions. This use of cutting-edge technology can drive innovation, increase productivity, and introduce novel methodologies across various sectors, including business, education, and research [7].

The significance of incorporating LLM into modern data processing and decision-making systems is underscored by its considerable advantages in speed, flexibility, and efficiency, compared to traditional approaches.

Utilising LLM further democratizes access to artificial intelligence, extending its benefits to small and medium-sized enterprises, start-ups, educational institutions, and individual users. This access was previously limited to large corporations.

In addition, LLM serves as a potent tool for enhancing the quality and precision of data analysis. This is especially vital in sectors like medicine, finance, and research, where decisions based on large data volumes must be both rapid and accurate.

Integrating LLM into automation and robotics opens new prospects for smart technology development, significantly bolstering the intellectual functions of robotic systems and offering more adaptable and flexible solutions.

LLM also paves the way for personalized and individualized customer interactions, creating new opportunities in marketing and CRM systems. This potential to improve customer satisfaction and communication efficacy is noteworthy.

In the realm of educational technologies, GenAI's role is transformative, providing adaptive learning systems that cater to individual student needs and preferences. Its deployment heralds a new era of innovation and enhanced efficiency across various fields, establishing it as a key component in an ever-evolving technological landscape.

Finally, the application of GenAI in web system development opens up new avenues for developers, enhancing productivity and resource efficiency. This marks a new phase in web technologies where innovation and optimal performance are paramount, emphasizing the need for ongoing research into LLMs' benefits over traditional programming methods, particularly in the development of complex web and information systems.

3 Approach to Employing GenAI in Replacement of Traditional Backend Systems for Enhanced Efficiency and Dynamism

Generative Artificial Intelligence, simultaneously possess the advantage and drawback of providing a multitude of response options based on queries. This necessitates the clear definition of tasks and response formats for their correct and consistent use. Addressing this challenge involves creating a query template or base message, structuring the query and defining parameters to be considered in generating a response. For instance, Fig. 1 illustrates a query template for searching countries based on specific parameters, where parameters are added in line after the task formulation, aiding the model in precisely understanding the required information and its desired format.

```
Chat, I need you to perform the following tasks without any interpretation.
I will give you a list of task parameters and you will return single array of ISO
countries with no other text or desctiptionalways excluding RU, RB. Give answer
only as array . If answer is null - return null or empty array. If you don't have data
for any of the parameters - use archived data.
Task: what are the countries
```

Fig. 1. Example of a basic template for searching countries by criteria

Developing an effective query template requires a deep understanding of both the model and the query context. It's crucial to articulate questions clearly and consider possible model limitations to ensure the appropriateness and accuracy of the responses. This may involve specifying categories, refining linguistic structures, or defining key terms to be used in the response.

Correct use of the message template is fundamental for effective GenAI operation, particularly in natural language processing. It not only ensures response accuracy but also helps avoid misunderstandings or ambiguities arising from the response variability offered by modern GenAI.

Figure 2 demonstrates a query for identifying countries near specified coordinates with temperatures above 50 degrees Celsius. The corresponding countries include Kazakhstan, Azerbaijan, Iran, Saudi Arabia, Iraq, Kuwait, and the United Arab Emirates, each meeting the set conditions.

```
Task: what are the countries
nearest to latitude: 47.90, longitude: 35.14.
and temperature is higher 50 Celsius

["KZ", "AZ", "IR", "SA", "IQ", "KW", "AE"]
```

Fig. 2. Example of a query for identifying countries by two parameters

Figure 3 shows a similar query but uses a different temperature range – from 20 to 50 degrees Celsius. According to this query, the identified countries are mostly located in Europe, the Middle East, and Africa.

```
Task: what countries are
nearest to latitude: 47.90, longitude: 35.14.
and temperature is between 20 and 50 Celsius

["UA", "BY", "MD", "RO", "TR", "GR", "AL", "MK", "RS", "BG", "HR", "SI", "HU", "SK",
"CZ", "PL", "LI", "AT", "CH", "DE", "LU", "BE", "NL", "FR", "MC", "ES", "PT", "MA",
"DZ", "TN", "LY"]
```

Fig. 3. Example of a query for identifying countries by two parameters

Figure 4 illustrates a query expanded with an additional condition of using the euro as the local currency. As a result, most European countries from the second list meet the new criterion, while countries from the Middle East, North Africa, and Ukraine are excluded.

```
Task: what countries are
nearest to latitude: 47.90, longitude: 35.14.
and temperature is between 20 and 50 Celsius
and currency is Euro

["DE", "FR", "BE", "LU", "NL", "ES", "PT", "IT", "GR", "AT", "SK", "SI", "ME", "EE", "LV",
"LT", "IE", "CY", "MT", "FI"]
```

Fig. 4. Example of a query for identifying countries by three parameters

These examples highlight the importance of precisely defining query parameters when working with GenAI, as even minor changes can significantly impact the outcome. They also underline the adaptability of GenAI, which allows for flexible responses to various queries, catering to specific user needs.

Figure 5 depicts a basic query with a single criterion – temperature between 20 and 50 degrees Celsius. The absence of a condition for identifying nearby countries leads to a significantly expanded list, including countries worldwide. This demonstrates how omitting or modifying individual criteria can broaden the scope and variety of data responses.

```
Task: what countries are
temperature is between 20 and 50 Celsius

["AF", "DZ", "AO", "AR", "AU", "BD", "BF", "BJ", "BO", "BR", "BT", "BW", "CF", "CG",
"CH", "CI", "CL", "CM", "CN", "CO", "CR", "CU", "DJ", "DO", "DZ", "EC", "EG", "ER",
"ET", "GA", "GH", "GM", "GN", "GQ", "GT", "GW", "GY", "HN", "ID", "IN", "IQ", "IR",
"JM", "JO", "KE", "KH", "KM", "KW", "LA", "LB", "LR", "LY", "MA", "MG", "ML", "MM",
"MR", "MT", "MX", "MY", "MZ", "NA", "NE", "NG", "NI", "NP", "OM", "PA", "PE", "PH",
"PK", "PR", "PS", "PY", "QA", "SA", "SD", "SG", "SL", "SN", "SO", "SS", "SV", "SY",
"TD", "TG", "TH", "TN", "TT", "TW", "TZ", "UG", "US", "VE", "VN", "YE", "ZA", "ZM",
"ZW"]
```

Fig. 5. Example of a query for identifying countries by one parameter

Figure 6 shows the introduction of a new parameter – climate type. Integrating such a parameter into an existing system would require adding this new parameter to the database and updating each record, either through gathering information from other sources or manually [8]. Typically, in programming situations, a programmer could create not just a Boolean field for tropical climate but also a climate field for entering specific textual values, potentially leading to errors due to the lack of strict typing.

```
Task: what countries are
temperature is between 20 and 50 Celsius
and climate is tropical

["BR", "ID", "CD", "PH", "CO", "MY", "VE", "PE", "MM", "NG", "TH", "VN", "MG", "EC",
"GH", "CI", "PG", "CM", "UG", "TM", "LA", "PY", "SN", "MZ", "LK", "KH", "BD", "LR",
"DO", "HT", "CR", "BS", "FJ", "GY", "PA", "TT", "JM", "MU", "BW", "SL", "GW", "GM",
"ET", "ML", "GA", "NE", "MR", "BF", "DJ", "GN", "SZ", "SO", "GW", "TL", "CV", "SC",
"ST", "BB", "AG", "LC", "VC", "GD", "DM", "BN", "SG", "MU", "GY", "BB", "GD", "VC",
"DM", "LC", "AG", "KN", "TT", "SR", "BZ", "JM"]
```

Fig. 6. Example of a query for identifying countries with a new parameter

The application of GPT offers the ability to add new parameters in any order, forming an easily modifiable and adaptable system that does not require extensive and complex changes, enhancements, or error corrections. This simplifies the data handling process, ensuring flexibility and efficiency in response to changing conditions and needs. Thus, GPT opens up new possibilities for the development and improvement of information systems that readily adapt to new requirements and circumstances.

Figures 7 and 8 illustrate the example of introducing a parameter unforeseen at the system design stage. In this case, it pertains to a range of years during which terrorist attacks did or did not occur.

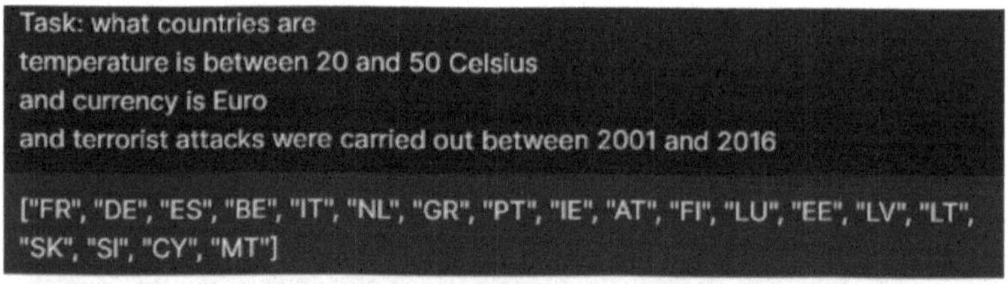

Fig. 7. Example of a query for identifying countries with unpredictable parameters

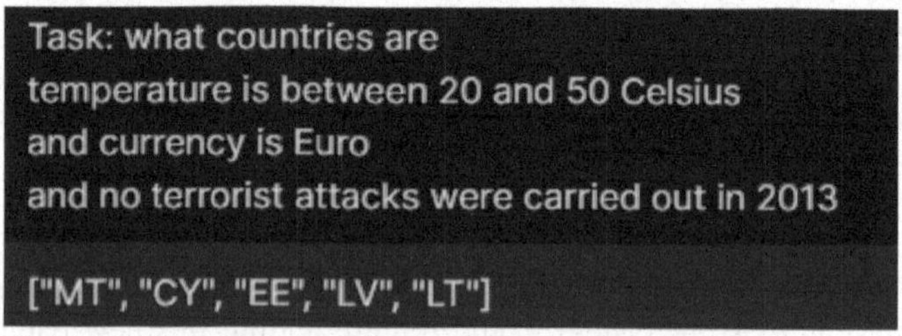

Fig. 8. Example of a query for identifying countries with unpredictable parameters

Analysing the model's responses reveals its aptitude to adequately react to the introduction of such a parameter, returning a list of countries that indeed meet these criteria. This demonstrates the flexibility and adaptability of systems based on GenAI in responding to unforeseen demands or parameters.

The application of such an approach can have significant advantages in various fields, particularly in security, data analytics, social research, and many others. It allows systems to quickly adapt to new challenges and changes in requirements without necessitating deep redesign or complete reconfiguration. This capability makes GenAI, like GPT, indispensable tools for modern dynamic and changing conditions, where response speed and system flexibility are key success factors.

This also opens avenues for deeper analysis and understanding of complex issues, as the ability to include diverse and specific parameters can help uncover new connections and insights previously inaccessible with more traditional data processing approaches. Thus, the use of GenAI like GPT not only ensures adaptability and flexibility but also depth and richness in analytical capabilities.

4 Analysis of Generative Artificial Intelligence for Providing User Responses

In the previous examples, an approach for end-user interaction with a GenAI through well-formulated automated queries was demonstrated. A key feature of this interaction is the use of unique parameters in each query, indicating the impracticality of predicting and recording all potential responses in a database. This feature underscores that the queries were aimed at satisfying end-users' needs rather than collecting and analysing data for internal use.

This leads to the conclusion that a feasible system can operate without a large content storage database, relying only on systematic user information. Such an approach significantly simplifies and accelerates system deployment, as all necessary data are provided by an external service, acting as the server-side with all required data.

This GenAI operation model is crucial for developing and implementing modern information systems oriented towards fulfilling specific user requests. It ensures a high level of flexibility and adaptability, allowing the system to quickly respond to changing user needs without the constant updating of a large database. Thus, using GenAI in this context opens new possibilities for efficient data management and user interaction, offering solutions that can be quickly adapted to various usage scenarios.

To estimate the man-hours required to create a standard web-based tourist guide system with the described functionalities, one must consider a comprehensive development approach encompassing requirement analysis, system design, development, testing, implementation, and database compilation.

Before presenting the detailed breakdown of the man-hour estimates for the various stages of system development, it is pertinent to acknowledge that these figures are approximate. They have been derived from an analytical study of the development timelines of analogous web systems. This estimation process involved a meticulous examination of comparable projects, allowing for a nuanced understanding of the time commitments typically required at each stage. By leveraging this empirical data, the estimates provided offer a comprehensive, albeit approximate, guide to the expected man-hour investment. It is crucial to consider these figures as indicative, providing a framework for planning and resource allocation rather than definitive metrics. Such an approach ensures a pragmatic and informed basis for anticipating the developmental demands of similar web-based systems.

- The requirement analysis process involves defining the system's functional and non-functional requirements, including the collection of requirements for each query and analysis of potential data sources, requiring approximately 40–60 man-hours.
- System design entails developing the architecture, choosing technologies, and designing the user interface, which can take 80 to 120 man-hours.
- Development, including creating the database, server logic, and client part, demands significant effort, estimated at 400–600 man-hours.
- Testing ensures system reliability and correctness, requiring around 100–150 man-hours.
- Implementation and maintenance, including server deployment and monitoring, take about 50–100 man-hours.
- Separately, database compilation, which may involve manual data entry or automated parsing from other sources, can vary widely depending on the data volume and availability, requiring an additional 200–400 man-hours.

Thus, the estimated time frame for the creation of such a system ranges between 1000 to 1400 man-hours, conditionally, contingent upon a variety of factors including the complexity of requirements, the team's experience, and the technologies employed.

Utilising GenAI as the cornerstone for a web-based tourist guide system, as delineated previously, can markedly revolutionise the methodology employed in the development and operation of such systems. The primary advantage of this approach lies in the diminution of the structure and quantum of work conventionally required.

Incorporating a GenAI bestows upon the system an enhanced capability for the dynamic processing of user queries, thereby considerably streamlining the process of requirements analysis. Notably, the conditions under which users formulate their queries become increasingly malleable. This automated flexibility can potentially reduce the time allocated to requirements analysis by half, as the majority of queries and their processing are dynamically adapted by the system.

Furthermore, the system design phase experiences substantial alterations. The necessity to develop only the 'systemic' components of the system could lead to a reduction in design time by a factor of 3 to 4, enabling a more concentrated effort on the implementation of pivotal functionalities and saving considerable time on the intricacies of content processing.

The expeditious development of a comprehensive system is a further benefit. As the GenAI undertakes the bulk of the content-related tasks, the time required for development may be curtailed by approximately 12 to 15 times in comparison to traditional methodologies. This implies that the most labour-intensive elements of content work, such as analysis, sorting, and presentation, are executed through automated systems underpinned by artificial intelligence.

While testing remains a crucial aspect of the development process, it too can undergo simplification. Given that 80 to 95 percent of the functionality is furnished by an external service, the primary focus of testing pivots towards the integration of the system with the GenAI and the interaction with users.

In terms of implementation and maintenance, although the overall effort may not be substantially diminished, the incorporation of dynamic parameters and the reliance on the LLM's knowledge base, as opposed to a conventional database, reduces the necessity for continuous content updates and modifications.

A particular time-saving is achieved by eliminating the need to populate a proprietary database. In this case, the content database is part of the LLM's knowledge base, which already contains a large amount of information. All these indicators are summarized in Table 1.

Having analysed Table 1, one can extrapolate time expenditure comparison graphs over time for both approaches using TableCurve2D. These graphs are depicted in Fig. 9.

Consequently, the total time required for the development of a full-fledged system can decrease from the usual 900–1400 man-hours required using standard programmed logic to 150–300 man-hours by utilising the capabilities of Generative Artificial Intelligence. This corresponds to an acceleration of the system development process by 5–7 times.

Table 1. Comparative characteristics of design stages for two approaches (in man-hours).

Stages of System Development	Standard approach	Utilising Generative AI
Requirement Analysis	40–60	20–30
System Design	80–120	20–40
System Development	400–600	50–100
Testing	100–150	30–60
Implementation and Support	50–100	30–50
Database Compilation	200–400	0
Total	900–1400	150–300

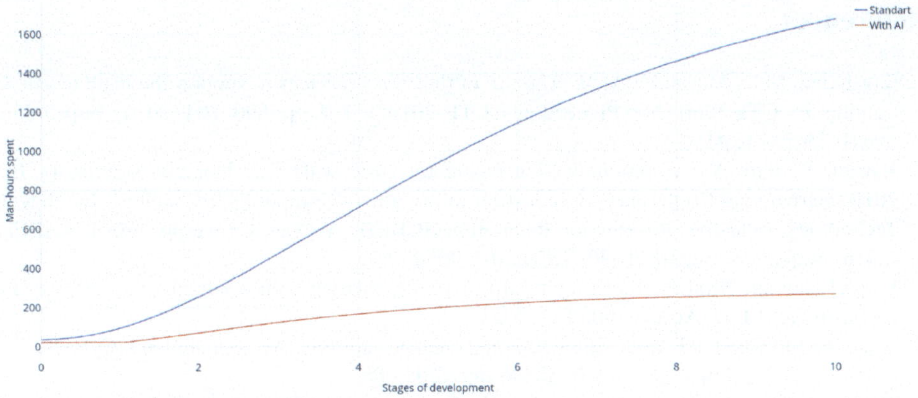

Fig. 9. Extrapolated Graphs Comparing Time Expenditure Over Time for Both Approaches

5 Conclusions and Further Researches

This research emphasises the crucial role of Generative Artificial Intelligence, especially Large Language models such as GPT, in projects where data confidentiality is not of primary importance. Utilising this technology significantly improves the efficiency of query processing and accelerates the provision of relevant responses, thereby enhancing system productivity and facilitating the swift dissemination of vital information to users.

Incorporating GenAI into information system development enhances both the dynamism and adaptability of these systems, leading to an improved level of response efficiency. The thorough analysis and application of GenAI open new avenues in system interactivity and personalisation. This approach not only ensures precise and pertinent responses to user queries but also notably diminishes the timeframe necessary for system design and development.

The research findings underscore the significant potential of GenAI in developing and implementing advanced information systems. They are instrumental in optimising resource utilisation, accelerating development processes, and creating more adaptable and intelligent systems that address a broad spectrum of user requirements.

The innovation in this research is encapsulated in the proposed methodology, which prioritises the use of GenAI over conventional programmed logic in systems with open datasets. This strategy notably enhances development speed, simplifies maintenance, and bolsters system dynamism.

The practical implementation of these findings has been demonstrated in the development approach, shown to increase the efficiency of web systems by 5–7 times compared to analogous systems. This method is particularly effective in managing systems with dynamic parameters within open data environments. Further research is needed to fully explore the range of applications for GenAI across different sectors, with the goal of continually improving system responsiveness and user experience. This may also extend to the development of commerce systems to observe their operational efficacy in real-world scenarios.

References

1. Shyshkina, M.P., Marienko, M.V.: The use of the cloud services to support the math teachers training. In: CTE Workshop Proceedings, CTE-2019, vol. 7, pp. 690–704 (2020). https://doi.org/10.55056/cte.419
2. Jiawan, T., Jian, A.: A communication model for large scale ship handling simulators. In: 2016 International Conference on Industrial Informatics - Computing Technology, Intelligent Technology, Industrial Information Integration (ICIICII), Wuhan, China, pp. 170–173. IEEE (2016). https://doi.org/10.1109/ICIICII.2016.0050
3. Large Language Models, TechTarget. https://www.techtarget.com/whatis/definition/large-language-model-LLM. Accessed 02 Dec 2023
4. Aljanabi, M.: ChatGPT: future directions and open possibilities. Mesopotamian J. CyberSecur. **1**, 16–17 (2023). https://doi.org/10.58496/mjcs/2023/003
5. Tregub, V., Korobiichuk, I., Klymenko, O., Byrchenko, A., Rzeplińska-Rykała, K.: Neural network control systems for objects of periodic action with non-linear time programs. In: Szewczyk, R., Zieliński, C., Kaliczyńska, M. (eds.) AUTOMATION 2019. AISC, vol. 920, pp. 155–164. Springer, Cham (2020). https://doi.org/10.1007/978-3-030-13273-6_16
6. Mou, N., et al.: Personalized tourist route recommendation model with a trajectory understanding via neural networks. Int. J. Digit. Earth **15**(1), 1738–1759 (2022). https://doi.org/10.1080/17538947.2022.2130456
7. Let the devil speak for itself: Should ChatGPT be allowed or banned in hospitality and tourism schools?. https://digitalcommons.usf.edu/jght/vol2/iss1/1/. Accessed 02 Dec 2023
8. Korobiichuk, I., Mel'nick, V., Kosova, V., Ostapenko, Z., Pavlenko, V., Wrona, J.: Modeling a dynamic object with distributed parameters. In: Proceedings of the 2023 Conference on Measurement, Modelling and Automation, pp. 388–393. IEEE (2023). https://doi.org/10.1109/MMAR58394.2023.10242400

The Modified Genetic Algorithm for Solving the Traveling Salesman Problem

Illia Solohubov(✉), Artur Moroz, Andrii Oliinyk, Sergey Subbotin, and Stepan Skrupsky

National University Zaporizhzhya Politechnic, Zaporizhzhya, Ukraine
illia.solohubov@gmail.com

Abstract. The Traveling Salesman Problem is one of the most relevant challenges today, as the need to solve it arises in various fields. However, there is no clear algorithm for its solution. It falls within the realm of NP-complete problems and is optimization-oriented. While it is assumed that the solution to these problems involves exhaustive search, in practice, various algorithms and approaches are often used to reduce the volume of analyzed solutions, aiming to find a way to solve the problem without exhaustive search. This includes methods like branch and bound to expedite the search or artificial intelligence approaches such as the ant colony algorithm and genetic algorithms to accelerate the process while achieving the closest possible solution.

Genetic methods are one of the most powerful tools for solving such tasks. These algorithms model evolutionary processes, which, thanks to the basic principles of evolution, allow finding better solutions. This approach allows representing the task as a certain evolutionary process. It is based on the fact that from the very beginning, a certain set of solutions is chosen, and they start competing with each other. Each solution represents an individual, and a set of its solutions represents a "gene". Each of these genes during the algorithm's operation can evolve, adapt, or mutate according to the rules of the environment. In the end, only the strongest genes should remain, which will be the approximate optimal solution.

A characteristic of these algorithms is that each implementation can be unique. An important task in building algorithms is precisely the optimization of their parameters, as well as choosing the right approach to some of its stages, where only the best of them will remain in the end. This methodology not only allows finding approximately optimal solutions but also provides flexibility and adaptability for various requirements and constraints.

Keywords: Algorithmic Modification · Genetic algorithm · Graph algorithms · Mutation Coefficient · Optimization · TSP · Traveling Salesman Problem

1 Introduction

As described above, optimization problems are addressed using methods that aim to avoid exhaustive search, reducing the volume of possible options or analyzed data, while at the same time achieving results that are as close as possible to the optimal

R. Szewczyk et al. (Eds.): AUTOMATION 2024, LNNS 1219, pp. 59–68, 2024.
https://doi.org/10.1007/978-3-031-78266-4_6

(best) outcome. This makes them highly intriguing in the field of computer science, especially in algorithms. They remain relevant because there is no single algorithm that can solve them maximally effectively, and there is always room to try and find ways to optimize them. Their universality and flexibility are also appealing; optimization tasks and solutions can be applied in various scientific fields such as combinatorics, graph problems, game theory, artificial intelligence, etc. [1, 6, 9].

The Traveling Salesman Problem (TSP) is particularly interesting in areas such as logistics, network design, genetic testing, etc. One of the challenges of this problem is the absence of a definitive algorithm capable of finding the optimal path, leading to the development of numerous optimization approaches and algorithms to find a more or less optimal route in a shorter time. For solving such a task, algorithms belonging to artificial intelligence methods – genetic and swarm, the latter being represented in the form of an ant algorithm, are considered among the best [3, 8, 11].

However, these algorithms are merely approaches without concrete implementations, allowing for their full utilization in combination with others to achieve maximum efficiency. From this perspective, they can take any form, and each task can provide its own opportunities for optimizing the problem [10]. In some cases, using conditions from the existing domain in these algorithms can result in acceleration of up to 20% [7]. Nearly in every task, such conditions or restrictions can be found. A more detailed approach can also help overcome certain drawbacks and uncertainties in genetic algorithms [5], providing clearer stopping criteria, directions for constructing the fitness function, or the selection of individuals.

The aim of this paper is to demonstrate, using the example of a genetic algorithm for solving the Traveling Salesman Problem, how interesting optimization paths can be found and how a specialized approach to the problem using other graph algorithms can help solve the problem more quickly than universal implementations. This will allow us to assess the possibilities of a combined approach and demonstrate that the use of non-trivial approaches can be entirely justified, even though it brings additional complexity in terms of configuring the genetic algorithm.

2 Modification of the Genetic Algorithm

There are many graph algorithms, and some of them can be used to obtain useful data. Speaking of the Traveling Salesman Problem (TSP), it would be useful to have an idea of the boundaries within which the correct solution needs to be sought or some initial path that appears optimal. For example, the human mind allows a person to almost instantly construct a solution for such a problem that would be in the top 10% of the best solutions. One algorithm that can help establish the boundaries for searching for the best solutions and also allow the construction of a first potentially good solution is the algorithm for a minimum spanning tree.

The idea of using this algorithm is simple. The shortest possible solution for the TSP in terms of path length would be a path that passes through the N-1 shortest edges, where N is the number of vertices. A minimum spanning tree is a graph that connects all the vertices of the graph without cycles, and the length of all edges is minimal with their number being N-1. Therefore, in the best case, this tree could be the optimal solution, and in the worst case, if one tries to traverse the graph through all its points, the path length will be at most twice as long, since in the worst case, each edge might have to be traversed twice [2]. This statement works for any spanning tree. Thanks to this data, the boundaries of the search for solutions become clear, from the shortest to the longest, which can be used in a genetic algorithm.

However, the tree can be easily transformed into a path. Performing any simple depth-first traversal of the graph, one can always obtain a path that will be within the necessary boundaries. This is proven by the fact that the worst path in terms of length from vertex A to vertex B will be the path through the tree. Thus, the distance between points is always minimal, and the depth-first traversal does not allow jumping from the tree without traversing its branches to the end.

The provided description outlines a scenario where Fig. 1 depicts a simple graph example. In this figure, the minimum spanning tree is highlighted with a thick line, and a short path is indicated in red. It's important to note that this example is illustrative and meant to enhance understanding. In reality, such small graphs are rare in real-world tasks, and more often, they are complete graphs (where each vertex is connected to every other vertex).

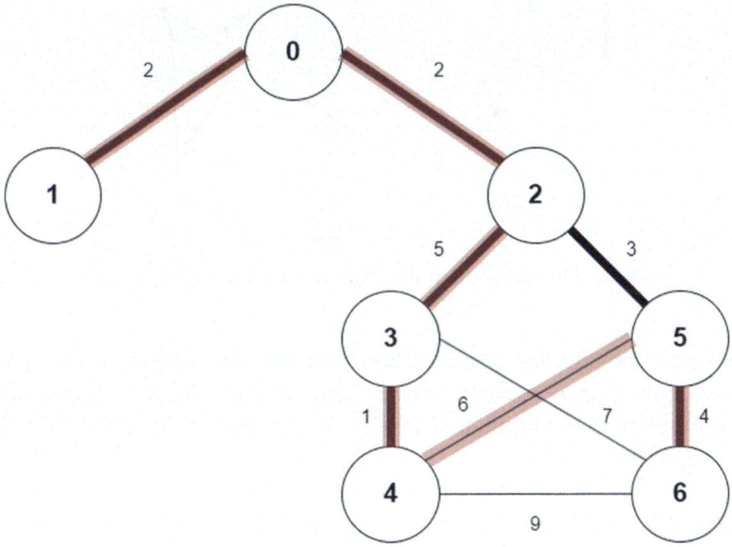

Fig. 1. Obtaining the Initial Path from a Minimum Spanning Tree Algorithm

As can be seen, the minimum spanning tree is highlighted in black, while the path constructed using depth-first search is marked in semi-transparent red. If one searches for the optimal path, it can be observed that the shorter path is 1-0-2-5-6-3-4, as its total gives a distance of 19, while the illustrated path has a distance of 20. However, this path still falls within the range of "best" values.

The main interest and utility of this path lie in the fact that it is not only potentially optimal and can be used as an individual in a colony, but also in that this path can serve as a basis for constructing other individuals in the population.

The idea of building the first colony from such a base consists of the following observation. Figure 2 demonstrates the initial path that can be obtained from the minimum spanning tree. From this, it is evident that some subgraphs (the collection of points between which a path is laid) are already optimal paths. The obvious problems include the presence of a loop in the path and an impractical reversal, which, obviously, cannot exist in the final solution. Based on this, it may be sensible to generate the first colony not from a random sequence of points, but from pieces of various lengths of the current graph, as shuffling by pieces can retain the optimal pieces and only differently construct the so-called bridges between them. A bridge in a graph is an edge that connects two subgraphs, and its absence breaks the graph itself.

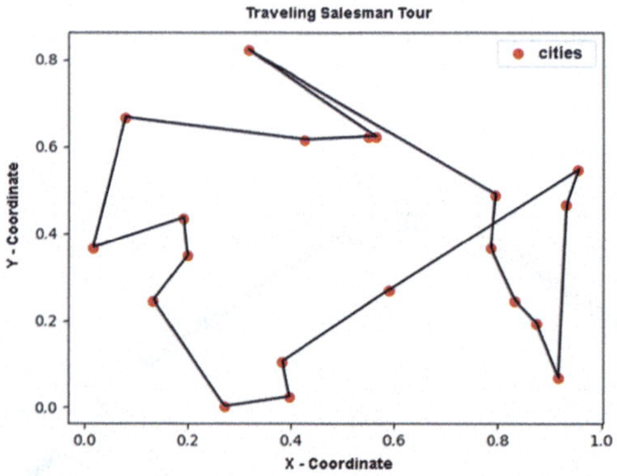

Fig. 2. Obtaining the Initial Path from a Tour of 20 Vertex

Figure 3 conceptually demonstrates how dividing into subgraphs of 5 points each and constructing bridges between them can help optimize the path. It can be seen that this method might help to eliminate the loop. At this stage, one of the individuals will already have a decent solution.

In such a shuffle, it's also prudent to perform certain manipulations with the initial path. For example, it is useful to conduct a cyclic shift of the path to constantly take different pieces and randomly reverse pieces to generate more varied options.

From this generation, one can immediately see one of the potential but fundamental drawbacks that can significantly affect the correctness of the solution – insufficient genetic diversity. In biology, this concept means that a population has insufficiently diverse genes, which slows down their evolution and can even stop at some point because each individual will be similar to another.

In genetic algorithms, as in nature, there is a way to partially avoid such a situation – mutation. Mutation is a random, unpredictable change in the genes of an individual. The boundaries obtained from the minimum spanning tree, as mentioned above, can be used for this.

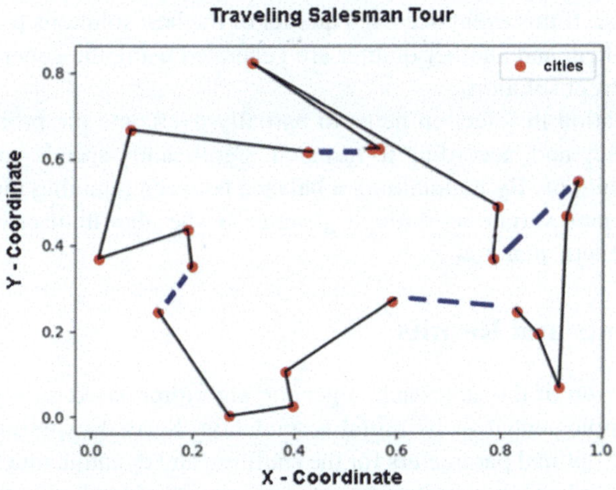

Fig. 3. Possible Individual after Building Bridges in Subgraphs of 5 Vertices

The main idea of using two boundaries and mutation is as follows and is visually represented in Fig. 4. The algorithm will gradually shift the right boundary, which corresponds to the longest length, to the left – the shortest length, gradually eliminating individuals that fall outside the shifted boundary. The mutation algorithm, in turn, will accelerate the mutation of those individuals who remain in the system as the boundary itself advances.

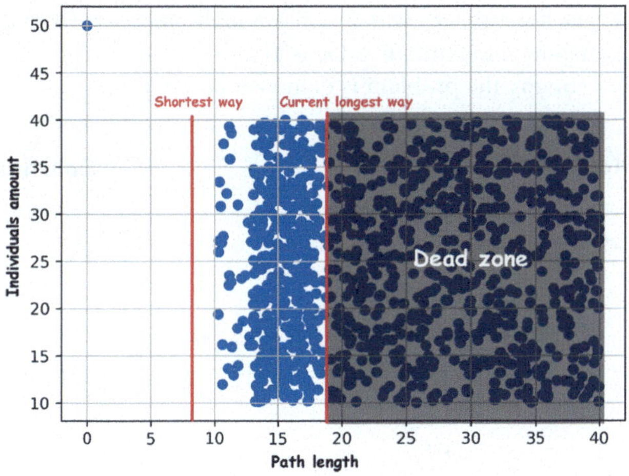

Fig. 4. Visualization of Colony Behavior with Boundary Shift

The final part of the algorithm that is modified is selection. A selection algorithm is proposed, which follows the idea of cleansing based on a boundary. The suggestion is to clear the population down to no less than half of the value set by the algorithm at the initialization stage. If this event occurs, a quarter of the best solutions from the previous population are taken, and another quarter are generated using the generation algorithm from the best current solution.

This modification in selection helps to partially overcome the problem of genetic diversity deficiency and, according to research, significantly speeds up the process of finding better solutions. By maintaining a balance between retaining the best solutions and introducing new variations through generation, the algorithm ensures a constant influx of fresh genetic material.

3 Experiments and Results

For the investigation of this approach, a genetic algorithm based on it was developed. Selection was implemented in its initial formulation. Several experiments were conducted to achieve optimal parameters for the shuffling and dynamic mutation algorithm. The crossover (breeding) stage is implemented in one of the standard forms - the remaining solutions in pairs next to each other are crossed through a single-point crossover [4].

The first unique feature of this algorithm is that the parameter for generating the initial solution is not a certain coefficient or number, but an array of pieces. This array corresponds to how many pieces the obtained path should be divided into for shuffling. In each iteration, the algorithm will run through the array of pieces, shuffle according to these pieces, and then cyclically shift the path to the left at the end of each iteration.

The mutation coefficient is not unique, and its algorithm is similar to a trivial approach. It will also be tied to the stopping criterion of the algorithm, which is considered to be the absence of a new better solution for an extended period. Experimentally, a coefficient of 0.1 was determined, which will gradually increase from the moment the algorithm stops, reaching a maximum value of 0.2.

An example of solving the problem is demonstrated in Fig. 5.

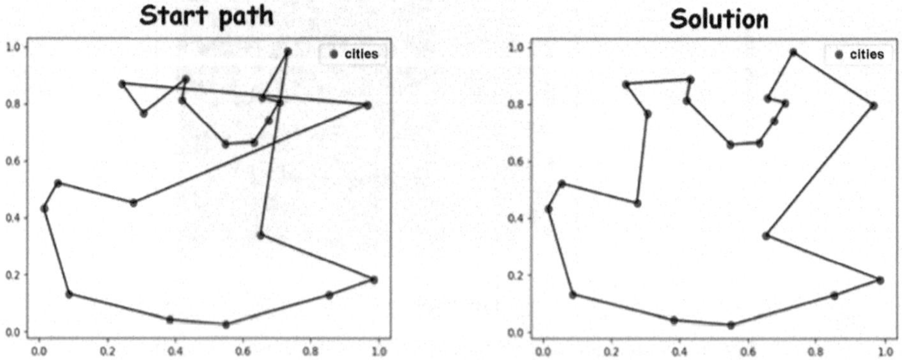

Fig. 5. Finding the Optimal Path

In Table 1, the results in terms of time and length for the algorithm using population generation through the use of a basis as the solution found via a minimum spanning tree and without it are presented. Additionally, a special parameter Δ is introduced, which determines the degree of difference between the solutions of the two algorithms. A positive value indicates that the algorithm using the basis obtained a better result, while a negative value indicates a worse outcome.

Table 1. Test Results with and without the Base.

№	Algorithm with a basis		Algorithm without a basis		Δ (length)	Δ (time)
	Length	Time (sec)	Length	Time (sec)		
1	3.985	8.592	4.041	10.595	0.014	0.189
2	3.592	6.129	3.592	9.666	0	0.366
3	4.169	8.636	4.169	11.103	0	0.222
4	3.753	7.627	3.769	10.849	0.004	0.297
5	4.546	7.291	4.429	11.486	−0.026	0.365
6	3.617	10.345	3.798	13.586	0.048	0.239
7	4.330	9.531	4.236	5.301	−0.022	−0.798
8	3.316	9.337	3.305	11.837	−0.003	0.211
9	3.259	9.884	3.146	9.342	−0.036	−0.058
10	4.099	7.290	4.188	11.954	0.021	0.390
11	3.697	7.061	3.682	9.527	−0.004	0.259
12	4.074	6.811	4.101	8.235	0.007	0.173
13	3.925	5.367	3.925	9.502	0	0.435
14	3.557	9.505	3.486	9.563	−0.021	0.006
15	3.486	9.563	4.587	9.969	0.240	0.041
16	3.592	9.969	3.591	10.268	0	0.029
17	4.327	9.573	4.411	5.246	0.019	−0.825

From the data in the table, it can be observed that the algorithm without a basis generally operates more slowly, and in some cases, it takes as much time as the algorithm with a basis. Both algorithms provide values that do not differ from each other by more than 5%, with the basis-utilizing algorithm more frequently achieving a better result. However, it should be noted that occasionally the random factor plays a role, and the algorithm without the use of a basis randomly attains a better solution during the value formation stage. Thus, there are instances when it operates faster or obtains a superior value compared to the designed algorithm. Experimentally, such cases account for no more than 10%.

Figure 6 presents a graph showing the relationship between the runtime of genetic algorithms and the number of vertices. For testing, a standard algorithm using a fitness function as a criterion for conducting mutations and selecting parents, with a stopping criterion based on checking for population change, was chosen. This algorithm is standard and universal for any problem. As another algorithm, the best implementations with ready-made coefficients were reviewed.

These observations point to the advantage of using fundamental elements such as minimum spanning trees in genetic algorithms to solve optimization problems. A comparison of standard and modified algorithms demonstrates the potential benefits of such algorithm adjustments and parameter optimizations to achieve better performance and results.

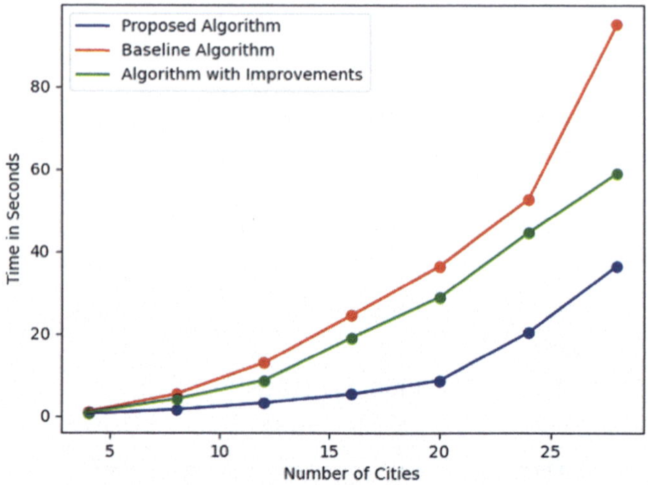

Fig. 6. Dependency of Running Time on the Number of Vertices in Different Algorithms

The graphs are displayed from top to bottom – the basic algorithm, the improved one, and the proposed one, respectively. From the graph, it can be seen that the algorithm with a basis for solving the Traveling Salesman Problem works faster as well. However, it should be noted that with the increase in the number of vertices, the likelihood of a calculation error by the algorithm with a basis also increases. The algorithm with improvements more often produces a more optimal route option for a large number of vertices. In worse cases, the algorithm with a basis starts to have an error of about 4%, which is likely due to insufficient training of the algorithm.

In Table 2, these test results are presented in terms of time and accuracy of the response. Accuracy is determined by the delta parameter, which defines how many times the solution is worse compared to the real solution. As a study of the delta parameter, some of the worst indicators were taken to determine potential error.

From the table, it is evident that as the number of cities increases, the margin of error in the solution may also increase, but at 28 vertices, it does not exceed 3%. Furthermore, even a less trained algorithm may start introducing a slight error into the solution. The key factor here lies in selecting optimal parameters that directly impact accuracy and execution time. Incorrect settings are the primary reason for the diminished accuracy of both algorithms – the basic one and the one with a basis – as the number of vertices

Table 2. The results of the work of three algorithms.

N cities	Execution time (t, seconds)			Solution error (Δ, %)		
	$t_{proposed}$	$t_{baseline}$	$t_{improvements}$	$\Delta_{proposed}$	$\Delta_{baseline}$	$\Delta_{improvements}$
4	0.672	1.154	0.924	0	0	0
8	1.655	5.510	4.243	0	0	0
12	3.257	13.037	8.735	0	0	0
16	5.468	24.615	22.153	0	0.002	0
20	7.767	36.551	29.972	0.002	0.008	0
24	17.511	52.904	40.736	0.01	0.012	0
28	30.645	95.227	52.096	0.019	0.016	0.01

increases, in contrast to the well-trained algorithm. These settings also contribute to the prolonged execution of the algorithm, despite achieving comparable results.

Analyzing the specific data obtained, it can be seen that this modified algorithm allows for a significant speedup in finding solutions by several times for small numbers of vertices and an average improvement of around 30% when the number of vertices becomes sufficiently high. However, it should be noted that this increased speed comes at the cost of some loss of accuracy. If one attempts to fine-tune the parameters, for example, using machine learning methods, it may be possible to achieve the same level of accuracy as a well-trained universal algorithm. In this case, there may be a reduction in speed, but considering the current growth dynamics and figures, the acceleration can still remain at the 30.

This approach to creating a genetic algorithm using auxiliary methods and ideas specific to the particular problem, such as generating an optimal initial colony, can indeed expedite the algorithm's execution. However, the cost of using this approach is a loss of the universality of the genetic algorithm. Considering this implementation, it becomes clear that the algorithm has become specialized for the Traveling Salesman Problem and can no longer be used for other tasks.

The loss of universality can become a critical factor and complicate the setting up of the algorithm. An algorithm that uses a specialized approach may have additional, unique parameters for it, or a unique approach for any of its stages, which in turn requires additional adjustments, more complex testing, and training of the algorithm.

4 Conclusions

The article proposes a modification of the algorithm using additional graph algorithms, allowing for a preliminary analysis of the problem. Thanks to this approach, the first "good" path is identified, which is chosen as the basis for forming the initial colony. This leads to obtaining better initial characteristics for the algorithm compared to the traditional approach. Modifications to the selection, with the introduction of additional individuals through modified generation, and mutations with dynamic acceleration, are suggested.

Research has demonstrated that using this approach in genetic algorithm for solving the Traveling Salesman Problem allows for a 30% improvement in algorithm efficiency on average compared to commonly accepted universal implementations.

In the proposed algorithm modification, there is one new parameter that can be adjusted directly - the array of fragments, and one parameter that has significantly more weight than in the trivial version of the algorithm - the mutation coefficient. Thus, it can be concluded that by specifying the algorithm, it can be accelerated for specific tasks, although at the expense of complicating its learning process due to the introduction of additional criteria and changing the weight of existing ones.

References

1. Alridha, A., Salman, A.M., Al-Jilawi, A.S.: The applications of NP-hardness optimizations problem. J. Phys.: Conf. Ser. **1818**, 012179 (2021). https://doi.org/10.1088/1742-6596/1818/1/012179
2. Baidoo, E.: A preliminary study on minimum spanning tree algorithm approach for travelling salesman problem. Commun. Appl. Electron. **5**(10), 29–36 (2016). https://doi.org/10.5120/cae2016652386
3. Bajpai, P., Kumar M.: Genetic algorithm – an approach to solve global optimization problems. Indian J. Comput. Sci. Eng. **1**(3), 199–206 (2010). https://www.researchgate.net/publication/283361244_Genetic_Algorithm_-_an_Approach_to_Solve_Global_Optimization_Problems
4. Crossover Operators in Genetic Algorithm. https://medium.com/geekculture/crossover-operators-in-ga-cffa77cdd0c8. Accessed 29 Nov 2023
5. Genetic Algorithms; Summary & Limitations. https://medium.com/@arbistarkillerllaveshi/genetic-algorithms-summary-limitations-3da8df3e1138. Accessed 02 Dec 2023
6. Kochkarov, R.: Research of NP-complete problems in the class of prefractal graphs. Mathematics **9**, 2764 (2021). https://doi.org/10.3390/math9212764
7. Korobiichuk, I., Hrybkov, S., Seidykh, O., et al.: Development of a modified ant colony algorithm for order scheduling in food processing plants. J. Autom. Mob. Robot. Intell. Syst. 53–61 (2023). https://doi.org/10.14313/jamris/1-2022/6
8. Kulenović, F., Hošić, A.: Application of genetic algorithm in solving the travelling salesman problem. In: IOP Conference Series: Materials Science and Engineering, vol. 1208, no. 1, p. 012032 (2021). https://doi.org/10.1088/1757-899x/1208/1/012032
9. Reiling, A.J.: Optimization of convolutional neural network accuracy using genetic algorithm. Master's thesis, University of Dayton (2017). http://rave.ohiolink.edu/etdc/view?acc_num=dayton1512662981172387
10. Subbotin, S., Oleynik, A.: Modifications of ant colony optimization method for feature selection. In: Proceedings of the 9th International Conference on the Experience of Designing and Applications of CAD Systems in Microelectronics, 19–24 February 2007, pp. 1–6 (2007). https://doi.org/10.1109/cadsm.2007.4297628
11. Yao, X.-S., Ou, Y., Zhou, K.-Q.: TSP solving utilizing improved ant colony algorithm. J. Phys.: Conf. Ser. **2129**(1), 012026 (2021). https://doi.org/10.1088/1742-6596/2129/1/012026

Failure Modeling of Industrial Electric Motors Using Unsupervised Learning Methods

Nataliia Lutska[1(✉)], Nataliia Zaiets[2], and Lidiia Vlasenko[3]

[1] National University of Food Technologies, Kyiv 01033, Ukraine
lutskanm2017@gmail.com

[2] National University of Life and Environmental Sciences of Ukraine, Kyiv 03041, Ukraine
z-n@ukr.net

[3] State National University of Trade and Economics, Kyiv 02156, Ukraine

Abstract. The task of detecting breakdowns helps to improve the reliability of industrial devices through early identification of problems and their timely resolution, which helps to improve the resource and energy efficiency of production as a whole. When modeling breakdowns of industrial electric motors, the following characteristics are primarily taken into account: average load, rotational frequency, power supply voltage, current consumption, leakage current, collected from data from a traditional automated control system and breakdown logs of repair teams of an industrial enterprise. The model for detecting reliability and breakdowns of electric motors in the work is built on an anomaly detection algorithm using unsupervised machine learning methods. Its effectiveness has been successfully tested using many evaluation metrics. It can be used for a range of tasks and applications in equipment management and maintenance, visualization, forecasting and management decision support.

Keywords: unsupervised learning · Gaussian Mixture Model · motor · modelling · breakdown · reliability

1 Introduction

Recently, artificial intelligence has been widely used to diagnose and predict the occurrence of faults and manage production performance [1]. The use of machine learning algorithms is actively used to monitor the condition of industrial equipment [2], in particular, industrial engines. Data characterizing the normal operation of the engine and data indicating the occurrence of malfunctions can be obtained from a traditional automation system and from a specially introduced engine condition monitoring system and from breakdown logs. Traditional automation systems can usually provide information about engine stalls, or problems with engine stalling or speed. Failure logs are a mandatory form of documentation of all malfunctions of industrial equipment, including electric motors, pumps, etc., by the repair team of the production department. Construction of an industrial enterprise in accordance with the principles of Industry 4.0 involves the introduction of intelligent systems for data processing and construction of appropriate

R. Szewczyk et al. (Eds.): AUTOMATION 2024, LNNS 1219, pp. 69–80, 2024.
https://doi.org/10.1007/978-3-031-78266-4_7

models [3]. Such systems can automatically analyze the characteristics of input data to predict the occurrence of faults in industrial equipment [4], and carry out effective maintenance planning, which is critical for ensuring the efficiency of industrial equipment, in particular motors. The ability to implement IoT [5] and cloud technologies allows the use of more complex fault prediction methods with minimal investment of time and resources.

Object of the study is the modeling of breakdowns and reliability of industrial electric motors to improve the control and maintenance system.

Subject of the study is a model of breakdowns of industrial electric motors based on data obtained from traditional automated control systems and breakdown logs of repair teams of an industrial enterprise.

The purpose of the work is the development of breakdown models of industrial electric motors using unsupervised learning methods in order to effectively identify anomalies and predict possible faults.

2 Review of the Literature References

Effective fault detection is important in any service area because recognizing problems early can prevent major failures and downtime, saving time and financial resources. Conventional failure detection methods usually require large amounts of investment, including both time and human resources. This has contributed to growing interest in the development of machine learning methods, in particular, unsupervised learning algorithms to optimize diagnostic processes.

Fault detection based on machine learning principles is a common approach that uses data to automatically learn diagnostic knowledge without relying on expert experience or engineering knowledge. This method is based on machine learning theory to automatically learn the relationships between collected data and equipment conditions to create a corresponding model.

For the efficient functioning of production, it is necessary to ensure reliable and uninterrupted operation of electric motors [6]. The reliability of an electric motor is assessed by its ability to operate uninterruptedly for a significant period of time under various operating conditions, taking into account resistance to various influences, such as mechanical loads, temperature changes, humidity, stress fluctuations and other factors. To assess reliability, the following are taken into account: mean time between failures (MTBF – Mean Time Between Failures), probability of failure, degree of reliability, reservoir for eliminating failures.

The most common machine learning methods used to analyze and predict the condition of engines can be divided into three groups [7, 8]:

1. Shallow machine learning (SML). Shallow machine learning (SML). SML-based fault modeling is a pattern identification process that uses signal processing techniques to extract fault parameters and generate feature vectors. To complete fault identification, classifiers such as artificial neural network (ANN) and support vector machine (SVM) are used [9].
2. Deep learning (DL). DL-based fault models contain more layers in neural networks, aiming to make full use of large amounts of data to obtain more accurate results [10].

3. Transfer learning (TL). Transfer learning consists of transferring existing knowledge to solve learning problems in a target domain with few or even no designated examples. Moreover, the more factors are divided into two different domains, the easier it is to carry out training [11].

Gaussian Mixture Model (GMM) – is a probability function that has parameters and is represented as a weighted sum of the densities of the Gaussian components. Typically, GMM is used as a parametric model to distribute the probabilities of continuous measurements or characteristics in technical systems, for example, in determining the operating characteristics of engines to reduce the occurrence of anomalies during their operation. The GMM parameters are determined through an iterative Expectation-Maximization (EM) algorithm or Maximum A Posteriori (MAP) algorithm based on the data produced by a well-trained model.

There are different approaches to using GMM to monitor the occurrence of engine failures. In [12], the results of developing a model for clustering, identification and diagnosis of six different faults in electric motors based on a combination of GMM and Mahalanobis distance are presented. There are a number of works confirming the effectiveness of using GMM to improve the reliability and accuracy of assessment, detection and identification of faults in motor bearings [13, 14].

In general, the range of application of machine learning methods is quite wide: this includes identifying short circuits in synchronous motors [15], and various types of failures for asynchronous motors of centrifugal pumps [7], and multiphase drive systems [16], and inverter-fed induction motors [17] etc.

3 Methods

To model the reliability of electric motors based on current measurements of their characteristics using machine learning models, we used the algorithm Fig. 1.

The first step is to obtain data representative of the reliability of electric motors, such as temperature, current consumption, voltage, speed, etc. Different food processing plants may measure different parameters that may reflect engine reliability. Most often, the following parameters include:

- average load, % – x_1;
- rotational frequency, rpm – x_2;
- power supply voltage, V – x_3;
- current consumption, A – x_4;
- leakage current, mA – x_5.

A feature of the data set is a large amount of data that is marked as reliable and a small amount of data that is marked as faulty. The number of the latter, as a rule, is $<$ 15% of the total. Considering the variety of engine breakdowns and the small amount of data for each type of breakdown, the use of anomaly detection methods will be effective. Such methods are used to identify data that is unexpected or too deviating from patterns for normal system operation.

After obtaining a data set, checking it and pre-cleaning it, the data should be divided into reliable data and breakdown data and considered separately for each set.

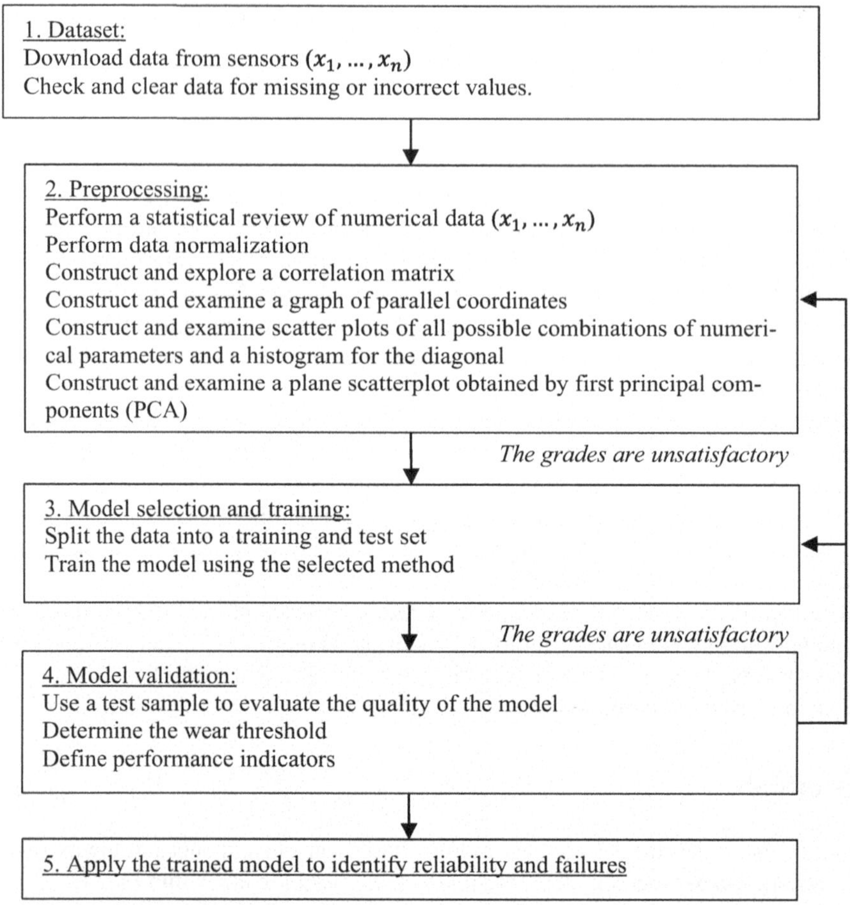

Fig. 1. Algorithm for modeling the process of detecting faults in electric motors.

The second step is to review the data and transform it. This step involves performing a statistical review of the numerical data for each distribution and normalizing the data. The latter will ensure stability and rapid matching of the model training algorithm and unify the influence of parameters on the model.

Studying the correlation matrix helps confirm the influence of parameters on the output indicator, however, parameters with a low correlation coefficient should not be discarded. A low correlation coefficient does not necessarily indicate that there is no relationship or that the parameter has no effect on the original measure. There may also be a situation where there is a relationship, but it is not apparent through linear correlation (measuring only linear relationships). Parameters with low correlation coefficients can also have an important influence on the model, especially if they interact with other parameters or do not exert their influence through a linear relationship. Therefore, it is important to carefully analyze all possible relationships and take into account all aspects of the influence of parameters on the initial indicator. If a parameter is considered important in terms of the task context (such as those given above), even in the absence of a strong correlation, its influence should be examined more carefully or other methods of identifying the influence of parameters on the original measure should be considered.

Examining a parallel coordinates graph allows you to visualize objects in a multi-dimensional parameter space, where each line represents one object and the coordinate axes represent different parameters. A first principal components area (PCA) scatterplot provides a visual representation of the structure of the data in its own coordinate system, where it varies most. Together, these two visualizations will help reveal the overall data structure and distribution of observations in high-dimensional data, allowing for simplified interpretation and visualization.

Creating scatter plots for all possible combinations of numerical parameters and their histograms will allow you to conveniently visualize the relationships between pairs of variables in a data set, and will also help you understand relationships in multivariate data. In addition, when working with histograms, you can decide on subsequent methods for building models, in particular, choose the metric that best describes the data. It will also help to understand whether the normalization algorithm chosen is correct and whether additional transformation of parameters is needed.

The next step is selecting and training the model on the training set. Considering the parameters of the data that were indicated earlier, the training set should contain only data related to the reliable operation of the engines. The test set must contain both reliable equipment data and equipment failures. The model is trained according to all ML guidelines.

After training the model, it is necessary to carefully check it on test data and select the wear threshold for electric motor failure. To do this, you can apply the following performance indicators:

- Accuracy is a general measure of how well the model classifies normal and abnormal (broken) samples. Recommended for use only with others;
- Recall or True Positive Rate – measures the model's ability to detect anomalies among all truly abnormal cases or the model's ability to correctly classify normal cases. It is recommended to achieve high values for both indicators;
- F1-score is a balanced measure between Recall and Precision. It is calculated as the harmonic mean between them. Given the uneven distribution of classes, it is useful to use the F1-score for a class of breakdowns to evaluate the effectiveness of the model in detecting them;
- AUC-ROC – measures the ability of the model to separate classes, namely, the probability that an anomalous sample will receive a higher score than a normal one;
- Likelihood Ratio – particularly useful for determining the limit of acceptance of an anomaly. For this task, a high probability can serve as an indicator of high reliability of the electric motor.

If the model efficiency assessments are unsatisfactory, you must return to the previous paragraphs and review the selected metric for obtaining model parameters and/or data preprocessing. The resulting model can be used to optimize the operation of electric motors, increase their reliability and reduce the risk of emergency situations.

4 Result and Discussion

Let us carry out the process of modeling the reliability of electric motors with a power of 55 kW, 1500 rpm, data on which were obtained at a sugar factory. The modeling process will be performed using the Python programming language, using the pandas, matplotlib, seaborn and sklearn libraries. Colaboratory (Google) was chosen as the programming environment.

The peculiarity of this data set is a large amount of data with high reliability ($y = 1$) and a small amount with low reliability ($y = 0$) – 2857/42. Figure 2 shows the first five data points and the main statistical characteristics of the resulting parameters. There is likely to be significant variability in the data associated with failure identification data. It can be concluded that further rationing is necessary.

	x1	x2	x3	x4	x5	y
0	93	978	379	8.9	3.0	1
1	83	971	360	7.8	2.6	1
2	89	968	368	10.8	4.9	1
3	80	964	370	7.0	3.2	1
4	81	960	364	9.4	2.7	1

a)

	x1	x2	x3	x4	x5
count	2899.0	2899.0	2899.0	2899.0	2899.0
mean	88.1	966.6	368.2	9.3	3.4
std	5.7	45.4	22.3	4.1	0.9
min	79.0	121.0	57.0	7.0	2.0
25%	83.0	965.0	365.0	7.9	2.8
50%	88.0	970.0	370.0	9.0	3.3
75%	93.0	975.0	375.0	10.0	3.9
max	116.0	980.0	380.0	73.0	11.7

b)

Fig. 2. First values in the data set (a) and statistical parameter estimates (b).

Let's consider the correlation matrix and the parallel coordinates graph of normalized parameter values (Fig. 3). All parameters, except the first, have a high correlation

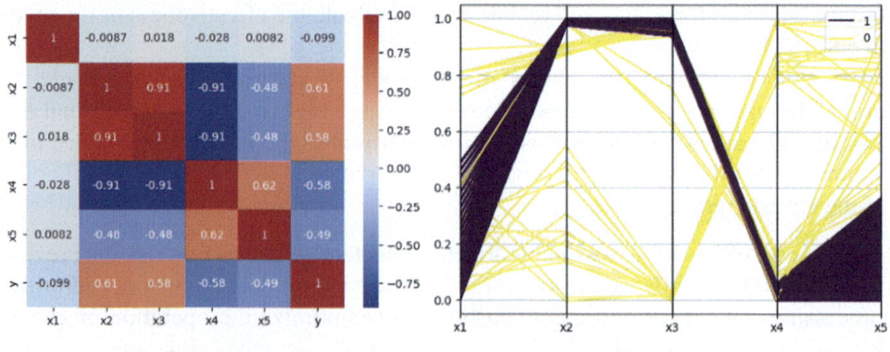

a) b)

Fig. 3. Data heat map and parallel coordinates plot for normalized parameter values.

coefficient relative to the original variable, also x_2 and x_3 have a high linear relationship, and x_4 is also correlated with x_5. However, in the parallel coordinates plot for reliable devices ($y = 1$), one can see similarities that confirm the existence of a single cluster of normal engine operation. And vice versa, during malfunction ($y = 0$), we observe several clusters (at least two), but we cannot draw a conclusion about the number of clusters during anomalies, because there is too little data.

After constructing a pairwise scatter plot for data characterized by reliable operation (Fig. 4, a), characteristic patterns can be observed on the interdiagonal plots. On the histograms (diagonal graphs), it can be assumed that the probability distribution of each feature is approximated by a Gaussian Mixture with the number of components 4 or 5. Thus, you can choose a method for modeling the reliability of electric motors – Gaussian mixture model (GMM), which is a parametric model and is based on distribution parameters Gaussian mixtures. Figure 4, b also demonstrates the probability of the existence of two clusters for data describing engine failures.

Figure 5 shows a scatterplot plotted on the plane of Principal Component 1 and 2. The main directions of variation in the data can be observed. Normal objects tend to cluster together, whereas anomalies stand out as individual points, presumably forming three clusters. One cluster comes fairly close to the normal data points.

After constructing a GMM model simulating the reliability of electric drives, the following parameters were obtained:

$$p(x) = \sum_{i=1}^{K=4} \pi_i \mathcal{N}(x; \mu_i; \sigma_i), \tag{1}$$

where K is the number of components in the mixture of Gaussians;

π_i – weight of the i-th component (probability of choosing the i-th component):

$$\pi_{i=1\ldots K} = [0.39; 0.28; 0.16; 0.17]; \tag{2}$$

$\mathcal{N}(x; \mu_i; \sigma_i)$ – Gaussian probability density function for the i-th component with parameters μ_i (mean value) and σ_i (covariance matrix) of the corresponding dimensions for the multidimensional case.

To test the resulting model, we used a scatter plot for the test data (Fig. 6). The probability threshold for detecting abnormal behavior of electric motors was selected so that all failures were identified by GMM. Thus, the following expression was obtained for making decisions on equipment breakdowns:

$$if\ p(x) < threshold\ then\, y = 0, threshold = 5.0. \tag{3}$$

As can be seen from the scatter plot (Fig. 6), a 100% fit to the data on the test set was obtained. This is also proven by the analysis of performance indicators given in Table 1. In particular, False Positive Rate (FPR) = 0.0, that is, there are no false positive cases – all objects that are not actually positive are correctly classified as negative. The accuracy on the training data does not reach 1.0 and this is normal because we tried to avoid overfitting and ensure that the model generalizes to new data.

In the anomaly plot on the Log Likelihood scale (Fig. 7), two clusters can again be observed that exhibit an anomaly. One cluster is concentrated in the region of low logarithmic probability, the other is closer to the cluster of normal engine operation.

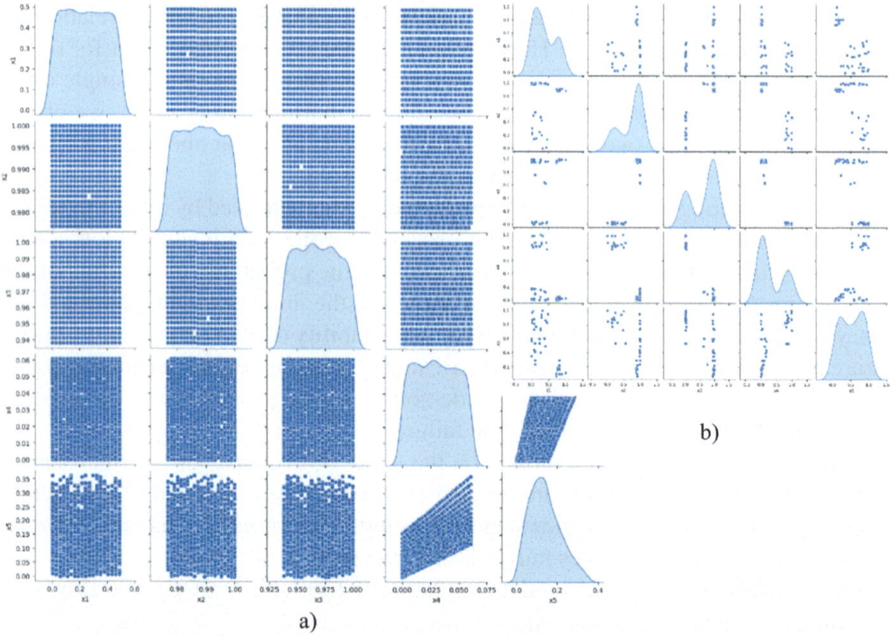

Fig. 4. Plot of the pairwise relationships in a dataset for $y = 1$ (a) and $y = 0$ (b).

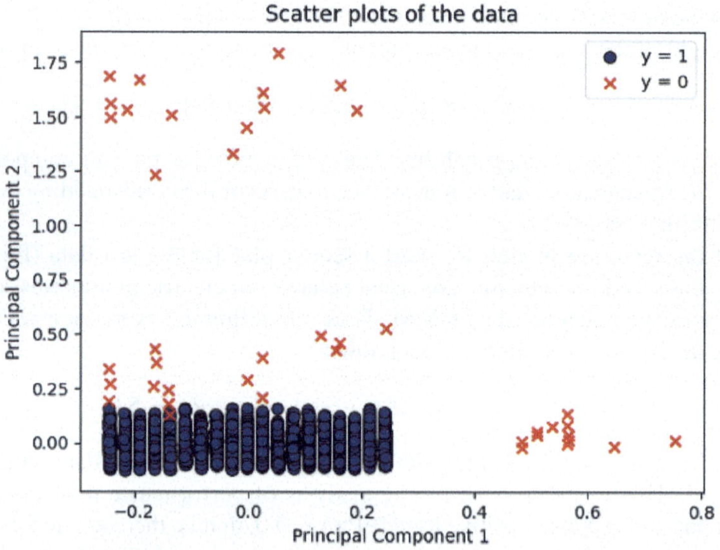

Fig. 5. Scatter plot of the training data on the principal components plane.

There is likely a third cluster of low item reliability that overlaps with the $y = 0$ cluster. That is, there are two local anomalies (small clusters of low probability) and one global

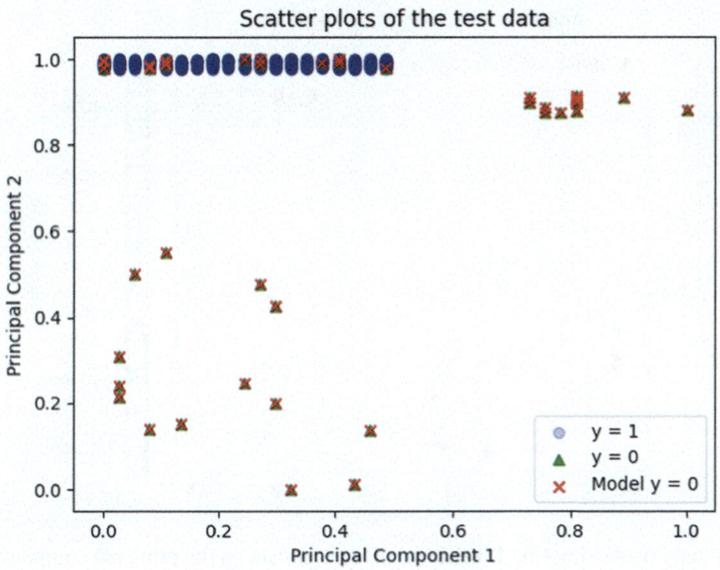

Fig. 6. Scatter plot of the test data on the principal components plane.

Table 1. Evaluation metrics.

Metric	Train data	Test data
Precision	0.99	1.0
Recall	1.0	1.0
F1-score	0.99	1.0
Accuracy	0.99	1.0
AUC-ROC	0.99	1.0
False Positive Rate	0.0	0.0

anomaly (large area of low probability). This all points to different types of anomalies in the data. Additional research into the nature of anomalous clusters and the circumstances of their occurrence is necessary.

Figure 7 provides an opportunity to visually indicate the wear threshold of anomalies. However, the authors used a tool to evaluate the effectiveness of detection error tradeoff (DET) anomaly detection systems, which found $threshold = 6.47$, not significantly different from the obtained threshold above.

Thus, using the resulting model, it is possible to predict engine failures, and further analysis of the resulting patterns will help identify the causes of electric drive failures.

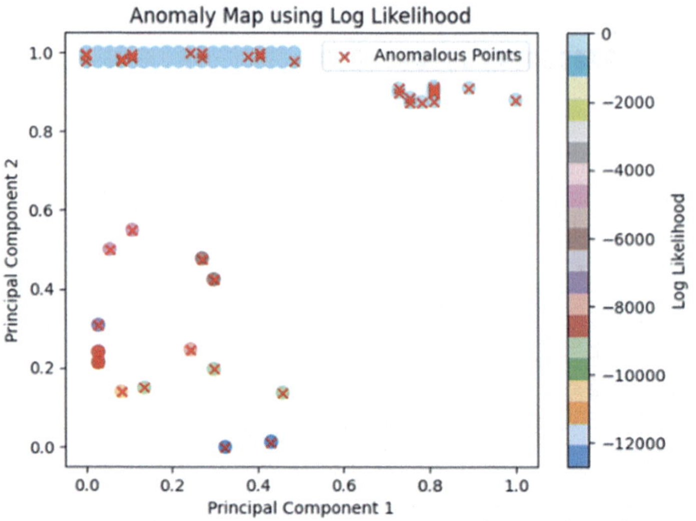

Fig. 7. Anomaly map using Log Likelihood of the test data on the principal components plane.

5 Conclusion

A model for detecting reliability and breakdowns of electric motors, built using an anomaly detection algorithm using machine learning methods, can be used for a number of tasks and applications in the field of management and maintenance, in particular:

1. Production monitoring of equipment and detection of anomalies – the model can be used to promptly identify anomalies and inconsistencies, which may indicate possible breakdowns or problems with the operation of equipment to avoid emergency situations.
2. Maintenance and repair planning – anomalies detected based on the model will help plan maintenance and repair work. By analyzing the resulting patterns, the model can provide information about which parts or components of electric motors may require attention or replacement.
3. Optimization of energy costs – the model can be used to optimize energy costs and improve efficiency.
4. Improving the operation of technical systems - timely anomalies detected based on the model can be used for continuous improvement of technical systems, taking into account the problems found when developing new models or control strategies.
5. Interaction with control systems – control feedback can be built when anomalies are detected, taking into account redundant connections, devices, etc.
6. Visualization and reporting – the model can form the basis for an interface for visualizing the results of electric drives and generating reports for operators or other interested parties.

In general, the application of the approach described in the paper and the developed model will help optimize the operation of electric motors, increase their reliability and reduce the risk of emergency situations.

References

1. Vlasenko, L.O., Lutska, N.M., Zaiets, N.A., Shyshak, A.V., Savchuk, O.V.: Domain ontology development for condition monitoring system of industrial control equipment and devices. Radio Electron. Comput. Sci. Control (1), 157 (2022). https://doi.org/10.15588/1607-3274-2022-1-16
2. Lutska, N., Vlasenko, L., Zaiets, N., Lysenko, V.: Modeling the productivity of a sugar factory using machine learning methods. In: 2022 IEEE 17th International Conference on Computer Sciences and Information Technologies (CSIT), pp. 353–356. IEEE (2022). https://doi.org/10.1109/CSIT56902.2022.10000571
3. Korobiichuk, I., Mel'nick, V., Kosova, V., Ostapenko, Z., Gnateiko, N., Rzeplinska-Rykala, K.: Mathematical model of the approximate function as the result of identification of the object of automatic control. In: Szewczyk, R., Zieliński, C., Kaliczyńska, M., Bučinskas, V. (eds.) AUTOMATION 2023. LNNS, vol. 630, pp. 173–182. Springer, Cham (2023). https://doi.org/10.1007/978-3-031-25844-2_16
4. Drakaki, M., Karnavas, Y.L., Tziafettas, I.A., Linardos, V., Tzionas, P.: Machine learning and deep learning based methods toward industry 4.0 predictive maintenance in induction motors: state of the art survey. J. Ind. Eng. Manag. (JIEM) **15**(1), 31–57 (2022). https://doi.org/10.1109/TLA.2022.9757372
5. Tran, M.Q., et al.: Robust fault recognition and correction scheme for induction motors using an effective IoT with deep learning approach. Measurement **207**, 112398 (2023). https://doi.org/10.1016/j.measurement.2022.112398
6. Korobiichuk, I., Kachniarz, M., Bezvesilna, O., Nowicki, M., Ilchenko, A., Szewczyk, R.: Calorimetrie flow meter of motor fuel with inlet temperature regulation. In: 2017 4th International Conference on Control, Decision and Information Technologies (CoDIT), pp. 0975–0979. IEEE (2017). https://doi.org/10.1109/CoDIT.2017.8102725
7. Sunal, C.E., Dyo, V., Velisavljevic, V.: Review of machine learning based fault detection for centrifugal pump induction motors. IEEE Access **10**, 71344–71355 (2022). https://doi.org/10.1109/ACCESS.2022.3187718
8. Cen, J., Yang, Z., Liu, X., Xiong, J., Chen, H.: A review of data-driven machinery fault diagnosis using machine learning algorithms. J. Vibr. Eng. Technol. **10**(7), 2481–2507 (2022). https://doi.org/10.1007/s42417-022-00498-9
9. Lei, Y., et al.: Applications of machine learning to machine fault diagnosis: A review and roadmap. Mech. Syst. Signal Process. **138**, 106587 (2020). https://doi.org/10.1016/j.ymssp.2019.106587
10. Mushtaq, S., Islam, M.M., Sohaib, M.: Deep learning aided data-driven fault diagnosis of rotatory machine: a comprehensive review. Energies **14**(16), 5150 (2021). https://doi.org/10.3390/en14165150
11. Li, C., Zhang, S., Qin, Y., Estupinan, E.: A systematic review of deep transfer learning for machinery fault diagnosis. Neurocomputing **407**, 121–135 (2020). https://doi.org/10.1016/j.neucom.2020.04.045
12. Ribeiro Junior, R.F., de Almeida, F.A., Jorge, A.B., et al.: On the use of the Gaussian mixture model and the Mahalanobis distance for fault diagnosis in dynamic components of electric motors. J. Braz. Soc. Mech. Sci. Eng. **45**, 139 (2023). https://doi.org/10.1007/s40430-023-04056-6a
13. Lin, S.-L.: Application of machine learning to a medium Gaussian support vector machine in the diagnosis of motor bearing faults. Electronics **10**(18), 2266 (2021). https://doi.org/10.3390/electronics10182266
14. Toma, R.N., Kim, J.: Bearing fault classification of induction motor using statistical features and machine learning algorithms. In: Abraham, A., Gandhi, N., Hanne, T., Hong, TP.,

Nogueira Rios, T., Ding, W. (eds.) ISDA 2021. LNNS, vol. 418, pp. 243–254. Springer, Cham (2021). https://doi.org/10.1007/978-3-030-96308-8_22
15. Shih, K.-J., et al.: Machine learning for inter-turn short-circuit fault diagnosis in permanent magnet synchronous motors. IEEE Trans. Magn. **58**(8), 1–7 (2022). https://doi.org/10.1109/TMAG.2022.3169173
16. Liu, Z., et al.: A machine-learning-based fault diagnosis method with adaptive secondary sampling for multiphase drive systems. IEEE Trans. Power Electron. **37**(8), 8767–8772 (2022). https://doi.org/10.1109/TPEL.2022/3153797
17. Cherif, B.D.E., et al.: Machine-learning-based diagnosis of an inverter-fed induction motor. IEEE Latin Am. Trans. **20**(6), 901–911 (2022). https://doi.org/10.1109/TLA.2022.9757372

Using Node-Red to Visualize Dairy Production Data via Modbus

Yuliia Samoilenko[1(✉)], Yaroslav Smitiukh[1], Yuliia Kostiuk[2], Kateryna Stepashkina[2], Dmytro Hnatchenko[2], and Valentyn Yaremych[2]

[1] National University of Food Technologies, Kyiv 01033, Ukraine
juliyasamoil@gmail.com
[2] State University of Trade and Economics, Kyiv 02156, Ukraine

Abstract. The dairy industry is under consideration, namely the possibility of collecting data from sensors of the technological process of whipping cream into butter to assess the quality of the finished product. In today's conditions of general digitalization and the Industry 4.0 strategy, the industrial sector is gradually integrating industrial IoT trends that combine automation, big data, IoT, Cloud. In this work, we will try to make it clear how easy it is to store, analyze and visualize data from certain industrial systems with the help of innovative IT developments using the advantages of the Node-RED inter-face. At industrial plants, the set of data on technological parameters is avail-able only to the equipment operator through the HMI. The integration of Node-RED with the cloud application Google Sheet (spreadsheet) is shown for storing data in the cloud, analyzing and visualizing data of technological parameters. Node-RED, along with the Modbus and Dashboard modules, can easily create a web dashboard that can be shared via a web link and viewed in any web browser on the network.

Keywords: industrial IoT · automation · big data · cloud applications · Node-Red

1 Introduction

Historically, production systems developed in parallel with a rigid hierarchical structure and were limited to the organization and use of resources related to production. The ISA951 model classifies production functions into five logically separated levels. The first three lower-level systems are secondarily responsible for operational technology (OT) operations and include hardware and software that monitor and control physical processes. The functions of the higher levels of the third and fourth are related to the field of information technology (IT), covering the management of manufacturing operations (MOM), enterprise resource planning (ERP). Because IT manufacturing and OT applications are managed distributedly within the enterprise, their integration is difficult because traditional manufacturing systems are slow to respond to changes in market demands. In addition, the rigid architecture of traditional production systems leads to difficulties in the reuse or modernization of existing production capacities. The new

R. Szewczyk et al. (Eds.): AUTOMATION 2024, LNNS 1219, pp. 81–90, 2024.
https://doi.org/10.1007/978-3-031-78266-4_8

trend of smart manufacturing (SM) (see Fig. 1) allows transforming production systems into flexible and efficient cyber-physical systems. SM integrates the latest information technologies and communication technologies into production systems, which allows responding in real time to the variability of requirements and operating conditions of the enterprise. In this, a new paradigm, the Internet of Things (IoT), the digital factory, and cloud computing technologies play an important role in transforming a rigid hierarchical structure into a flexible one.

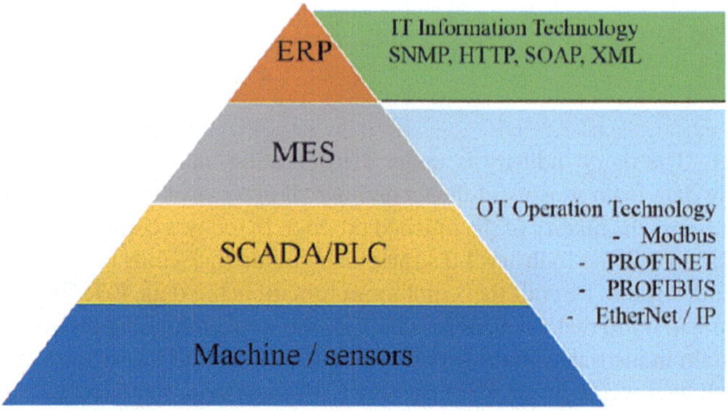

Fig. 1. Cyber-Physical System in industry

Extensive research on the Internet of Things (IoT) with cloud technologies makes it possible to accumulate huge data and turn it into knowledge with the help of intelligent data analysis [1–17], form optimal solutions, predict future situations and properly manage resources the enterprises themselves.

In the article [18], a program was developed that transmits data coming from temperature and humidity sensors connected to Raspberry Pi via the Node-RED interface to the IBM Bluemix Cloud with the possibility of access via a mobile device.

The article [19] shows a home automation system based on IoT using the Node-RED interface and the MQTT broker.

Articles [20, 21] show the possibility of applying Node-RED for smart factories using the traditional Modbus industrial network and/or the OPC UA protocol, where large volumes of data are generated that need to be stored and processed.

Building on this and focusing on data, we explore the dairy industry, namely improving butter quality through statistical diagnostics and intelligent prediction, emphasizing Modbus TCP/IP communication protocol, advanced database systems and Node-RED integration node. Based on research and articles, the goal of the work is to develop flows through the Node-RED interface to ensure the collection, analysis and processing of data from sensors and actuators of the oil production process.

2 Material and Methods

Industry is a complex, heterogeneous system consisting of various subsystems that are interconnected to ensure optimal productivity of the plant as a whole. The improvement of coordination between industrial subsystems is covered by the concept of the Internet of Things (IoT) due to the interconnection between the physical and digital worlds, which significantly improves the management of processes as a whole.

As smart manufacturing advances, I/O devices, sensors, and actuators become smarter by incorporating new functions. Similarly, the communication between these devices has improved due to the improvement of communication standards and protocols, such as Modbus, Profibus/Profinet, OPC at the lower levels, making the devices more versatile in connection through standard communications. With the growth of Ethernet-based industrial networks, it is possible to control field devices directly from the network.

The development of such trends leads to the emergence of new information architectures for industrial automation. Vogel-Heuser et al. [22] presented a new architecture as a global information architecture (see Fig. 2) for industry. It consists of two main layers, represented by two cones placed between business processes and technological processes. The bottom layer is the field and control levels in the traditional pyramid, including devices and functionality, while the top layer is represented by the process control and management level, which sits on top of the control level in the automation pyramid. The automation cloud (AT-Cloud) provides functions and services at the lower levels, while the IT-Cloud provides applications and services at the upper levels of automation. The integration between these infrastructures is based on an information model through the direct interaction of services and functions in delivery standards and presented as everything as a service (XaaS) for automation. Custom functions and automation services can be provided as SaaS with IT-Cloud and AT-Cloud. Also, it is possible to use PaaS for specific integration needs, such as process logs.

The advantage of the cloud is not only in providing data storage services, but also in supporting software and giving users remote access to data through available infrastructures such as web browsers.

Also, it is worth noting that the possible ways of integrating components of management systems with cloud platforms primarily depend on the necessary task to be solved, the choice of cloud platform, the services it provides, and, accordingly, the protocols supported by these services. In some cases, it is enough to ensure the transfer of archival data with a certain frequency for the execution of reports, in others - data must be transferred in real time with the possibility of online analysis and decision-making based on cloud computing using Data Mining.

As an integration tool, you can use the streaming programming tool – Node-RED. Node-RED is an open source environment based on JavaScript and Node.js that creates "data streams" from sensors to the cloud by connecting hardware and software. Also, this environment provides easy compilation of data processing logic and transfer of processed data to higher-level systems (SQL server, enterprise management system, centralized data collection, cloud service). Node-RED consists of main elements: node panel, flow panel, information panel and debug panel. It is great for prototyping because of its flexibility and fast turnaround time. The Node-RED platform provides opportunities

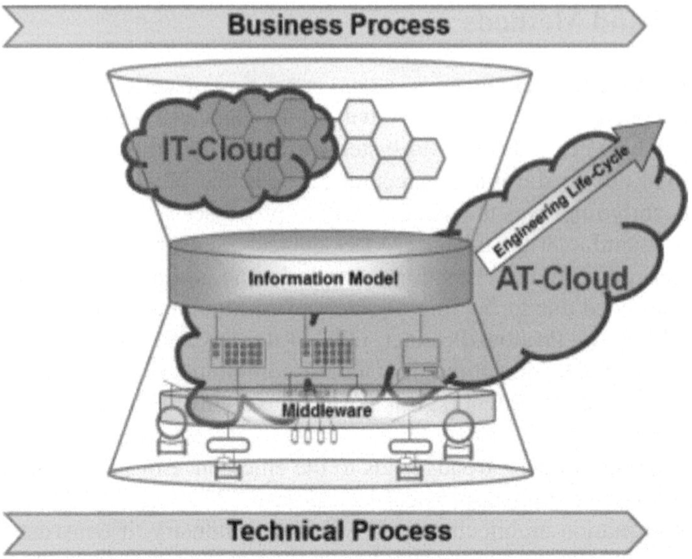

Fig. 2. Global Information Architecture with use of Cloud Computing

to create and test new prototypes embedded in existing technologies or to combine them [23–29]. Node-RED has built-in support for many protocols, such as OPC/UA, Modbus RTU/TCP/IP, S7, HTTP, MC on node-red-contrib-opcua, node-red-contrib-modbus, node-red-contrib-s7comm, node-red-contrib-omron-fins, node-red-contrib-mcprotocol, node-red-contrib-df1, node-red-contrib-cip-st-ethernet-ip.

The limits of the development of industrial automation, the use of industrial IoT, the transition of services to the cloud, the saturation of industrial networks with field-level digital devices create additional entry points for potential industrial cyber threats. Broad opportunities for increasing the level of cyber security of automation systems networks at the OT level are presented by Indegy, Ciberbit, TrapX Security solutions in the portfolio of one of the largest distributors in the field of IT security - Softprom by ERC. Solutions in combination with other measures make it possible to significantly increase the level of security in accordance with NERC CIP, NIST 800-82, IEC 62443 standards. At the IT level, existing solutions are used to ensure the protection of networks from possible cyber-attacks.

3 Functional Overview of Our Solution

Based on the above theoretical basis, an important stage is the creation of the structure of the information system for the collection, transmission and processing of technological data of dairy production, using the example of evaluating the quality of butter and water content, using modern tools in combination with cloud technologies. The hardware and software structure of such a system is shown in Fig. 3 and Fig. 4. Such systems make it possible to remotely control processes, collect, exchange and analyze data and use all the information obtained for constant process correction.

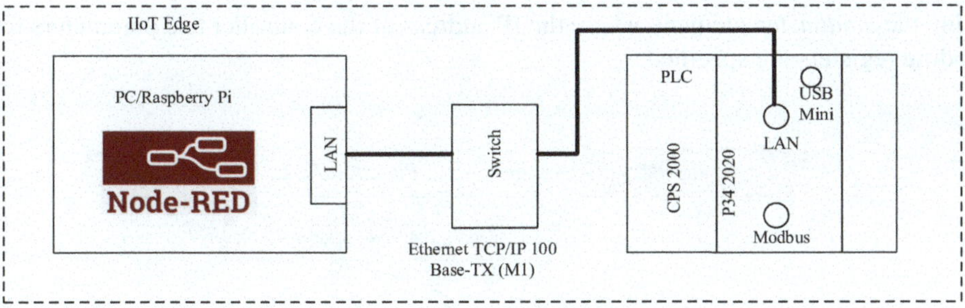

Fig. 3. Hardware architecture of the dairy production data collection system

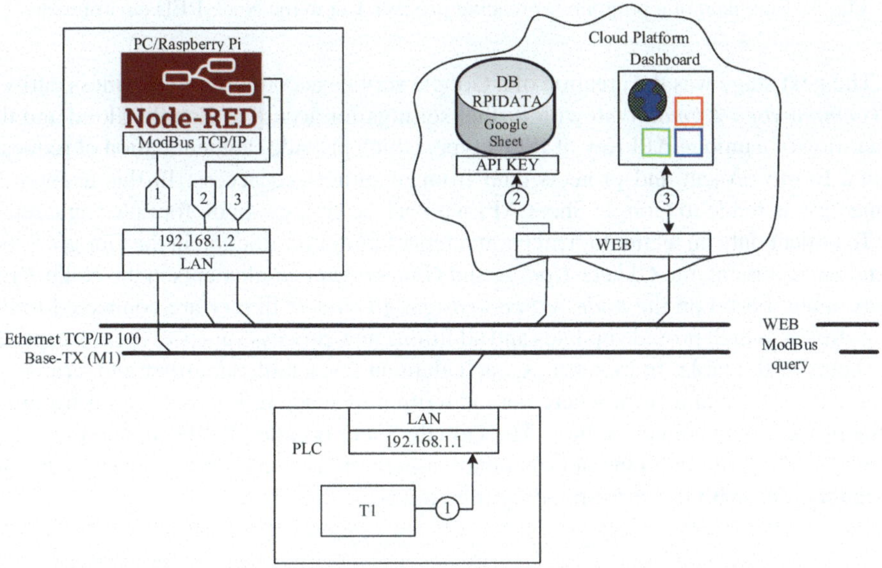

Fig. 4. Software architecture of information flows for the dairy production data collection system

All sensors together with the entire infrastructure for integration (see Fig. 3) with the level of event processing via the Internet form the Edge area. Node-Red is typically used on the Edge side as an application, for example on a PC or Raspberry PI hardware platform, thus providing data collection from various devices using industrial networking protocols. Today, the most common protocol used is Modbus TCP/IP, which works on an Ethernet network. Thus, as a PC/Raspberry PI option, it will interact with the PLC controller using the Modbus TCP/IP protocol, and on the other hand, interact with cloud applications and services (see Fig. 4). A number of IoT protocols are used to transfer data – systems along with traditional HTTP protocols and the new MQTT, AMPQ and CoAP are capable of easily sending data to the cloud.

To ensure data collection of technological process parameters for the dairy industry using the ModBus TCP/IP protocol, flows based on the *node-red-contrib-modbustcp* library were created in the Node-RED environment (see Fig. 5). Direct reading occurs using the *modbustcp* element, where the IP address of the controller and parameters for reading registers are specified.

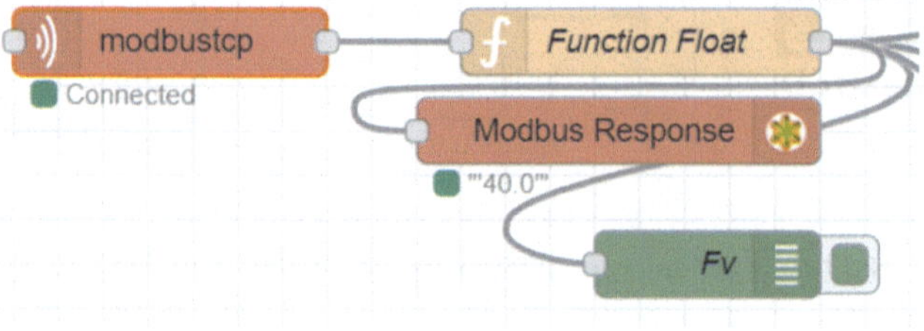

Fig. 5. Fragment of a program for reading process data in the Node-RED environment

The next stage was the creation of a Google service account for the business entity – *milkcompanynew@gmail.com* with further settings for access to Google Cloud and the generation of a unique API Key of JSON type, which provides authentication of requests related to the project and protects data from unauthorized access. In this account, a connection is made to Google Sheet API with access to the e-book for entering data.

To collect data on technological parameters of butter production in the Google Sheet Cloud environment, the *GSheet-Update* and *GSheet-Append* elements in the Node-RED environment based on the *node-red-contrib-google-sheets* library are connected to the main stream, which provide updates and additions new process values in the corresponding sheets of the table. In its settings, each element has a table identifier and a range of sheet cells, where and from where data is written or read. In Fig. 6 shows a fragment of the process data collection flow. The Google Sheet is called RPIData, and the corresponding values of cream fat and ripening temperature are recorded in the sheet F_v and t_d are recorded with the corresponding time interval.

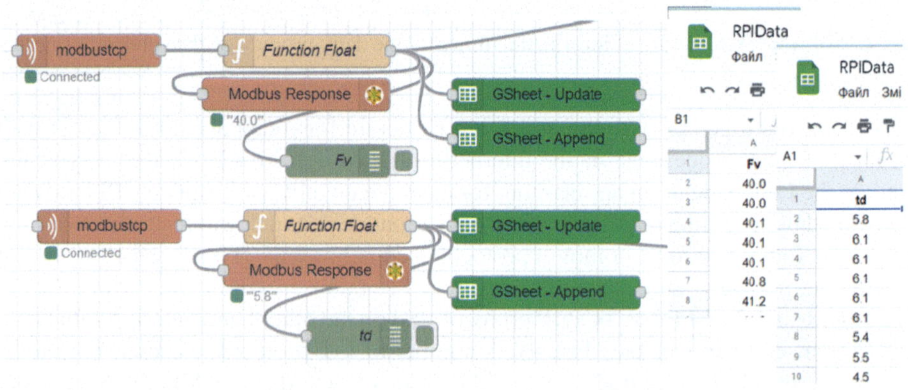

Fig. 6. Fragment of the program for connecting to the Google Sheet Cloud service

A dashboard is presented as a dashboard consisting of collections of visualization elements for presenting process data through a user interface (UI). Each visualization is a graphical display of information based on categories of attributes and their types. Other visualization characteristics can be data sets, headings and optional description, interaction elements in the form of scaling, detailing, style and semantic information. The WEB user interface in the Node-RED environment is developed based on the node-red-dashboard module. The created user interface nodes are available at *http://localhost:1880/ui* and are shown in Fig. 7. The general dashboard is called Process Monitoring, consisting of two tabs: Parameters and Charts.

The Parameters tab contains visualization elements and text fields: current date and time, parameter indicator widgets, setting settings.

Fig. 7. Information panel of the butter churning process of the dairy production

The qualitative process parameters are indicated using the Gauge node with the names of the corresponding parameters, such as: cream fatness F_v, cream ripening temperature t_d, cream supply temperature t_p, frequency of rotation of the stirrer of the whipping device S_z and consumption of the normalizing component F_{nk}. A fragment of the flow is shown in Fig. 8. This widget converts the value of the msg.payload [0] – msg.payload [4] parameters into a numeric value selected according to the *Value* format and the *Color gradient*, the *Units* display units. In this case, a yellow color was chosen for the variable cream fatness F_v, and green for the cream ripening temperature t_d with the indication of the corresponding measurement units and display range.

Also, the user interface is supplemented with fields for outputting textual information using the Text node based on the given format of the Value Format value. This widget is used to display the current date and time, as well as to display parameter limits: normal, below normal, above normal.

It is possible to directly set settings (see Fig. 8) for qualitative process parameters and record them in the corresponding registers via ModBus TCP/IP. The flow branch is implemented using the Text input node, the input mode is Mode – number. The numerical value entered in this field is sent as msg.payload with a delay of 300 ms to the corresponding register using the *modbustcp* element of the *FC6: Write Single Holding Register* function code, the input float value is previously converted to the buffer and vice versa based on the Function element JavaScript type code.

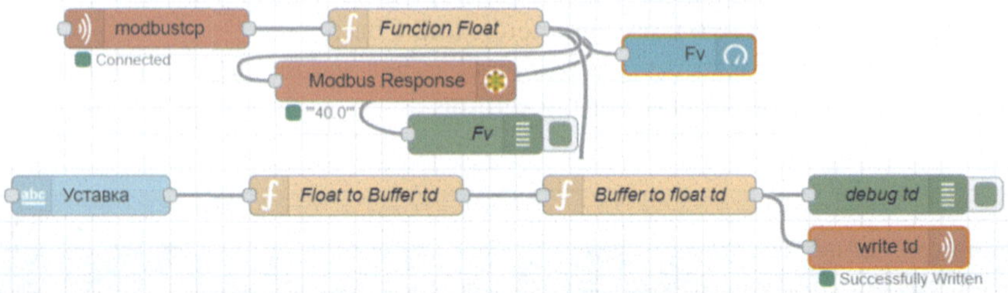

Fig. 8. A fragment of the flow in the Node-RED environment for the Gauge indicator widget and for recording the settings of quality parameters by the user

For better monitoring of the quality parameters of the process, the Charts tab (see Fig. 9) contains linear diagrams of changes in the values of the cream supply temperature t_p, frequency of rotation of the stirrer of the whipping device S_z and consumption of the normalizing component F_{nk}, which are implemented according to the chart element.

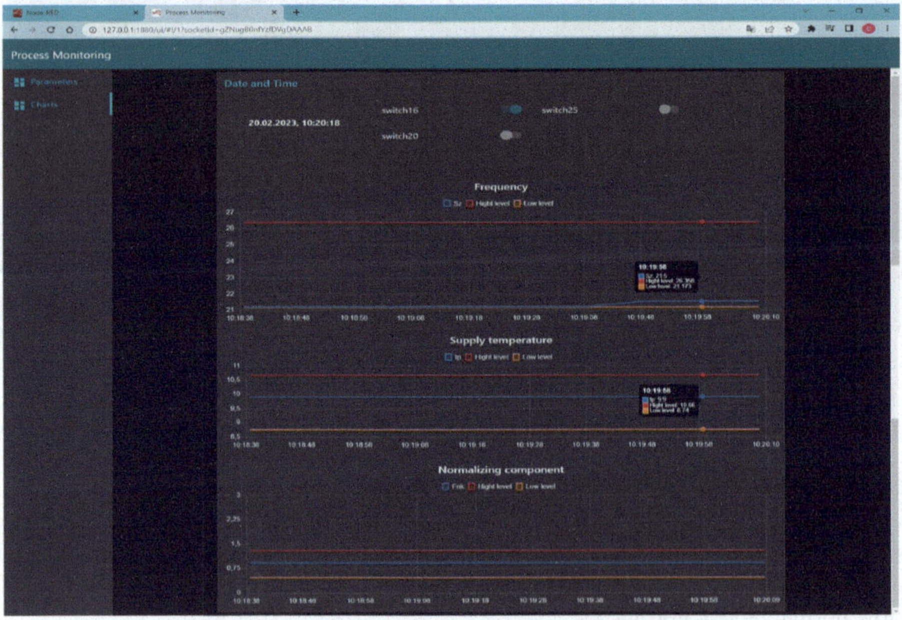

Fig. 9. Process variable dashboard with control limits

Each line graph has $\pm 2\sigma$ process limits, which warn of approaching the limits, which can lead to deterioration of the process and the quality of the final product.

4 Conclusion

This work shows the possibility of using Node-RED for the development of IIoT prototypes for dairy production, in particular for the process of whipping cream into butter. Significant benefits have been demonstrated in using Node-RED to integrate software and hardware data collection from sensors and actuators into a cloud-based process data storage and visualization environment. Thus, the combination of the two technologies with minimal development work allows the development and testing of prototypes and proof-of-concepts, optimizing industrial processes through proper data analysis.

References

1. Zaiets, N., Vlasenko, L., Lutska, N., Shtepa, V.: Resource efficiency forecasting neural network model for the sugar plant diffusion station. In: Conference on Automation, pp. 151–161. Springer, Cham (2022)
2. Lutska, N., Zaiets, N., Vlasenko, L., Shtepa, V., Savchuk, O.: Forecasting the efficiency of the control system of the technological object on the basis of neural networks. In: IEEE 20th International Conference on Modern Electrical and Energy Systems (2021)
3. Boccaletti, S., et al.: The structure and dynamics of multilayer networks. Phys. Rep. **544**(1), 1–122 (2014)
4. Korobiichuk, I., Shevchuk, D., Prokhorenko, I., Tymoshenko, N., Smityuh, Y., Boyko, R.: Synthesis of an intelligent UAV control system based on fuzzy logic in external disturbance conditions. J. Autom. Mob. Robot. Intell. Syst. **14**(3), 3–9 (2020)
5. Oliinyk, A., Skrupsky, S., Subbotin, S., Korobiichuk, I.: Parallel method of production rules extraction based on computational intelligence. Autom. Control. Comput. Sci. **51**(4), 215–223 (2017)
6. Agatonovic-Kustrin, S., Beresford, R.: Basic concepts of artificial neural network (ANN) modeling and its application in pharmaceutical research. J. Pharm. Biomed. Anal. **22**, 717–727 (2000)
7. Fellner, M., Delgado, A., Becker, T.: Functional nodes in dynamic neural networks for bioprocess modeling. Bioprocess Biosyst. Eng. **25**, 263–270 (2003)
8. Walia, N., Harsukhpreet, S., Sharma, A.: ANFIS: adaptiv neuro-fuzzy inference system – a survey. Int. J. Comput. Appl. **123**(13), 32–38 (2015)
9. Hanai, T., Iwata, N., Furuhashi, T., Honda, H., Kobayashi, T.: Proposal of reliability index in search for reliable solutions of reverse calculation based on fuzzy neural network modeling. J. Chem. Eng. Jpn. **37**, 523–530 (2004)
10. Horiuchi, J., Shimada, T., Funahashi, H., Tada, K., Kobayashi, M., Kanno, T.: Artificial neural network model with a culture database for prediction of acidification step in cheese production. J. Food Eng. **63**, 459–465 (2004)
11. Cabrera, A., Prieto, J.: Application of artificial neural networks to the prediction of the antioxidant activity of essential oils in two experimental in vitro models. Food Chem. **118**, 141–146 (2010)
12. Martinez, O.: Artificial neural network in response of mass transfer parameters predictions (moisture loss and solid gain) during osmotic dehydration of fruits. Acta Agronómica **65**(4), 318–325 (2016)

13. Jimenez-Marquez, S., Thibault, J., Lacroix, C.: Prediction of moisture in cheese of commercial production using neural networks. Int. Dairy J. **15**, 1156–1174 (2005)
14. Paquet, J., Lacroix, C., Thibault, J.: Modeling of pH and acidity for industrial cheese production. J. Dairy Sci. **83**, 2393–2409 (2000)
15. Funahashi, H., Horiuchi, J.: Characteristics of the churning process in continuous butter manufacturing and modeling using an artificial neural network. Int. Dairy J. **18**, 323–328 (2008)
16. Funahashi, H., Horiuchi, J., Nakamura, Y.: Predicting manipulated variables to control water content in continuous butter manufacture by an artificial neural network. Milchwissenschaft **63**, 41–44 (2008)
17. Sakairi, T., Palachi, E., Cohen, C., Hatsutori, Y., Shimizu, J., Miyashita, H.: Designing a control system using SysML and Simulink. In: 2012 Proceedings of SICE Annual Conference (SICE), pp. 2011–2017 (2012)
18. Kodali, R., Anjum, A.: IoT based home automation using node-red. In: Second International Conference on Green Computing and Internet of Things (ICGCIoT) (2018)
19. Sicari, S., Rizzardi, A., Coen-Porisini, A.: Smart transport and logistics: a node-RED implementation. Internet Technol. Lett. **2**(2) (2019)
20. Toc, S.-I., Korodi, A.: Modbus-OPC UA wrapper using node-RED and IoT-2040 with application in the water industry. In: 2018 IEEE 16th International Symposium on Intelligent Systems and Informatics (SISY). IEEE (2018)
21. Tabaa, M., et al.: Industrial communication based on modbus and node-RED. Procedia Comput. Sci. **130**, 583–588 (2018)
22. Vogel-Heuser, B., Kegel, G., Bender, K., Wucherer, K.: Global information architecture for industrial automation. Automatisierungstechnische Praxis (atp), Oldenbourg-Verlag, Muenchen, Bd. 51, Nr. 01–02, pp. 108–115 (2009)
23. Smitiukh, Y., Samoilenko, Y., Kostiuk, Y., Kryvoruchko, O., Stepashkina, K.: Development of a prototype of an intelligent system for predicting the quality of dairy manufacture. In: 2022 IEEE 11th International Conference on Intelligent Systems (IS), Warsaw, Poland, pp. 1–6 (2022)
24. Dikaiakos, M., Katsaros, D., Mehra, P., Pallis, G., Vakali, A.: Cloud computing: distributed internet computing for it and scientific research. IEEE Internet Comput. **13**(5), 10–13 (2009)
25. Jadeja, Y., Modi, K.: Cloud computing – concepts, architecture and challenges. In: International Conference on Computing, Electronics and Electrical Technologies (ICCEET) (2012)
26. Shaikh, F., Haider, S.: Security threats in cloud computing. In: International Conference for Internet Technology and Secured Transactions (ICITST) (2011)
27. Yu, W., Chen, J.: Semantic Service in Cloud Computing. Springer (2011)
28. Givehchi, O., Trsek, H., Jasperneite, J.: Cloud computing for industrial automation systems a comprehensive overview. In: 18th IEEE Conference on Emerging Technologies Factory Automation (ETFA) (2013)
29. Dickey, N., Banks, D., Sukittanon, S.: Home automation using cloud network and mobile devices. In: Proceedings of IEEE Southeastcon (2012)

Automatic Functional Tests of Cash Registers

Agnieszka Hys[1] and Paweł Drapikowski[2(✉)]

[1] Central Office of Measures, ul. Elektoralna 2, 00-139 Warszawa, Poland
agnieszka.hys@doctorate.put.poznan.pl

[2] Poznan University of Technology, pl. Marii Sklodowskiej-Curie 5, 60-965 Poznan, Poland
pawel.drapikowski@put.poznan.pl

Abstract. The paper presents formal requirements and a prototype of a stand for automatic functional tests of cash registers. The authors are unaware of any automatic cash register test stands offered on the market and operating. An OCR system was used to detect digits and inscriptions on keys, and their spatial coordinates are determined using stereo vision. Deep neural networks were used to detect pictograms from outside the OCR database. Due to the different size of the cash registers, the stereo vision system must be reconfigured in terms of the height of the cameras and their base. Fast stereo calibration based on only a pair of images was used, which provides sufficient measurement properties. The calculated key coordinates are sent to an industrial robot equipped with a flexible pointer adapted to pressing the keys. The key sequence enabling individual tests is entered by the operator based on the cash register documentation. The advantage of the proposed stand is its versatility, which allows for testing of cash registers of different manufacturer and other devices with similar functionality.

The stand is intended for implementation at Central Office of Measures.

Keywords: Cach register · Functional tests · Stereovision · Convolutional neural network

1 Introduction

1.1 Purpose and Scope of Study

The aim of the research was to automate the process of functional testing of cash registers performed manually by the operator to the extent possible to automate. Automation covers typical general-purpose hardware cash registers, but does not include special purposes cash registers, e.g. pharmacy, taxi drivers, ticket or software cash registers. Time-consuming and monotonous activities performed by the operator were eliminated by using a vision perception system and an industrial robot. The perceptual system is responsible for recognizing the keys necessary to perform the tests and determining the coordinates of their centers in the local coordinate system. It is also responsible for recognizing the content displayed on the display and paper printouts. Key coordinates are determined using stereo vision, and a dedicated camera is used to analyze printouts.

R. Szewczyk et al. (Eds.): AUTOMATION 2024, LNNS 1219, pp. 91–98, 2024.
https://doi.org/10.1007/978-3-031-78266-4_9

1.2 Test Stands

The authors are unaware of any automatic cash register test stands offered on the market and operating. The closest solution is the robot offered by Abrantix [1] for testing payment terminals. It is a Cartesian robot that presses keys and has the ability to insert and remove payment cards. The latest version is designed to test four terminals simultaneously. Tests of cash registers for operational safety, flammability and mechanical properties are offered by the House of Testing laboratory [2].

The advantage of the proposed stand is its versatility, which allows for testing of cash registers of different manufacturer and other devices with similar functionality. Universality is ensured by easy adaptation of the perception system to detect features of various devices available on market.

A little drawback is a need of stereo vision system calibration each time, when the base and height of cameras is changed.

1.3 Formal Requirements

One of the basic tasks of the Central Office of Measures is to perform legal metrological control of measuring instruments in order to ensure the protection of the interests of the State and citizens of the Republic of Poland. This protection involves checking whether measuring instruments meet the technical requirements specified in legal regulations both before being placed on the market and during their use. Relevant legal acts specify the scope and type of devices covered by legal metrological control and appropriate tests.

Tests of the cash register functions include tests to ensure compliance with the correct operation of the sales service and cash register reporting functions, the correctness of calculating the tax due, the format and content of fiscal printouts, and the correct operation of the system for saving electronic copies of printouts for cash registers with electronic recording of copies of documents. The research includes:

- checking the formal correctness of printouts and getting to know how to operate the cash register in non-fiscal mode,
- fiscalization of the cash register and verification of zeroing of the cash register counters during the cash register in order to check the correctness of calculating the tax due,
- introduction of sales operations including functionalities specified in the regulation in order to check the correct operation of the sales function, the content and format of fiscal printouts and the correct operation of the system for saving electronic copies of printouts,
- closing the cash register's fiscal mode,
- readings of fiscal memory and the contents of the electronic copy in order to check the correctness of the cash register reporting function and the operation of the system for saving electronic copies of printouts,
- assessment of the computational correctness of the registration of turnover and VAT due by the cash register program based on the content of fiscal printouts,
- verification of documents contained in the electronic copy in order to check the correctness of the cash register reporting function and the operation of the system for saving electronic copies of printouts for cash registers with an electronic copy.

A typical test procedure involves the operator pressing a sequence of keys and checking the correct results on the display and printouts in terms of numerical values and printout features.

To sum up the above, the design and introduction of an innovative methodology for testing cash registers using a perception system and a robot will free the operator from the need to perform repetitive and monotonous activities to the extent possible.

2 Features of Cash Registers and the Perception System

2.1 Features of Cash Registers

General-purpose hardware cash registers are characterized by different sizes, different types of keys and displays that make recognition of digits and inscriptions by OCR (Optical Character Recognition) systems difficult. An additional difficulty is the reflective keyboards (or touch screen keyboard), which impede the operation of the vision system. There are also markings that the OCR system cannot recognize, for example, the decimal separator. In this case, the key must be marked with a special code replacing the decimal point.

2.2 The Perception System

To carry out the tests introduced in 1.3, it is necessary to know the keys coordinates and read the content on paper printouts and displayed on the display. At the beginning of study, the RGBD SR300 camera was examined [3]. However, the obtained spatial model did not allow to identify of key coordinates with the required precision. It was decided to use a stereo vision system. Due to the different sizes of cash registers, it is necessary to change the distance of the cameras from the objects and rotate them. This reconfiguration of the camera system requires calibration each time. Due to the convenience of using the test stand, it was decided to perform calibration only on the basis of a pair of images of a checkerboard pattern. This method of calibration based on only a pair of images means that the identification of polynomial coefficients describing the distortions of the camera model may be insufficient. To avoid the significant impact of lens distortions, lenses with a longer focal length (12 mm) and cameras with high resolution (3840×2160 px) were selected (Basler a2A3840-4 cameras [4] and Basler C11-1220 lenses [5]). An interesting proposals to accelerate this procedure while maintaining changes in the location of the calibration pattern were presented by the authors of the work [6]. They used a computer screen to display the pattern and during the iterative calibration procedure corrected the parameters of the stereo vision system model.

A camera and lens identical to those used for the stereo vision system were used as a dedicated camera for reading the content of printouts and display.

3 Industrial Robot

Conducting functional tests of cash registers requires pressing appropriate key sequences. This can be achieved by building a Cartesian robot of appropriate dimensions or using a commercially available industrial robot. The second solution was chosen and in the initial testing phase an ABB IRB 120 robot was used. Ultimately, a robot SCARA Epson G6-651SW [7] will be used.

Interesting examples of robotic stations used for quality control and measurement of items manufactured for the aviation industry were presented by authors from the Rzeszow University of Technology [8, 9]. These stations combine the flexibility of using industrial robots with the measurement accuracy of optical sensors.

The vision system allows to recognize the coordinates of the keys, but does not allow to estimate the type of keyboard and the depth of pressing them. In order to overcome this inconvenience, a flexible tool for the robot (pointer) was designed, which, due to the resistance of the pressed key, withdraws without damaging the cash register or the robot. A certain value (offset) is added to the coordinates determined by the vision system in the direction of operation of the tool, enabling keystrokes to be pressed.

The movement sequence including time of key press duration is encoded by writing text file records corresponding to individual test phases and the type of cash register.

4 Software

The software for the perception system was written in Python using the OpenCV library for image processing and calibration of the stereovision system. An OCR system was used to recognize digits, keyboard inscriptions, displayed and printed content. Two OCR systems have been investigated: keras_ocr [10] and easyocr [11]. In terms of recognizing characters and inscriptions, their effectiveness is very similar, but the keras_ocr system was unable to recognize the content displayed on the displays. Hence, it was decided to use the easyocr system. The result of the ocr system is the recognized text and the rectangle bounding box.

Recognition of pictograms (arrows) enabling navigation between checkout menu items was performed using a pre trained YOLOv7 convolutional neural network [12]. Based on internet sources and cash registers owned by the authors, data sets were prepared: training 339 pictograms, validation (29) and test (27). The Open3D library [13] was used to graphically present the measurement results in 3D.

The program for the ABB IRB 120 robot was written in the RAPID language using socket communication between processes. The robot plays the role of a "server" waiting for coordinates to be sent from the perception system (client).

5 Results

Figure 1 shows a view of the test stand. Local lighting was used, which significantly improved the effectiveness of keyboard character recognition.

5.1 Calibration and Measurements Using a Stereo Vision System

For easy concatenation of the coordinate systems of the stereo vision and the robot, the origin of the coordinate system is placed at point p0 of the calibration pattern (the first point detected by the corner detector), and the axes directions correspond to the lines

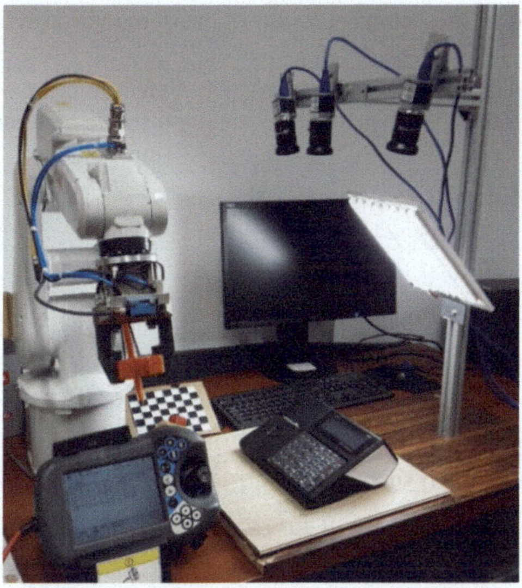

Fig. 1. Prototype test stand of cash registers.

between the fields of the chessboard. This coordinate system position is easily accessible to the robot and allows calibration of its local coordinate system. Robot local coordinate system is calibrated according to its implemented method mostly 3-points method. The coordinates of the points detected by the stereo vision system become arguments for the robot's motion commands.

Verification of the calibration process was carried out for each camera based on RMS of reprojection error for all corners of calibration pattern. RMS should be below 1 pix mostly is in range of 0.5–0.6 px.

Verification of stereo vision system measurement properties was carried out using a chessboard by moving it horizontally and vertically. The coordinates of 54 corners were detected and the 48 distances between corners in X and 45 distances in Y were calculated knowing real distance between them (the pattern size PS – 16.14 mm). The test board was moved horizontally and vertically relative to the calibration pattern. The parameters of the stereo vision system are:

- the cameras height – 575 mm,
- the cameras base – 330 mm.

The measurement results are included in Table 1.

Errors are calculated as follows:

$$EX = \frac{abs(PS - XMean)}{PS} * 100, EY = \frac{abs(PS - YMean)}{PS} * 100 \quad (1)$$

Records marked with an asterisk indicate measurements taken with image rectification. The results indicate that rectification has no significant effect on the measurements. Mean percentage error for 14 measuring position was calculated in X (1.14%) and Y (1.31%). Based on this errors one can evaluate measuring properties of the stereo vision

Table 1. Measurement results of the stereo vision system.

No.	Position [mm]	Height relative to the calibration pattern [mm]					Measurement of the distance between corners [mm]				Error [%]	
	X	Real	Mean	Dev. st.	Min.	Max.	X Mean	Dev. st.	Y Mean	Dev. st.	EX	EY
1	0,0	0	0,30	0,38	−0,48	1,01	16,14	0,15	16,10	0,17	0,00	0,25
2	−50,0	0	0,30	0,37	−0,53	1,04	16,12	0,14	16,12	0,17	0,12	0,12
3	50,0	0	0,42	0,28	−0,18	1,01	16,09	0,17	16,12	0,17	0,31	0,12
4	0,0	−22	−20,34	0,57	−21,56	−19,12	15,96	0,18	15,91	0,16	1,12	1,43
5	−50,0	−22	−20,91	0,61	−22,19	−19,72	15,97	0,17	15,95	0,13	1,05	1,18
6	50,0	−22	−19,19	0,59	−20,28	−18,04	15,79	0,18	15,84	0,17	2,17	1,86
7	0,0	22	20,45	0,57	19,40	21,87	16,29	0,14	16,32	0,15	0,93	1,12
8	0,0*	22	20,36	0,33	19,76	21,15	16,32	0,16	16,32	0,16	1,12	1,12
9	−50,0	22	19,68	0,74	18,39	21,27	16,23	0,14	16,30	0,15	0,56	0,99
10	50,0	22	20,96	0,48	20,12	22,26	16,25	0,15	16,28	0,17	0,68	0,87
11	0,0	−44	−38,67	0,92	−40,54	−36,97	15,81	0,19	15,74	0,18	2,04	2,48
12	0,0*	−44	−38,86	1,25	−41,21	−36,78	15,85	0,15	15,77	0,17	1,80	2,29
13	0,0	44	38,37	0,96	36,55	40,27	16,45	0,16	16,51	0,13	1,92	2,29
14	0,0*	44	38,37	0,65	37,07	40,05	16,48	0,19	16,51	0,13	2,11	2,29

Column ‘Position’ means position in relation to the base center. ‘X Mean’ and ‘Y Mean’ are the appropriate euclidean distances in X and Y direction of test board.

system as satisfactory. The measurement results confirm the correctness of one shot calibration method and correct selection of camera lenses.

Analysis of the results shows that the system tends to shorten the vertical distance compared to the real one (in relation to position of calibration pattern during calibration). This property will not be of major importance when testing cash registers (the robot presses keys), because the tool (pointer) has the ability to move backwards when is pressing keys. In addition, for cash registers whose keys have a certain click depth, an offset will be added to enable pressing the keys.

However, horizontal distances are reproduced with satisfactory accuracy. It can be concluded that the measurement properties of the stereo vision system meet the requirements to the extent necessary for measuring cash register keyboards.

5.2 Keyboard Coordinates Detection

The keyboard coordinates are determined based on the coordinates of the rectangles surrounding the text or digit so that they fall near their centers. The coordinates of the center of the numeric keys are calculated from the left side of the digit, while the coordinates of the inscription and arrow keys are calculated in the center of the rectangle.

Due to the difficulties of OCR systems in recognizing the characters ‘1’, ‘4’, ‘7’, it was assumed that the coordinates of these keys would be determined on the basis of the remaining numeric keys, omitting ‘0’. It also happens that the ocr system recognizes a character multiple times. Then the correct character is selected based on the distance relationship to other surrounding keys. The condition for proceeding to the test procedure is correct recognition and determination of the coordinates of the necessary keys, and confirmation by pressing the keys in the range <0;9> and their recognition on the display.

Figure 2 shows an example of character recognition for the POSNET Neo cash register, and Fig. 3 shows their spatial graphical representation. Figure 2 also shows the coordinate system created during calibration, which is also the base for the robot.

Fig. 2. POSNET Neo cash register keyboards (left and right camera) with recognized keys

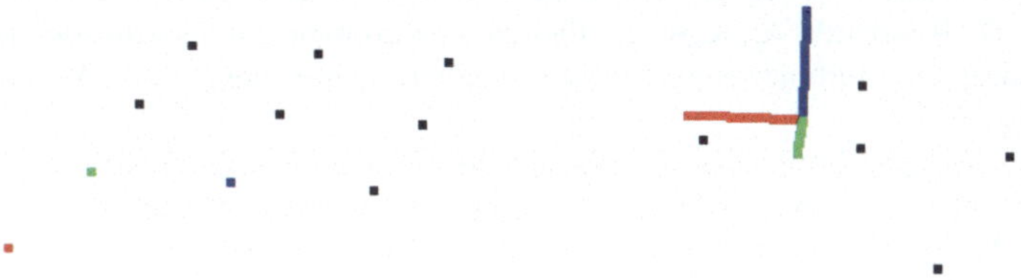

Fig. 3. Graphical representation of numeric keys, arrows and ‘RAZEM’ inscription key coordinates

6 Conclusion

The set up intended for functional tests of hardware cash registers is ready to carry out test procedures. At the same time, a cash register model will be created to enable a comparative procedure to be carried out with the model created by law regulation.

The key problem is keyboard recognition, which may seem like an easy task. However, the large variety of cash registers and the not entirely predictable operation of OCR systems make it not easy in reality. This is confirmed by the lack of test stands on the market with properties enabling testing the functionality of cash registers in accordance with law regulations. Especially since the test procedure is time-consuming and is still performed by the operator.

It is not possible to completely eliminate the operator during testing. Keystroke sequences for performing test operations are contained in the manufacturer's manual and must be entered to the system before testing begins.

To sum up, designing and introducing an innovative methodology for automatic testing hardware cash registers using a perception system and a robot will, free the operator from the need to perform repetitive and monotonous activities.

The stand is intended for implementation at Central Office of Measures.

References

1. Abrantix. https://www.abrantix.com/en/test-automation-products.html. Accessed 15 Dec 2023
2. House of Testing. https://www.houseoftesting.com/about-us. Accessed 15 Dec 2023
3. Intel. https://www.intel.com/content/www/us/en/products/sku/92329/intel-realsense-camera-sr300/specifications.html. Accessed 15 Dec 2023
4. Basler. https://www.baslerweb.com/en/shop/a2a3840-45ucbas/. Accessed 15 Dec 2023
5. Basler. https://www.baslerweb.com/en/shop/basler-lens-c11-1220-12m-p-f12mm/. Accessed 15 Dec 2023
6. Barone, S., Neri, P., Paoli, A., Razionale, A.V.: Flexible calibration of a stereo vision system by active display. In: 29th International Conference on Flexible Automation and Intelligent Manufacturing (2019). Procedia Manuf. **38**, 564–572
7. EPSON. https://neon.epson-europe.com/de/de/robots/products/product.php?id=12747&tab=downloads&content=547. Accessed 15 Dec 2023
8. Burghardt, A., Kurc, K., Szybicki, D., Muszyńska, M., Szczęch, T.: Monitoring the parameters of the robot-operated quality control process. Adv. Sci. Technol. **11**(1), 232–236 (2017)
9. Szybicki, D., Burghardt, A., Kurc, K., Pietruś, P.: Calibration and verification of an original module measuring turbojet engine blades geometric parameters. Arch. Mech. Eng. **66**(1), 97–109 (2019)
10. Keras-ocr Homepage. https://keras-ocr.readthedocs.io/en/latest/. Accessed 15 Dec 2023
11. Easyocr Homepage. https://www.jaided.ai/easyocr/. Accessed 15 Dec 2023
12. Wang, C.-Y., Bochkovskiy, A., Liao, H.-Y.M.: YOLOv7: trainable bag-of-freebies sets new state-of-the-art for real-time object detectors. In: CVPR (2022)
13. Open3D Homepage. http://www.open3d.org/. Accessed 15 Dec 2023

Sliding Mode Inventory Control with a Pregenerated Reference Trajectory Profile

Katarzyna Adamiak(✉) and Tymoteusz Zwierzchowski

Institute of Automatic Control, Łódź University of Technology, 18 Stefanowskiego St., 90-537 Łódź, Poland
{katarzyna.adamiak,tymoteusz.zwierzchowski}@p.lodz.pl

Abstract. This study tackles the problem of discrete time sliding mode control design for inventory management systems and proposes a new concept of demand structure and control. The paper considers a single product multi supplier inventory system. It is assumed that the demand divides into two sections: the unknown but bounded random demand and an a priori known contractual component. Therefore, it is proposed to firstly take into account the contractual demand only and generate the desired trajectory profile. In other words, the strategy aims to shape the chain of supplies according to the a priori known demand term. Next, the obtained trajectory profile is applied as a reference in the sliding mode control scheme, where an additional compensation of the random sales is introduced. This concept guarantees satisfaction of both contractual and random demand within their bound at any time instant. Moreover, the proposed strategy greatly simplifies the control design process.

Keywords: Control Design · Discrete-Time Systems · Inventory Systems · Sliding Mode Control · Trajectory Generator

1 Introduction

Sliding mode control is recently one of the most eagerly applied variable structure control strategies [1–5]. Its foundation is based on a selected sliding hyperplane in the state space, designed to guarantee a stable performance. Once bounded to the sliding hyperplane the order of the control system is reduced by one. The sliding mode controller's task becomes driving the representative point of the system onto the sliding hyperplane and ensuring it remains on the plane afterwards. Such control action utilizes two linear state feedback control laws applied alternately, depending on the system's position relative to the switching hyperplane, denoted with a so-called sliding variable.

With the sliding mode concept the state trajectory may be divided into two sections. The first section is when the representative point is driven towards the switching hyperplane and external disturbances may affect its performance. However, once the representative point is on the sliding hyperplane, the control action pushes it back onto it whenever any deviation occurs. Consequently, in the sliding phase the system becomes

R. Szewczyk et al. (Eds.): AUTOMATION 2024, LNNS 1219, pp. 99–110, 2024.
https://doi.org/10.1007/978-3-031-78266-4_10

robust to matched external disturbances and modelling uncertainties [5]. Time variant sliding planes expand this effect to the whole control process [6, 7]. In that case, the sliding hyperplane crosses the initial position of the system's representative point as well as the demand state. Therefore, the switching control action occurs from the very beginning of the control process, suppressing the disturbance influence.

Except for the reduction of system's order and their robustness, sliding mode control systems are also characterized by lower computational effort than other comparably effective strategies. Thanks to these features, sliding mode control schemes have found applications in electromechanics and power electronics [8–10], robotics [11, 12] and logistic systems [22, 23], which are the subject of this study as well.

The problem of resupply chain management for inventory systems has been widely discussed in the literature. Most conventional control strategies aim to fill the maximum warehouse capacity or satisfy the maximum demand. However, due to markets globalization and rapid changes in customer demand, more flexibility and robustness to changing economical situation is nowadays required. Although the task is not trivial, many control theory based methods have been successfully applied to the logistic systems [13–15]. They include PID controllers [15, 16], $H\infty$ norm minimizing strategies [17, 18] and optimal control [19, 20]. Most recent contributions apply neural networks to learn and predict market fluctuations [21]. Discrete time sliding mode algorithms have also been proposed in [22, 23]. These works are characterized by relatively high robustness and much lower computational difficulty than the aforementioned strategies. Therefore, this study proposes a novel reference trajectory based discrete time sliding mode control scheme for an inventory system with a redefined customer demand.

This paper considers a single product multi supplier based inventory system, where each supplier requires a certain lead time to deliver the goods. The system must handle the demand of dual nature: an a priori known component representing contractual obligations and random demand. Therefore, the system's primary goal is to fulfill the contract while the secondary goal is to provide spare product for random buyers. Having the knowledge of the contract, we propose to use it to generate a desired state trajectory that satisfies the contract only and then apply it as a reference in a sliding mode control scheme. We propose to define a time variant sliding plane, which shifts according to the reference trajectory profile and design a control action which keeps the system's representative point on this plane in spite of the presence of random demand. We prove that this approach ensures full demand satisfaction, both contractual and random.

2 Inventory System

This paper deals with the problem of control design for an inventory system based on a single product delivered by multiple suppliers with different delivery delays. All suppliers with the same delivery time are considered as one. The lead times in the system are within the range $i = 1, 2, \ldots, n$, where n is the maximum lead time. This yields that all orders placed at the time instant k, with the control $u(k)$, will have been delivered and available for sale latest after $n + 1$ discrete time instants. Therefore $n + 1$ is the order of the dynamical system. At each step k the controller calculates the required order volume and allocates its a_i part to the supplier with lead time i. The supplier with lead time 1 is allocated the a_1 part of the order, whereas the supplier with lead time n is allocated with a_n part of the total order volume. If a supplier with a certain lead time i does not exist then $a_i = 0$. Parameters a_i satisfy

$$0 \le a_i \le 1 \quad \text{and} \quad \sum_{i=1}^{n} a_i = 1. \tag{1}$$

The dynamics of the system are described with the following state equation

$$\boldsymbol{x}(k+1) = \boldsymbol{A}\boldsymbol{x}(k) + \boldsymbol{b}u(k) - \boldsymbol{f}h(k), \tag{2}$$

where the state vector $\boldsymbol{x}(k)$ contains the warehouse stock and all the deliveries being currently carried out. Therefore, it is of $n+1$ order and has the following structure

$$\boldsymbol{x}(k) = \left[x_1(k)\ x_2(k)\ \ldots\ x_n(k)\ x_{n+1}(k)\right]^T. \tag{3}$$

The state variables may be interpreted so that x_1 is the current warehouse stock available for sale, x_2 represents the goods that have been ordered and will arrive at the warehouse after one discrete time instant, x_3 denotes the goods that will arrive after two time instants and so on. The current stock of the warehouse, and at the same time the system's output, is the first state variable $y(k) = x_1(k)$. The initial stock is $y(0) = y_0$. The discrete time control signal is $u(k)$ and it represents the volume of product ordered at step k. In response to the customer demand, the system sells goods from the warehouse, $h(k)$ denoting the volume of product sold at instant k. The state matrix of the system $\boldsymbol{A}$, the input distribution vector $\boldsymbol{b}$ and the sales vector $\boldsymbol{f}$ have the following structure

$$\boldsymbol{A} = \begin{bmatrix} 1 & a_n & a_{n-1} & \cdots & a_2 & a_1 \\ 0 & 0 & 1 & \cdots & 0 & 0 \\ \vdots & \vdots & \vdots & \ddots & \vdots & \vdots \\ 0 & 0 & 0 & \cdots & 1 & 0 \\ 0 & 0 & 0 & \cdots & 0 & 1 \\ 0 & 0 & 0 & \cdots & 0 & 0 \end{bmatrix}_{n+1 \times n+1}, \quad \boldsymbol{b} = \begin{bmatrix} 0 \\ 0 \\ \vdots \\ 0 \\ 1 \end{bmatrix}_{n+1}, \quad \boldsymbol{f} = \begin{bmatrix} 1 \\ 0 \\ \vdots \\ 0 \\ 0 \end{bmatrix}_{n+1}. \tag{4}$$

The considered inventory system is subject to customer demand of dual nature: the contractual demand denoted with $d_c(k)$ and the random varying demand denoted with $d_r(k)$. Therefore, the total customer demand is $d(k) = d_c(k) + d_r(k)$. The contractual demand is a priori known, however, there may exist an instant when the customers do not purchase all contracted goods. Therefore, it holds that

$$0 \le d_c(k) \le \hat{d}_c(k), \tag{5}$$

where $\hat{d}_c$ is the a priori expected value. The random demand is bounded as follows

$$0 \le d_r(k) \le d_{\max}. \tag{6}$$

Due to the dual nature of the demand there are two kinds of sales in the system: $h(k) = h_c(k) + h_r(k)$, where $h_c(k)$ is the sale due to contractual demand and $h_r(k)$

corresponds with the random demand. The sales volume is limited by the amount of product in stock as well. The contractual demand has higher priority. Therefore, the contractual sales volume satisfies the demand up to the number of product in stock

$$h_c(k) = \min\left[y(k), d_c(k)\right]. \tag{7}$$

As follows, it holds that

$$h_c(k) \leq d_c(k). \tag{8}$$

If, the contract is satisfied and there exists any leftover product in stock, it may be used to satisfy the random demand. The leftover product amount $y_r(k)$ is obtained from

$$y_r(k) = y(k) - h_c(k). \tag{9}$$

As follows, the random sales is limited by the demand and the leftover stock

$$h_r(k) = \min\left[y_r(k), d_r(k)\right]. \tag{10}$$

Moreover, it holds that

$$h_r(k) \leq d_r(k). \tag{11}$$

The amount of product left in stock after the all sales commence may be obtained as

$$y(k) = y_0 + \sum_{i=1}^{n-1} a_i \sum_{j=0}^{k-i-1} u(j) - \sum_{j=0}^{k-1} h(j), \tag{12}$$

for any $k \leq n$ and for any $k \geq n + 1$ as

$$y(k) = y_0 + \sum_{j=0}^{k-n-1} u(j) + \sum_{i=1}^{n-1} a_i \sum_{j=k-n}^{k-i-1} u(j) - \sum_{j=0}^{k-1} h(j). \tag{13}$$

Finally, after both contractual and random sales have concluded, the product left in the warehouse is denoted with

$$y_s(k) = y(k) - h(k). \tag{14}$$

The next chapter proposes a novel control strategy, which utilizes the knowledge of the contractual demand in order to generate a reference trajectory for the system.

3 Reference Trajectory Based Sliding Mode Control

Partial knowledge of the demand allows to a priori generate the order sequence necessary to satisfy the contract. Therefore, the control design is divided into two steps. First, the contractual demand terms are used to generate the desired trajectory profile. As all suppliers must have commenced their deliveries after $n + 1$ time instants, the demand trajectory at each step incorporates the consequent $n + 1$ contractual demand terms. The reference will be denoted with $s_r(k)$, where

$$s_r(k) = \sum_{l=k}^{k+n} \hat{d}_c(l). \tag{15}$$

The value of $s_r(k)$ may represents the amount of product in stock necessary to fully satisfy the contractual demand for $n + 1$ time instants: k, $k + 1$, ..., $k + n$. For $k = 0$

$$s_r(0) = \hat{d}_c(0) + \hat{d}_c(1) + \ldots + \hat{d}_c(n). \tag{16}$$

The reference trajectory profile changes according to the contractual demand flow. As the contractual system obligations are known in advance it may be generated a priori and saved in a look-up table.

Next, the obtained trajectory profile $s_r(k)$ will be used to form a time varying sliding plane for the control of the system with the random demand acting as a bounded disturbance. We define the sliding hyperplane as $s(k) = 0$ and select the sliding variable so that it represents the distance of the system's representative point $\boldsymbol{cx}(k)$ from the reference position $s_r(k)$, which yields

$$s(k) = s_r(k) - \boldsymbol{cx}(k). \tag{17}$$

Vector $\boldsymbol{c}$ is chosen in a dead-beat manner, so that all poles of the closed-loop system are place at the origin of the complex plane, i.e. the characteristic polynomial satisfies $M(z) = z^{n+1}$. For that purpose vector $\boldsymbol{c}$ is defined as

$$\boldsymbol{c} = \left[1 \; a_n \; \sum_{i=n-1}^{n} a_i \ldots \sum_{i=2}^{n} a_i \; \sum_{i=1}^{n} a_i \right]_{n+1}. \tag{18}$$

The proposed sliding hyperplane may be understood as a time-varying hyperplane, as it shifts following $s_r(k)$. According to time-varying sliding plane concept, the system's initial position must belong to the hyperplane. Therefore, it must hold that $s(0) = 0$. As follows

$$y_0 = x_1(0) = s_r(0). \tag{19}$$

The condition (19) means that the initial stock of the warehouse must be sufficient to cover the contractual demand for steps $k = 0, 1, \ldots, n$, up to the moment when all goods ordered at $k = 0$ will have arrived at $k = n + 1$. However, it must be taken into account that from the very beginning of the control process, the random demand exists as well. Therefore, the initial condition (19) will be further modified by a compensation term.

Having introduced a time varying sliding hyperplane, we design a control law, which keeps the system's representative point on the sliding hyperplane for the whole control process. Therefore, we use a reaching law, which drives the representative point $\boldsymbol{cx}(k)$ to the reference position in each step

$$s(k + 1) = s_r(k + 1) - \boldsymbol{cx}(k + 1) = 0 \tag{20}$$

Applying the reaching law (20) to the system (2) with the sliding variable (17) results in the following control signal

$$u(k) = (\boldsymbol{cb})^{-1}\{s_r(k + 1) - \boldsymbol{cAx}(k) + \boldsymbol{cf}h(k)\}. \tag{21}$$

In order to make the control feasible, we must now compensate for the unknown sales volume $h(k)$. The upper and lower bounds for the demand may be applied. Therefore, considering (6), it holds that

$$h(k) \leq d_c(k) + d_r(k) \leq \hat{d}_c(k) + d_{\max}. \tag{22}$$

As the sales are upper bounded by (22) we substitute $h(k) = \hat{d}_c(k) + d_{\max}$ in order to obtain the control. However, the term $d_{\max}$ ordered by the controller at step k will only arrive at the warehouse after $n + 1$ time instants to compensate for the random demand $d_r(k + n + 1)$. In the meantime, there must be another compensation term to cover for the random demand at discrete time instants k, $k + 1$, ..., $k + n$. To solve this problem we introduce a compensation vector $\boldsymbol{D_{\max}}$, defined as follows

$$\boldsymbol{D_{max}} = \begin{bmatrix} 1 \; 1 \ldots 1 \; 1 \end{bmatrix}_{n+1}^{T} d_{\max}. \tag{23}$$

Compensating for the random sales results in the following control rule

$$u(k) = (\boldsymbol{cb})^{-1}\left\{s_r(k + 1) - \boldsymbol{cAx}(k) + \boldsymbol{cf}\left[\hat{d}_c(k) + d_{\max}\right] + \boldsymbol{cD_{max}}\right\}. \tag{24}$$

Considering the random sales, we modify the system's initial condition, so that it covers the maximum random demand up to step $k = n$. Therefore, (19) becomes

$$y_0 = x_1(0) = \sum_{j=0}^{n} \hat{d}_c(j) + \boldsymbol{cD_{max}}. \tag{25}$$

The above sliding mode control law (24) ensures that for any $k \geq 0$ the system's representative point satisfies

$$|s(k)| = |s_r(k) - \boldsymbol{cx}(k)| \leq \boldsymbol{cf} d_{\max} + \boldsymbol{cD_{max}}, \tag{26}$$

where $\boldsymbol{cf} d_{\max} + \boldsymbol{cD_{max}}$ is its maximum deviation from the reference. The following theorem shows that the proposed control ensures enough warehouse stock to satisfy both contractual and random demands within their bounds in the whole control process.

Theorem 1. The control law (24) proposed for the system (2) under the initial condition (25) for any $k \geq 0$ satisfies

$$u(k) \leq \hat{d}_c(k + n + 1) + d_{\max}. \tag{27}$$

Furthermore, for any $k \geq 0$ system output satisfies

$$y(k) \geq \hat{d}_c(k) + d_{\max}. \tag{28}$$

As a result the contractual sales matches the demand in each step, i.e. $h_c(k) = d_c(k)$ and the leftover product amount is sufficient to fulfill the random demand, i.e. $y_r(k) \geq d_{\max}$. Therefore, both demand components are satisfied for any $k \geq 0$.

Proof of Theorem 1. To prove the above, we consider the least favourable conditions of the system. We take into account a situation when the random demand is maximum, i.e. $d_r(k) = d_{\max}$. Moreover, it may be easily noticed that for system (2), with the matrix $\boldsymbol{A}$ and vectors $\boldsymbol{b}$ and $\boldsymbol{f}$ defined in (4) and vector $\boldsymbol{c}$ defined in (18) it holds that

$$\boldsymbol{cA} = \boldsymbol{c}, \qquad \boldsymbol{cb} = 1, \qquad \boldsymbol{cf} = 1. \tag{29}$$

This may be verified with a straightforward substitution. Taking into account (29), we may simplify the control law (24) into the following form

$$u(k) = s_r(k+1) - \boldsymbol{cx}(k) + \hat{d}_c(k) + d_{\max} + \boldsymbol{cD_{max}}. \tag{30}$$

Next, the structure of the compensation term $\boldsymbol{cD_{max}}$ shall be considered. The multiplication of the term gives

$$\begin{aligned}\boldsymbol{cD_{max}} &= \begin{bmatrix} 1 \; a_n \; \sum_{i=n-1}^{n} a_i \; \ldots \; \sum_{i=2}^{n} a_i \; \underbrace{\sum_{i=1}^{n} a_i}_{1} \end{bmatrix}_{n+1} \begin{bmatrix} 1 \; 1 \; \ldots \; 1 \; 1 \end{bmatrix}_{n+1}^{T} d_{\max} = \\ & d_{\max} + a_n d_{\max} + \sum_{i=n-1}^{n} a_i d_{\max} + \ldots + \sum_{i=2}^{n} a_i d_{\max} + d_{\max} \geq 2 d_{\max}.\end{aligned} \tag{31}$$

Taking into account the initial condition (25), the control signal at $k = 0$ becomes

$$\begin{aligned} u(0) &= \sum_{j=1}^{n+1} \hat{d}_c(j) - \sum_{j=0}^{n} \hat{d}_c(j) - \boldsymbol{cD_{max}} + \hat{d}_c(0) + d_{\max} + \boldsymbol{cD_{max}} \\ &= \hat{d}_c(0) - \hat{d}_c(0) + \hat{d}_c(n+1) + d_{\max} = \hat{d}_c(n+1) + d_{\max}. \end{aligned} \tag{32}$$

As the worst possible case is considered, the demand is maximum at each step. Hence, $\hat{d}_c(k) = d_c(k)$. The initial condition (25), ensures that at $k = 0$ the contractual sales are $h_c(0) = d_c(0)$ and the random sales are assumed to be maximum, i.e. $h_r(0) = d_{\max}$. As follows, we present the state vector at $k = 1$, according to (2)

$$\boldsymbol{x}(1) = \begin{bmatrix} \sum_{j=1}^{n} \hat{d}_c(j) + \boldsymbol{cD_{max}} - d_{\max} \; \underbrace{0 \ldots 0}_{n-1} \; \hat{d}_c(n+1) + d_{\max} \end{bmatrix}_{n+1} \tag{33}$$

where the control $u(0)$ has become the last state variable x_{n+1}. Moreover, it holds that

$$\boldsymbol{cD_{max}} - d_{\max} = a_n d_{\max} + \sum_{i=n-1}^{n} a_i d_{\max} + \ldots + \sum_{i=2}^{n} a_i d_{\max} + d_{\max} \geq d_{\max}. \tag{34}$$

We follow calculating the control $u(1)$

$$u(1) = \sum_{j=2}^{n+2} \hat{d}_c(j) - \sum_{j=1}^{n} \hat{d}_c(j) - \boldsymbol{cD_{max}} + d_{\max} - \hat{d}_c(n+1) - d_{\max} + \hat{d}_c(1) + d_{\max} + \boldsymbol{cD_{max}}$$
$$= \hat{d}_c(1) - \hat{d}_c(1) + \hat{d}_c(n+2) + d_{\max} = \hat{d}_c(n+2) + d_{\max}. \tag{35}$$

The stock at $k = 1$ satisfies the maximum demand, so $h_c(1) = d_c(1)$ and $h_r(1) = d_{\max}$. As follows, the state vector at $k = 2$ is

$$\boldsymbol{x}(2) = \begin{bmatrix} y(2) \underbrace{0 \dots 0}_{n-2} \hat{d}_c(n+1) + d_{\max} \; \hat{d}_c(n+2) + d_{\max} \end{bmatrix}_{n+1}, \tag{36}$$

where

$$y(2) = \sum_{j=2}^{n} \hat{d}_c(j) + \boldsymbol{cD_{max}} - 2d_{\max} + a_1\hat{d}_c(n+1) + a_1 d_{\max}. \tag{37}$$

We denote the terms compensating for random sales with $g(k)$. It may be noticed that

$$g(2) = \boldsymbol{cD_{max}} - 2d_{\max} + a_1 d_{\max} = a_n d_{\max} + \sum_{i=n-1}^{n} a_i d_{\max} + \dots + \sum_{i=2}^{n} a_i d_{\max} + a_1 d_{\max}. \tag{38}$$

Considering (1), we further obtain

$$g(2) = a_n d_{\max} + \sum_{i=n-1}^{n} a_i d_{\max} + \dots + \sum_{i=3}^{n} a_i d_{\max} + \underbrace{\sum_{i=2}^{n} a_i d_{\max} + a_1 d_{\max}}_{d_{\max}} \geq d_{\max}. \tag{39}$$

As the stock is sufficient the sales are $h_c(2) = d_c(2)$ and $h_r(2) = d_{\max}$. Next, $u(2)$ is

$$u(2) = \sum_{j=3}^{n+3} \hat{d}_c(j) - \sum_{j=2}^{n} \hat{d}_c(j) + 2d_{\max} - \underbrace{\left\{ a_1\left[\hat{d}_c(n+1) + d_{\max}\right] + \sum_{i=2}^{n} a_i\left[\hat{d}_c(n+1) + d_{\max}\right] \right\}}_{d_c(n+1)+d_{\max}}$$
$$-\hat{d}_c(n+2) + \hat{d}_c(2) = \hat{d}_c(n+3) + d_{\max}. \tag{40}$$

Furthermore, the state vector $\boldsymbol{x}(3)$ becomes

$$\boldsymbol{x}(3) = \begin{bmatrix} y(3) \underbrace{0 \dots 0}_{n-3} \hat{d}_c(n+1) + d_{\max} \; \hat{d}_c(n+2) + d_{\max} \; \hat{d}_c(n+3) + d_{\max} \end{bmatrix}_{n+1}, \tag{41}$$

where

$$y(3) = \sum_{j=3}^{n} \hat{d}_c(j) + \sum_{i=1}^{2} a_i \hat{d}_c(n+1) + a_1 \hat{d}_c(n+2) + g(3), \tag{42}$$

and, as follows from (1)

$$\begin{aligned} g(3) &= g(2) - d_{\max} + \sum_{i=1}^{2} a_i d_{\max} = \\ a_n d_{\max} &+ \sum_{i=n-1}^{n} a_i d_{\max} + \ldots + \sum_{i=4}^{n} a_i d_{\max} + \underbrace{\sum_{i=3}^{n} a_i d_{\max} + \sum_{i=1}^{2} a_i d_{\max}}_{d_{\max}} \geq d_{\max}. \end{aligned} \tag{43}$$

Therefore, sales at $k = 3$ are $h_c(3) = d_c(3)$ and $h_r(3) = d_{\max}$. The state vector, at any $0 \leq k \leq n$ becomes

$$\boldsymbol{x}(k) = \begin{bmatrix} y(k) \ \underbrace{0 \ldots 0}_{n-k} \ \hat{d}_c(n+1) + d_{\max} \ldots \hat{d}_c(n+k) + d_{\max} \end{bmatrix}_{n+1}. \tag{44}$$

Finally, for any $k \geq 0$ we obtain

$$u(k) = \hat{d}_c(k+n+1) + d_{\max}. \tag{45}$$

When the maximum demand values occur, for $k = n$ the state vector becomes

$$\boldsymbol{x}(n) = \left[y(n) \ \hat{d}_c(n+1) + d_{\max} \ldots \hat{d}_c(n+n) + d_{\max} \right]_{n+1}^{T}, \tag{46}$$

where $y(n) = \hat{d}_c(n) + \sum_{i=1}^{n-1} a_i \sum_{j=n+1}^{n+n-i} \hat{d}_c(j) + g(n)$ and $g(n) = a_n d_{\max} + \sum_{i=1}^{n-1} a_i d_{\max} = d_{\max}$.

From the above considerations we conclude that for any $0 \leq k \leq n$, it holds that

$$y(k) = \sum_{j=k}^{n} \hat{d}_c(j) + \sum_{i=1}^{n-1} a_i \sum_{j=n+1}^{k+n-i} \hat{d}_c(j) + g(k), \tag{47}$$

where

$$g(k) = \underbrace{\sum_{i=k}^{n} a_i d_{\max} + \sum_{i=1}^{k-1} a_i d_{\max}}_{d_{\max}} + \sum_{i=k+1}^{n} a_i d_{\max} + \ldots + \sum_{i=n-1}^{n} a_i d_{\max} + a_n d_{\max} \geq d_{\max}. \tag{48}$$

It may be concluded that for $0 \leq k \leq n$ both demand terms are fully satisfied. Next, we derive $u(n+1) = \hat{d}_c(n+n+1) + d_{\max}$. We follow with the state vector. From (44), with $k = n+1$, we get

$$\boldsymbol{x}(n+1) = \left[y(n+1) \ \hat{d}_c(n+2) + d_{\max} \ldots \hat{d}_c(n+n+1) + d_{\max} \right]_{n+1}^{T}, \tag{49}$$

where

$$y(n+1) = \sum_{i=1}^{n} a_i \hat{d}_c(n+1) + \sum_{i=1}^{n-1} a_i \sum_{j=n+2}^{n+n+1-i} \hat{d}_c(j) + g(n+1) \tag{50}$$

and

$$g(n+1) = \sum_{i=1}^{n} a_i d_{\max}. \tag{51}$$

Considering (1), for $k = n + 1$ the system output becomes

$$y(n+1) = \hat{d}_c(n+1) + \sum_{i=1}^{n-1} a_i \sum_{j=n+2}^{n+n+1-i} \hat{d}_c(j) + d_{\max}. \tag{52}$$

System output, for any $k \geq n + 1$, becomes

$$y(k) = \hat{d}_c(k) + \sum_{i=1}^{n-1} a_i \sum_{j=k+1}^{k+n-i} \hat{d}_c(j) + d_{\max} = u(k-n-1) + \sum_{i=1}^{n-1} a_i \sum_{j=k-n}^{k-1-i} u(j) + d_{\max}. \tag{53}$$

With $u(k)$ satisfying (45), we conclude that for any $k \geq 0$

$$y(k) \geq \hat{d}_c(k) + d_{\max}, \tag{54}$$

which ends the proof. ∎

4 Simulation Verification

We now verify the theoretical findings with a simulation example. We take into account an inventory system with 3 suppliers, whose lead times are 5, 4 and 2 time instants. They deliver 50%, 30% and 20% of the orders, respectively. Considering the above, the maximum lead time is $n = 5$ and the system is of the 6-th order. The system's dynamics are described in (2), where matrix $\boldsymbol{A}$ and vectors $\boldsymbol{b}$ and $\boldsymbol{f}$ satisfy (4). The values of the a_i parameters in the system are $a_1 = 0$, $a_2 = 0.2$, $a_3 = 0$, $a_4 = 0.3$, $a_5 = 0.5$. The inventory system is subject to two kinds of customer demand. The contractual demand is a sequence of values $d_c(k) = \{20, 10, 40, 30, 40\}$ that repeats every 5 time instants. For example, for $k = 0, 5, 10, \ldots$, the value will be 20, for $k = 1, 6, 11, \ldots$, – 10 etc. The value of the random demand changes every 15 time instants between its lower ($d_{min} = 1$) and upper ($d_{max} = 5$) bounds.

The control vector $\boldsymbol{c}$, chosen according to (18) is $\boldsymbol{c} = [1\ 0.5\ 0.8\ 0.8\ 1\ 1]$. With the compensation term $\boldsymbol{D}_{\mathbf{max}} = [5\,5\,5\,5\,5\,5]^T$, the initial point of the system is $x_1(0) = 160 + \boldsymbol{cD_{max}} = 185.5$. The generated reference trajectory for the system is depicted on Fig. 1. It consists of the repeated sequence of contractual demand values. The control signal resulting from (24) is pictured in Fig. 2. Figure 3 illustrates the amount of goods in the warehouse before sales, after selling the contracted product and after all the demand in the system has been fulfilled. Finally, Fig. 4 shows the amount of product sold from the warehouse, which, due to full demand satisfaction, corresponds to the demand definitions above.

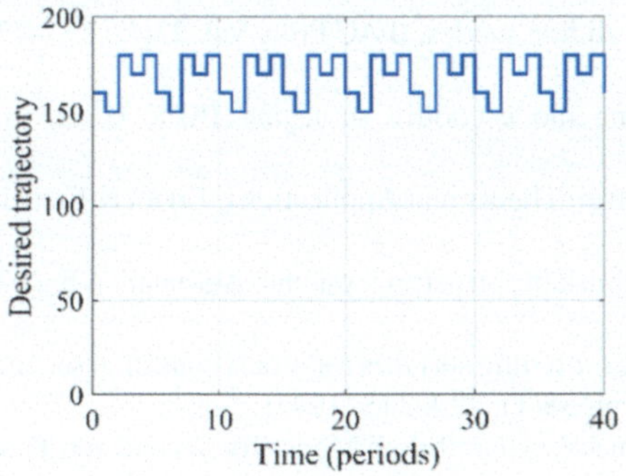

Fig. 1. Desired trajectory $s_d(k)$.

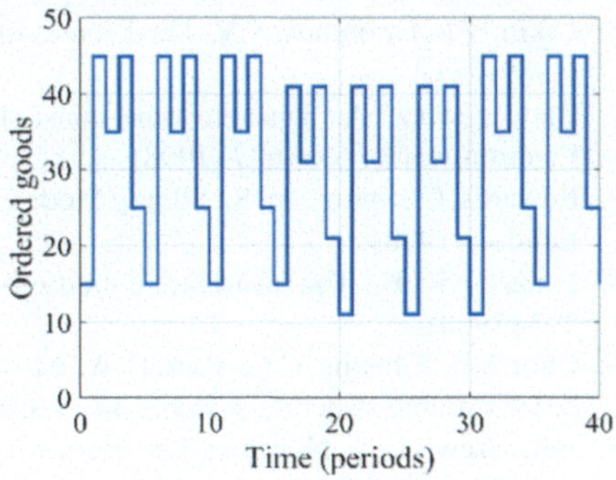

Fig. 2. Control signal $u(k)$.

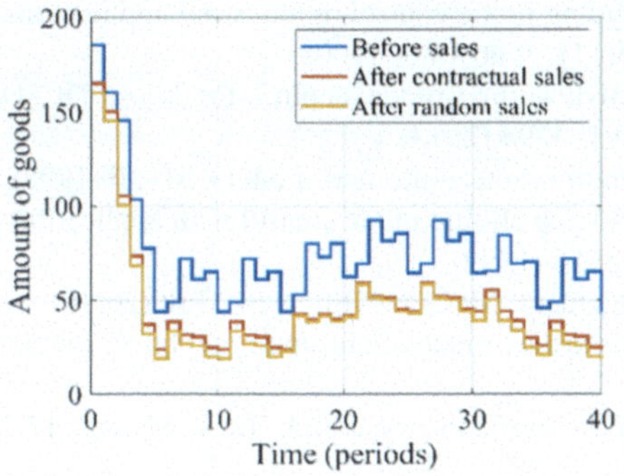

Fig. 3. Amount of product in the system.

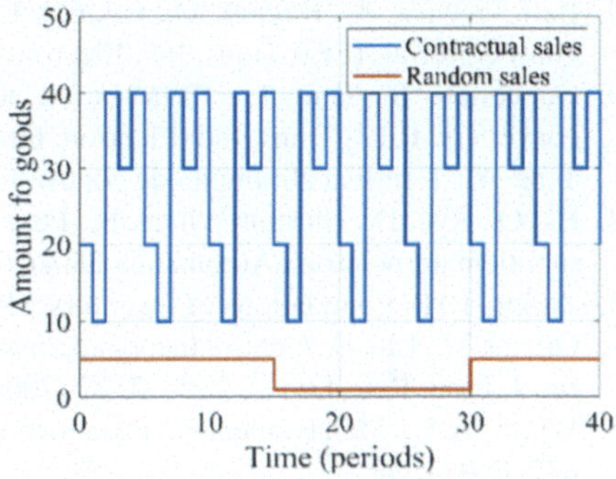

Fig. 4. Contractual and random sales.

From the above figures, it is clearly visible that the proposed control strategy ensures full demand satisfaction from the very beginning of the control process.

5 Conclusions

This study considered a novel control concept for a single product multiple supplier based inventory system, where each supplier requires a certain time to deliver the goods. Moreover, the demand is partially a priori known due to contractual obliga-tions. Therefore, the control design process utilizes the known demand term to form the required state trajectory profile. Next, the desired trajectory is used as a reference in a trajectory following quasi-sliding mode control scheme. As a result, the sliding mode control law drives the system to the demand position in each step. Finally, a compensation term is introduced to provide additional product reserved for the ran-dom buyers. The paper proves that the proposed control scheme satisfies both de-mand terms within their bounds. Theoretic results were verified with a simulation example.

References

1. Utkin, V.I.: Sliding Modes in Control and Optimization. Springer, Berlin (1992)

2. Utkin, V.I., Drakunov, S.V.: On discrete-time sliding modes. IFAC Proc. Vol. **22**(3), 273–278 (1989)
3. Bartoszewicz, A.: Discrete time quasi-sliding mode control strategies. IEEE Trans. Ind. Electron. **45**(4), 633–637 (1998)
4. Edwards, C., Spurgeon, S.: Sliding Mode Control: Theory and Applications. Taylor & Francis, London (1998)
5. Draženović, B.: The invariance conditions in variable structure systems. Automatica **5**, 287–295 (1969)
6. Choi, S.B., Cheong, C.C., Park, D.W.: Moving switching surfaces for robust control of second-order variable structure systems. Int. J. Control. **58**(1), 229–245 (1993)
7. Bartoszewicz, A., Nowacka-Leverton, A.: Time-Varying Sliding Modes for Second and Third Order Systems. Springer, Heidelberg (2009)
8. Utkin, V.I., Guldner, J., Shi, J.: Sliding Mode Control in Electromechanical Systems. Taylor & Francis, London (1999)
9. Bartoszewicz, A., Kaynak, O., Utkin, V.I.: Sliding mode control in industrial applications – guest editorial. IEEE Trans. Ind. Electron. **55**(11), 3806–3808 (2008)
10. Oucheriah, S., Guo, L.: PWM-based adaptive sliding-mode control for boost DC–DC converters. IEEE Trans. Ind. Electron. **60**, 3291–3294 (2013)
11. Tang, Y.: Terminal sliding mode control for rigid robots. Automatica **34**(1), 51–56 (1998)
12. Ha, Q., Rye, D., Durrant-Whyte, H.: Fuzzy moving sliding mode control with application to robotic manipulators. Automatica **35**(4), 607–616 (1999)
13. Zipkin, P.H.: Foundations of Inventory Management. McGraw-Hill, New York (2000)
14. Ortega, M., Lin, L.: Control theory applications to the production inventory problem: a review. Int. J. Prod. Res. **42**(11), 2303–2322 (2004)
15. White, A.S.: Management of inventory using control theory. Int. J. Tech. Manag. **17**(7), 847–860 (1999)
16. Chołodowicz, E., Orłowski, P.: Comparison of a perpetual and PD inventory control system with Smith predictor and different shipping delays using bicriterial optimization and SPEA2. Pomiary Automatyka Robotyka **20**(3), 5–12 (2016)
17. Rodrigues, L., Boukas, E.K.: Piecewise-linear H∞ controller synthesis with applications to inventory control of switched production systems. Automatica **42**(8), 1245–1254 (2006)
18. Boccadoro, M., Martinelli, F., Valigi, P.: Supply chain management by H-infinity control. IEEE Trans. Autom. Sci. Eng. **5**(4), 703–707 (2008)
19. Ignaciuk, P., Bartoszewicz, A.: Linear-quadratic optimal control of periodic-review perishable inventory systems. IEEE Trans. Control Sys. Tech. **20**(3), 1400–1407 (2012)
20. Wang, Y., Du, D., Huang, J.: Optimal replenishment strategy for inventory mechanism with step-shaped demand. J. Optim. Theory Appl. **190**, 841–860 (2021)
21. Cholodowicz, E., Orlowski, P.: Switching robust neural network control of perishable inventory with fixed shelf life products under time-varying uncertain demand. J. of Comp. Sci. **70**, 102035 (2023)
22. Bartoszewicz, A., Maciejewski, M.: Sliding mode control of periodic review perishable inventories with multiple suppliers and transportation losses. Bull. Pol. Acad. Sci. Tech. Sci. **61**(4), 885–892 (2013)
23. Bartoszewicz, A., Leśniewski, P.: Reaching law approach to the sliding mode control of periodic review inventory systems. IEEE Trans. Autom. Sci. Eng. **11**(3), 810–817 (2014)

The Concept of Wireless Energy Transmission Intended for Use in Two-Rotor Wind Turbines

Wojciech Miąskowski[1], Bartosz Moczulak[1], Wojciech Kiński[2](✉), Paweł Pietkiewicz[1], and Krzysztof Nalepa[1]

[1] Faculty of Technical Sciences, University of Warmia and Mazury in Olsztyn, ul. Oczapowskiego 11, 10-719 Olsztyn, Poland

[2] Łukasiewicz Research Network – Industrial Institute of Automation and Measurements PIAP, Al. Jerozolimskie 202, 02-486 Warsaw, Poland

wojciech.kinski@piap.lukasiewicz.gov.pl

Abstract. The process of designing wind turbines requires multi-threaded procedures when developing individual elements, taking into account their mutual influence. The WNT/PIAP research team, whose members include the authors of this article, proposes a solution that improves the efficiency of obtaining electricity by wind turbines, which requires the use of unconventional solutions in the field of energy transmission. The solution described in this article is a system for wireless transmission of electricity from the generator stator windings to the inverter. The article presents the effect of the team's work on solving the said design problem. As a result of the work, a concept of wireless transmission of electricity was proposed, aimed at its use in low-power two-rotor wind turbines.

Keywords: Renewable energy · wind turbine · wireless energy transmission

1 Introduction

Research on obtaining energy from renewable sources determines the current and future directions of energy development. There is therefore a need to use wind engines that use wind energy more effectively [1, 3, 7]. When designing new wind turbine solutions or improving existing structures, a multi-threaded approach is taken, taking into account the influence of individual structural nodes on each other [6].

Research is carried out in relation to the profiles of wind turbine blades, the number and shape of rotors as well as the entire wind turbine structures [8], in particular in order to improve the efficiency of wind turbines and their manufacturing technology [2, 6, 9, 10].

In line with the trend of increasing the use of alternative energy sources, solutions for a two-rotor micro wind turbine with a vertical and horizontal axis were proposed. [5] When developing design solutions for wind turbines, it was proposed to increase the efficiency of energy production by using two-rotor wind turbines with rotors rotating in opposite directions. This solution allows for obtaining the effect of summing the rotational speeds of two rotors, enabling the relative speed of the rotor and stator of the

R. Szewczyk et al. (Eds.): AUTOMATION 2024, LNNS 1219, pp. 111–119, 2024.
https://doi.org/10.1007/978-3-031-78266-4_11

electric energy generator to be twice as high as in the case of a single rotor. This allows you to use an electric machine with smaller dimensions. This means less copper, smaller permanent magnets, etc., which reduces the cost of producing an electricity generator. [2, 4, 5].

This solution requires the use of a system that allows energy to be extracted from the rotating stator of the electricity generator. The conventional method to accomplish this task is to use a rotating joint (Fig. 1).

Fig. 1. Rotary joint with mechanical contacts [11]

This solution was used in earlier structures of this type that the team worked on. One of them is shown below.

2 Contact Energy Transmission System – Used so Far

The presented solutions of two-rotor wind microturbines, developed by the team, require a complex system for transmitting power from the rotors to the generator and receiving the produced electricity (Fig. 2).

In order to specify the problem of receiving the generated energy in the generator, the solutions currently used in one of the assembly structures described in more detail, e.g. in [2, 3, 5], are briefly outlined. Rotation of rotors 1 and 2 (Fig. 2) in opposite directions increases the relative rotational speed of the rotor and generator stator. The operating characteristics of the electricity generator require the use of a drive with a rotational speed close to the rated rotational speed of the generator. Increasing the rotational speed allows you to obtain the expected power with a smaller size and price of the electric generator. The wind turbine structure presented below makes it possible to obtain a high rotor speed relative to the stator, which increases the useful energy yield.

This allows the use of a high-speed electric current generator in a wind turbine without the need to use multipliers. Unfortunately, this complicates the mechanical structure of the wind turbine, primarily causing problems with supporting the external shaft and receiving electricity through the system of slip rings 3 and carbon brushes 4 (Fig. 2).

3 Wireless Energy Transmission System

The disadvantage of the solution using mechanical rotary joints is the wear of the elements subjected to friction in the process of transmitting electricity through the joint. Therefore, a concept of a two-rotor wind power plant module was developed, using the

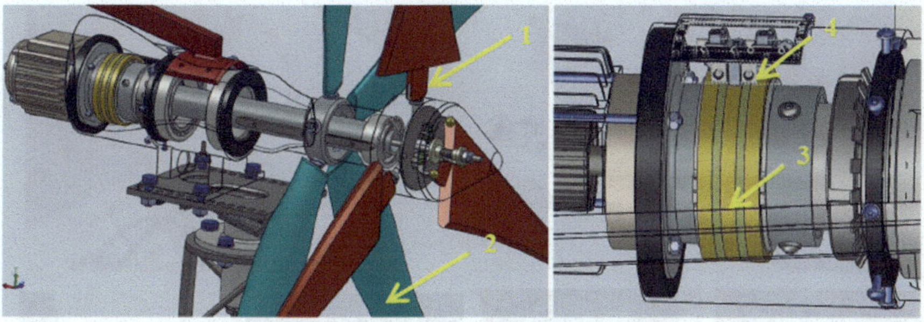

Fig. 2. System for transmitting power to the generator and receiving electricity from the generator [5]

effect of rotors rotating in opposite directions. One of the wind turbine rotors is mechanically connected to the rotor of the electricity generator, the other rotor is connected to a stator containing electromagnetic coils. A wireless electricity transmission system was used to transfer the energy generated in the rotating stator to a stationary supporting structure.

Stationary Module

The work was carried out in several stages. The first step was to build a simple stationary station for wireless energy transfer (Wireless Power Transfer). A system with resonant coupling was used for wireless energy transmission. The power of the set is 2.5 W and the efficiency achieved is approximately 59%. The energy transmitter and receiver coils were rigidly attached to both sides of a 5 mm thick polystyrene plate (Fig. 3). The transmitter was powered by a stabilized power supply, while the load was a set of resistors enabling various levels of electrical load.

The stationary module includes:

1. Transmitter (electronic system)
2. Transmitter coil
3. Receiver (electronic system)
4. Receiver coil
5. Load resistors
6. Load switches

Rotation Module

The next stage of the tests was the construction of a rotational station in which the energy transmitter and, together with the power supply system in the form of a lithium-ion battery and a converter increasing the voltage to the level of 5 V required by the transmitter, were placed in a rotating drum. The energy receiver coil is placed on a stationary support whose position in relation to the transmitter coil can be changed. A newer generation electronic system of the energy transmitter and coils made of high-frequency litz (multi-core insulated wires forming a bundle) allowed for an efficiency of approximately 70% with a coil distance between 5 mm and 20 mm (Fig. 4 and 5).

The rotation module includes:

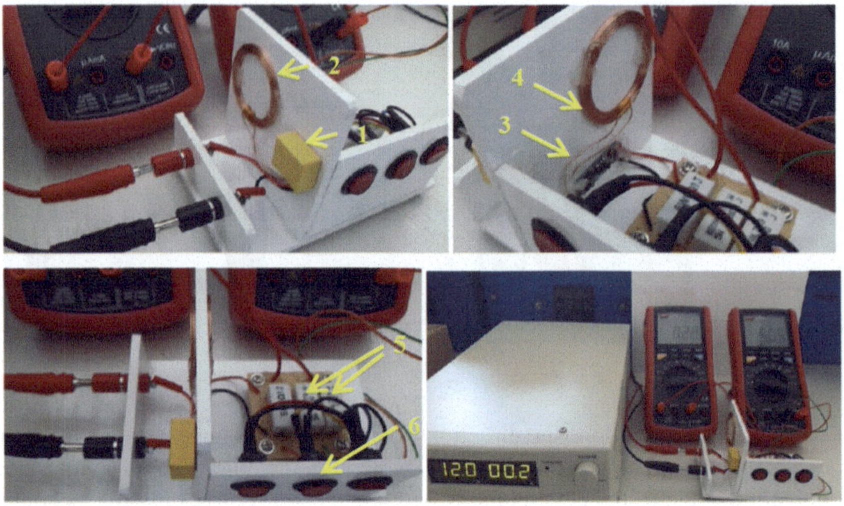

Fig. 3. Stationary station for wireless energy transfer [own work]

1. Transmitter (electronic system)
2. Transmitter coil
3. Lithium-ion battery (3.7 V)
4. DC/DC converter increasing voltage (3.7 V/5.0 V)
5. Receiver (electronic system)
6. Receiver coil
7. Load resistor
8. Drive motor
9. Regulated power supply to power the engine (speed controller)

Preliminary research allowed the selection of a set for wireless transmission of electricity, which can be used in the construction of a two-rotor wind turbine with counter-rotating rotors. For further work, the Proxi module 100 W set from Power by Proxi was used.

4 100 W Module for Use in a Wind Turbine Generator

The idea of the set's operation is shown in Fig. 6. 24 V direct current energy is transformed into high-frequency alternating current energy and then in the coil into the energy of a high-frequency alternating magnetic field. The receiver coil converts the alternating magnetic field into high-frequency electrical energy, which is converted into 24 V direct current energy in the receiver.

The transmitter and receiver are placed on a base that allows the transmitter and receiver coils to be moved relative to each other and the angular setting of the coils to be changed (Fig. 7). The efficiency of the energy transfer system reaches approximately 85% with a distance between the coils from 6mm to 35 mm and an angular position change from 0° to 30°.

Fig. 4. Rotary station for wireless energy transfer - energy transmitter elements [own work]

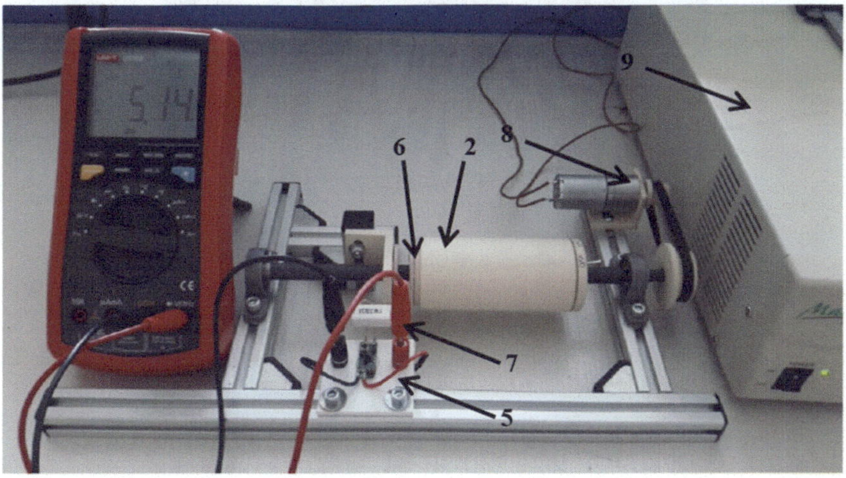

Fig. 5. Rotary station for wireless energy transfer [own work]

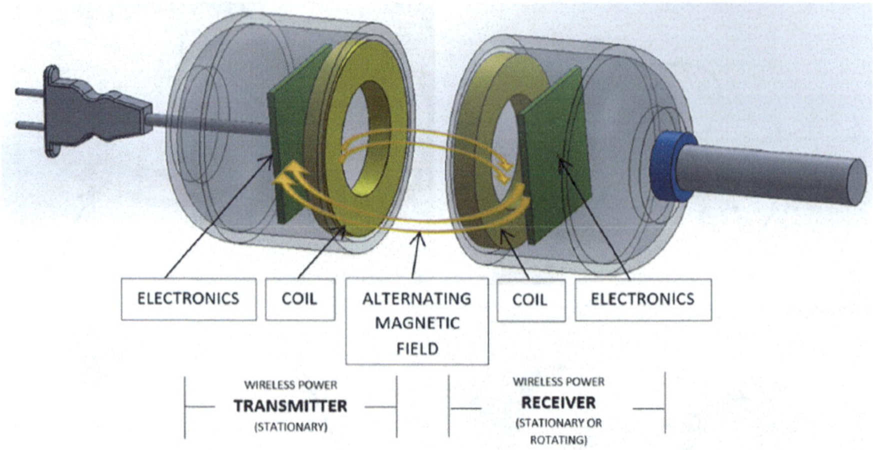

Fig. 6. The idea of energy transfer in a high-frequency electromagnetic field [own work]

Fig. 7. Station for testing energy transfer systems in a high-frequency electromagnetic field

5 Model of a Two-Rotor Turbine with a Vertical Axis of Rotation with Wireless Energy Transmission

Based on research on wireless energy transmission systems, a concept and design of a two-rotor wind power plant using the relative motion of rotors with a wireless energy transmission system was developed. The visualization of the concept is shown in Fig. 8.

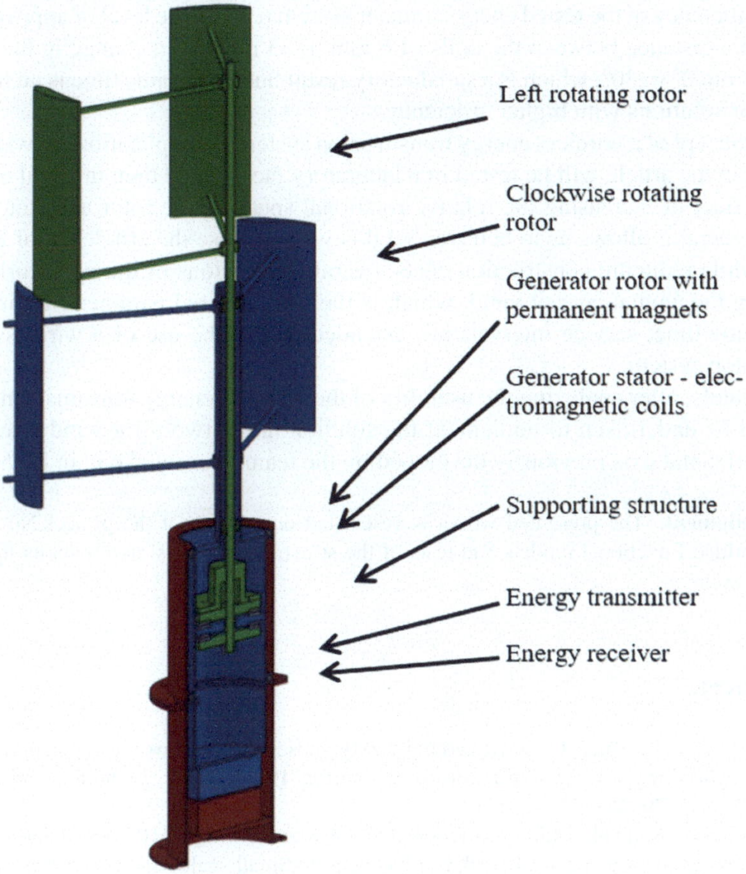

Fig. 8. Two-rotor wind power plant using energy transfer systems in a high-frequency electromagnetic field

It is a wind power plant that uses the counter-rotating movement of two rotors to increase the efficiency of the process of generating electricity from wind energy. Increasing the share of energy production from renewable sources and improving the efficiency of such processes is consistent with the assumptions of the EU energy policy.

Summary

The disadvantage of the solution using mechanical rotary joints is the wear of elements subject to friction in the process of transmitting electricity through the joint, which the proposed solution does not have.

Preliminary research allowed the selection of a set for wireless transmission of electricity, which can be used in the construction of a two-rotor wind turbine with counter-rotating rotors after necessary modifications.

The efficiency of the tested energy transfer system reaches the level of approximately 85% with a distance between the coils of 6 mm to 35 mm and a change in the angular position from 0° to 30°, which is a satisfactory result but at the same time is an incentive to look for solutions with higher efficiency.

The concept of a wireless energy transmission system for applications in wind farms proposed in the article will be tested in a laboratory facility and then in a real one.

The effect of increasing the relative rotational speed of the rotor and stator of the electric generator allows us to believe that this will increase the efficiency of the wind turbine while reducing construction costs, overall dimensions of the wind turbine and improving the annual energy yield, which is the most desired parameter by investors. At the same time, service intervals are extended due to the use of a wireless energy transmission system.

Ultimately, after analyzing the usability of the wireless energy transmission system, work will be undertaken to implement the solution in the two-rotor wind power plant with a horizontal axis previously developed by the team, presented e.g. in [2, 5].

Acknowledgment. The presented works were carried out as part of the project No. N N502 511940 entitled Functional models and tests of the structure of a quasi-autonomous lighting or signal-ing point

References

1. Kasperowicz, Ł., Kurz, D.: Zwiększenie efektywności turbiny Savoniusa pracującej w systemie hybrydowym w wyniku zastosowania kurtyn. Poznan Univ. Technol. Acad. J. Electr. Eng. Poznań, **70**, 268–275 (2012)
2. Miąskowski, W., et al.: Mikrogeneracja energii z wiatru i wody, IMP PAN, Gdańsk (2014)
3. Miąskowski, W., Nalepa, K., Paweł, P., Piechocki, J.: Small-scale wind power energy systems for use in agriculture and similar applications, Geothermal, Wind and Solar Energy Applications in Agriculture and Aquaculture / Editors: Jochen Bundschuh, Guangnan Chen, D. Chandrasekharam, Janusz Piechocki, Leiden : CRC Press Taylor& Francis Group, s. 259–288, Seria: Sustainable Energy Developments, 2164–0645; vol. 13 (2017).https://doi.org/10.1201/9781315158969,
4. Kłos, P., Miąskowski, W., Nalepa, K., Pietkiewicz, P.: Analiza koncepcyjna dwuwirnikowej turbiny wiatrowej z osią pionową typu H-Darrieus, Przegląd Mechaniczny, **2017**(6), 24–26 (2017)
5. Miąskowski, W., Nalepa, K., Pietkiewicz, P., Komar, W.: Dwuwirnikowa mikrosiłownia z osią poziomą. Mechanik, **7**(2012), 485–492 (2012)
6. Moczulak, B., Miąskowski, W., Kiński, W., Cader, M.: Application use of the HP multijet fusion technology in the rapid prototyping of wind turbines with a vertical axis of rotation. Agric. Eng. **26**(1), 253–263 (2022)
7. Li, Q., Maeda, T., Kamada, Y., Hiromori, Y., Kasuya, T.: Study on stall behavior of a straight-bladed vertical axis wind turbine with numerical and experimental investigations. J. Wind Eng. Ind. Aerodyn. **164**, 1–12 (2017)
8. Seifert, A., Shtendel, T., Dolgopyat, D.: From lab to full scale active flow control drag reduction: how to bridge the gap? J. Wind Eng. Ind. Aerodyn. **147**, 262–272 (2015)
9. Bottasso, C.L., Cacciola, S., Iriarte, X.: Calibration of wind turbine lifting line models from rotor loads. J. Wind Eng. Ind. Aerodyn.odyn. **124**, 29–45 (2014)

10. Wong, K.H., Chong, W.T., Sukiman, N.L., Poh, S.C., Wang, C.-T.: Performance enhancements on vertical axis wind turbines using flow augmentation systems: a review. Renew. Sustain. Energy Rev. **73**, 904–921 (2017)
11. https://sempolska.com.pl/files/zlacza_slizgowe.pdf. Accessed 15 Dec 2023

Robotics

Literature-Based Analysis of Lower Extremity Kinematics and Dynamics During Task-Oriented Physiotherapy for Rehabilitation Robot Design

Tomasz Osiak[1,2,3], Natalia Osiak[3], Piotr Falkowski[1,4](✉), Mehmet Emin Aktan[5], Piotr Czerechowicz[6], and Vasfi Emre Ömürlü[7,8]

[1] Łukasiewicz Research Network–Industrial Research Institute of Automation and Measurements PIAP, Warsaw, Poland
piotr.falkowski@piap.lukasiewicz.gov.pl
[2] Department of Rehabilitation, Józef Piłsudski University of Physical Education, Warsaw, Poland
[3] Z Miłości Do Kości, Warsaw, Poland
[4] Institute of Aeronautics and Applied Mechanics, Faculty of Power and Aeronautical Engineering, Warsaw University of Technology, Warsaw, Poland
[5] Bartin University, Bartin, Turkey
[6] Elmiko Biosignals, Warsaw, Poland
[7] 3LS R&D Consulting Industry and Trade Inc., Istanbul, Turkey
[8] Yildiz Technical University, Istanbul, Turkey

Abstract. Robot-assisted rehabilitation is becoming an increasingly common method and allows using previously unavailable solutions. Designers of rehabilitation robots must ensure that their devices' exercises lead to the most functional movements. This review aimed to collect biomechanical, kinematic, and dynamic data on lower limb movement that can assist designers of rehabilitation robots. Seventeen studies were retrieved from Scopus and reviewed. The partial information regarding ranges of motion, angular velocities and joint torques during gait and daily life activities. It was found that the gait involves mainly motion in the parasagittal plane, while the activities of daily living often require full activation of the hip joint. The research proves that more experimental trials for modelling task-oriented rehabilitation movements are needed.

Keywords: additive manufacturing · Industry 4.0 · process modelling · robotic welding · WAAM

1 Introduction

Rehabilitation robots recently raised the attention of physiotherapy specialists. It is due to their capabilities to substitute humans in physically tiring and repetitive work [1]. Therefore, the specialists have more time for more advanced and unique tasks. In rehabilitation, repetitive movements should be applied with a certain standard within

R. Szewczyk et al. (Eds.): AUTOMATION 2024, LNNS 1219, pp. 123–138, 2024.
https://doi.org/10.1007/978-3-031-78266-4_12

the framework of a program. With their sensors and actuators, robots can perform such tasks with higher accuracy and precision than physiotherapists and physicians [2]. In addition, these machines can constantly measure parameters such as the patient's range of motion and joint strength. Based on these, the therapy adjustments can be modified. At the top, the physicians can analyse the measurements. Hence, they are supported in further diagnoses.

The physiotherapy robots can be either end-effector type or exoskeletons. Regarding rehabilitation of the lower extremities, the latter are often used for gait training. Such devices are designed and programmed to follow healthy gait patterns. Thanks to the force support provided to the patient, it is possible to assess the lower limbs' kinematics when the patient cannot move or even stand. This also allows better diagnosing [3]. This contributes to recovering immobile patients by enabling them to perform gait movements passively or at low resistance levels. Another significant advantage is that the entire diagnosis and treatment process, exercise parameters and patient performance during each session can be recorded and accessed at any time.

Technological advances enabled the development of telerehabilitation. In such an approach, a patient can follow the therapist's guidelines from their home, while the specialist can continuously monitor the therapy effects and the patient's performance. Artificial intelligence (AI) based solutions can also be introduced to increase the effectiveness of such applications [4–6].

An exoskeleton of a lower extremity with its digital twin in VR for its teleoperation and biomedical signals registration is developed within a "*SmartEx-Twin*" project. The primary focus of this project revolves around researching the implementation potential of integrating advanced Information and Communication Technologies (ICT) into robot-assisted telerehabilitation processes. These advancements include tracking biomedical signals, such as Electromyography (EMG) and Electroencephalography (EEG), to monitor treatment effects or automatically adapt exercise routines based on individual performance. Additionally, the project aims to introduce immersive telerehabilitation using digital twins within virtual reality (VR) environments.

Given the pressing healthcare needs across Europe, the envisioned system holds the potential to elevate the quality of rehabilitation services. This becomes particularly critical for individuals residing in rural areas and for parents caring for children with illnesses. Thanks to this, they will not spend unnecessary time travelling to medical facilities but can exercise in a home-like environment.

This innovative approach hinges on the fusion of various ICT technologies to leverage their combined strengths in line with the principles of Industry 4.0. An integral aspect involves the development of an exoskeleton tailored specifically for the project, intended to facilitate highly efficient remote and personalised treatment for patients recovering from post-stroke complications or other orthopaedic conditions. This includes supporting passive and active neuromotor rehabilitation in standing, sitting or lying position.

While designing the mechanics of an exoskeleton, required torques and appearing resultant forces must be computed. This has to be based on the kinematics and dynamics of motions realised by patients. However, there is very little data on which rehabilitation robot designers can rely. Not many papers with experimental data on the biomechanics, kinematics, and kinetics of the human body are being published that can be used to design high-quality physiotherapy assistive devices. Many designers do not involve data from living human bodies in computing required mechanical parameters for their devices but copy the parameters from already existing ones. However, such an approach does not enable mass reduction of the designs, which is critical for home-environment applications [7].

This study reviews the literature from the past five years. Biomechanical data on the lower limb movement of healthy subjects were sought to be used to create a database that would be useful for the designers of rehabilitation robots.

2 Methodology

A literature search was performed on 26th November 2023 with Scopus by one author. The search parameters used for database were:

("range of motion" OR "angular acceleration" OR "angular velocities" OR "angular speeds" OR "joint moments" OR "joint reaction force") AND (hip OR knee OR ankle) AND (gait OR jump OR "walking upstairs" OR "walking down stairs" OR "sideways walking" OR running OR "activities of daily living" OR adls OR adl) AND "healthy human").

The selected keywords had to appear in the title, abstract or keywords of the articles. Only the articles in English from the last 5 years were considered. This was due to the requirement of remaining up-to-date with the physiotherapeutic standards and well-assessed by the international community. Search and exclusion schematic is presented in Fig. 1.

Selected literature was reviewed to determine the activities performed during task-oriented kinesiotherapy of the lower extremity. Moreover, the presented data was analysed to find maximal values of this limb's kinematic and dynamic joint parameters. The parameters of interest include range of motion, angular velocities, angular accelerations, and torques. Every parameter was analysed in regard to the anatomical motion. Therefore, it can be interpreted as a single degree of freedom.

If an article was rated by a division or several values for one activity, the authors used the extremal ones. When no numerical data was given explicitly, the values of the parameters were retrieved based on their graphical representation, including time-series charts.

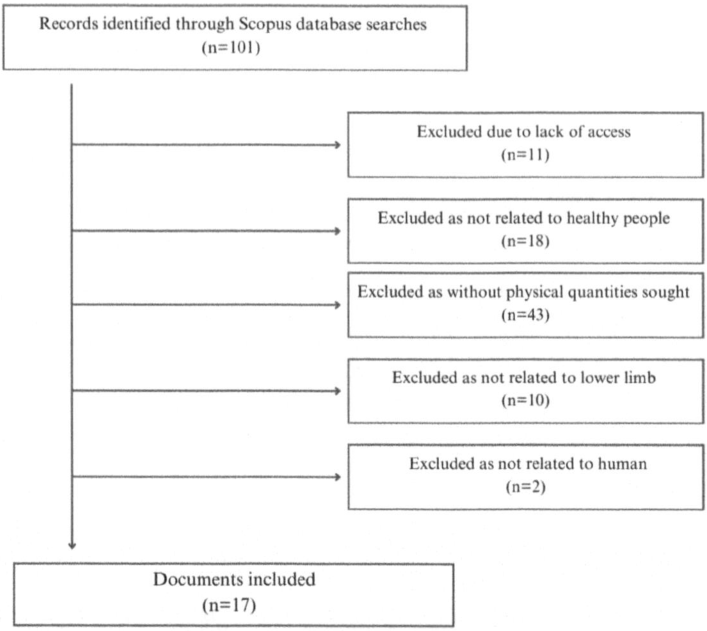

Fig. 1. Search schematic

3 Results

The data found in the articles are presented in Tables 1, 2, 3, 4 and 5. Table 1 shows how dynamic and kinematic joint values were measured for specific groups of subjects. This includes a description of the measured movements, a description of the trial groups, and measuring instruments.

Table 1. Key information from the documents analysed (felx.-felxion, ext-extension)

Author (Year)	Parameters tested	Activity tested	Participants	Measuring system
Das et al. (2023) [8]	Knee Flex./Ext. Joint angle	active range in the forward position	n = 23; male 23; age [yr] 20.45 ± 2.18; Height [cm] 160.00 ± 5.66; weight [kg] 59.80 ± 5.81	software Kinovea® version 0.9.5
Slade et al. (2022) [9]	Hip, knee, ankle joint angle	Gait (1,25m/s); Run (4m/s)	n = 5; male 3; female 2; age [yr] 25.0 ± 1.8; weight [kg] 63.0 ± 10.3; height [m] 1.70 ± 0.07	Open-Source System and Experimental Comparison to Optical Motion Capture

(*continued*)

Table 1. (*continued*)

Author (Year)	Parameters tested	Activity tested	Participants	Measuring system
Filli et al. (2020) [10]	Hip, knee, ankle Flex./Ext	Gait (1km/h; 2,5km/h)	n = 19; male 11 female 8; age [yr] 44.7 ± 12.3; height [cm] 172.1 ± 8.2; weight [kg] 69.1 ± 8.4	Vicon
Martinez-Hernandez et al. (2018) [11]	Hip, knee, ankle angular velocity	Level-ground walking, Ramp ascent, Ramp descent	n = 8; male 8; age [yr] 24 - 34; height [m] 1.74 - 1.79; weight [kg] 77.6 - 85	freedom IMUs, from Shimmer Inc.,
Langeard et al. (2021) [12]	Hip, Ankle joint angle	Gait	n = 14; male 7 female 7; age [yr] 73.6 ± 4.9; BMI 25.9 ± 3.2	VICON ® system and 8 Bonita B10 cameras (1 million pixels, 250fps) and one Bonita 720c camera (1280 x 720, 120Hz)
Cheatham et al. (2019) [13]	Knee Flex./Ext. Joint angle	Passive knee flexion ROM	n1 = 12 age [yr] 24.33 ± 3.47; height [cm] 169.54 ± 9.88; weight [kg] 77.34 ± 12.56; n2 = 12 age [yr] 25.25 ± 3.60; height [cm] 173.15 ± 10.13; weight [kg] 81.68 ± 14.96; n3 = 12 age [yr] 26.08 ± 6.02; height [cm] 167.61 ± 9.63; weight [kg] 67.98 ± 16.65	Baseline digital inclinometer
Taborri et al. (2022) [14]	Hip, knee, ankle joint angle	Gait	n = 12; male 4 female 8; age [yr] 26 ± 3; height [cm] 170.8 ± 4.7; weight [kg] 64.9 ± 9.6	Xsens Technologies (Enschede, The Netherland)
Rebeyrat et al. (2022) [15]	Hip, knee joint angle	Gait	n = 62; male = 24; female = 38; age[yr] 34 ± 13	Vicon Motion Systems
Song et al. (2022) [16]	Hip, knee joint torques	Gait	n = 1; female = 1; age [yr] 25; height [cm] 168; weight [kg] 52	NOKOV 3D optical motion capture system
Cash et al. (2022) [17]	Hip, knee joint angle	Gait (1m/s)	n = 12; male 6 female 6; age [yr] 20.8 ± 1.4; height [m] 1.7 ± 0.1; weight [kg] 75.2 ± 9.9	Vicon-Vantage System Oxford, UK

(*continued*)

Table 1. (*continued*)

Author (Year)	Parameters tested	Activity tested	Participants	Measuring system
Holcomb et al. (2022) [18]	Hip, knee, ankle joint angle, Vertical ground reaction force [N], EMG - Co-contraction of knee and ankle	Initial Contact and Loading response and Midstance and Stance Phase of Gait	n1 = 14; male 7 female 7; age [yr] 21.8 ± 2.1; height [cm] 174.6 ± 9.7; weight [kg] 73.0 ± 15.8; n2 = 14; male 7 female 7; age [yr] 70.1 ± 3.0; height [cm] 171.5 ± 12.0; weight [kg] 74.2 ± 17.9	Vantage, Vicon Motion Systems Ltd., Oxford, UK; force platform OR6, Advanced Mechanical Technology Inc., Watertown, MA; Visual3D (v6.0, C-Motion, Rockville, MD)
Riazati et al. (2022) [19]	Hip, knee, ankle joint angle	self-selected walking speed, fastest comfortable walking speed	n = 21; male 7 female 14; age [yr] 37.8 ± 18.8; height [m] 1.7 ± 0.1; weight [kg] 70.8 ± 11.0;	Eight video cameras with AccuPower 4.0 software
Schulte et al. (2023) [20]	Hip, knee, ankle joint angle, sEMG Normalized Amplitude (A.U.)	Walking, Stair Ascent, Stair Descent, Ramp Ascent, Ramp Descent	n1 = 10; male 7 female 3; age [yr] 24 ± 2; height [cm] 183 ± 9; weight [kg] 77 ± 10; n2 = 35; male 14 female 21; age [yr] 23 ± 2; height [cm] 179 ± 9; weight [kg] 73 ± 11; n3 = 10; male 4 female 6; age [yr] 24 ± 2; height [cm] 174 ± 6; weight [kg] 71 ± 9	IMU-based motion capture system; Trigno electrodes (Delsys, Boston, US) at a sample frequency of 1000 Hz; Wave electrodes (Cometa Systems, Bareggio, Italy) at a sampling frequency of 2000 Hz
Raghu et al. (2021) [21]	Knee, Ankle Flex./Ext. Joint angle	Gait (1,2,3 Mph)	n = 12; male 5 female 7; age [yr] 23.0 ± 3.6; height [cm] 1.7 ± 0.08; weight [kg] 72.3 ± 11.3;	five Vicon T40s cameras
Gutierrez-Coronado et al. (2023) [22]	Knee Ext. Joint angle	Knee extension with hip flexion to 90 deg	n = 62; male 52 female 10; age [yr] 21.1 ± 2.03; height [cm] 176 ± 0.09; weight [kg] 75.91 ± 17.04;	Posture Assessment Software (SAPO®)
Mentiplay et al. (2018) [23]	Hip, Knee, Ankle Flex./Ext.; Hip, Knee, Ankle angular velocity	Gait	n = 36; male 13 female 23; age [yr] 21.1 ± 2.03; height [cm] 169.7 ± 9.3; weight [kg] 69.3 ± 12.3	A 13-camera Optitrack three-dimensional motion analysis system

(*continued*)

Table 1. (*continued*)

Author (Year)	Parameters tested	Activity tested	Participants	Measuring system
Kashitaro et al. 2018) [24]	Hip, knee, ankle joint angle	Putting on pants while sitting, putting on pants while standing, putting on the first leg, putting on pants while standing, putting on the second leg, put on the shoes while sitting, by crossing our legs and placing the knee of one leg on top of the other knee, put on the shoes while sitting by crossing our legs and placing the ankle of one leg on top of the other knee, tying shoelaces in the right shoe, tying shoelaces in the left shoe, wiping buttocks with the right hand, wiping buttocks with the left hand, getting in the bath, getting out of the bath, picking up object, trunk rotation during crouching	n = 26; male 13 female 13; age [yr] 20 ± 1; height [cm] 168 ± 7; weight [kg] 60 ± 7	3SPACE FASTRAK, Polhemus, Colchester, VT, USA

Many articles have focused on analysing the lower limb range of motion during gait. The results of these studies are presented in Table 2. Empty cells indicate no data in the study on the discussed values. If the author gave only one value for a single plane movement (e.g. the value in the parasagittal plane as the sum of flexion and extension) – one cell contains the shared value for two directions of a single movement. The results of studies on lower limb range of motion in activities other than walking are presented in Table 3.

Table 2. Lower limb joints range of motion (degrees) during gait (* values estimated based on the graphical representation, flex.-flexion, ext.-extension, add.-adduction, abd.-abduction, int.rot.-internal rotation, ex.rot.-external rotation, inv.-inversion, ev.-eversion)

Author	Gait speed/ phase	Range of motion (deg)											
		Hip						Knee		Ankle			
		Flex.	Ext.	Add.	Abd.	Int. Rot.	Ex. Rot.	Flex.	Ext.	Flex.	Ext.	Inv.	Ev.
Slade et al. [9]		30*	10*	8*	8*	8*	8*	60*	0*	20*	10*		
Filli et al. [10]	1km/h	27.1 ± 1.0						45.0 ± 1.5		18.3 ± 1.0			
	2,5km/h	30.6 ± 1.1						51.0 ± 1.0		24.7 ± 1.0			
Langeard et al. [12]		40.35	14.04	12.06	3.93					34.28	9.72	0.57	0.16
Taborri et al. [14]		38*	2*					70*	0*	20*	10*		
Rebeyrat et al. [15]		35*	8*					62*	-5*				
Cash et al. [17]	Stance Phase	28.58 ± 6.57	6.21 ± 6.58	5.66 ± 4.62	6.26 ± 6.93	3.33 ± 7.19	12.41 ± 6.25	40.61 ± 5.12	4.30 ± 5.22				
	Swing Phase	31.62 ± 7.56	1.67 ± 6.17	6.56 ± 3.72	-0.73 ± 4.47	4.63 ± 7.00	13.08 ± 5.63	57.15 ± 4.73	0.85 ± 4.48				
Holcomb et al. [18]	Initial Contact (n1)	25.2(7.0)						8.2(6.2)		1.3(4.6)			
	Loading response (n1)	14.2(7.5)						14.4(5.8)		3.4(4.0)			
	Midstance (n1)	0.9(6.5)						8.5(5.1)		6.4(3.3)			
	Stance Phase (n1)	37.8(4.6)						36.7(7.1)		19.5(4.8)			
	Initial Contact (n2)	29.4(5.4)						12.8(7.7)		1.1(5.5)			
	Loading response (n2)	19.3(5.2)						22.2(7.2)		5.6(5.5)			
	Midstance (n2)	6.1(5.0)						15.0(7.1)		8.7(6.0)			
	Stance Phase (n2)	39.3(4.8)						33.1(8.0)		21.7(3.6)			

(continued)

Table 2. (*continued*)

Riazati et al. [19]	self-selected walking speed	26.9 ± 2.8	19.7 ± 2.9	6.7 ± 2.3	0.45 ± 2.1	6.4 ± 4.3	5.2 ± 4.2	64.9 ± 3.9	2.8 ± 3.4	31.0 ± 4.4	10.6 ± 4.2	10.9 ± 3.6	4.3 ± 3.5
		47.6 ± 4.5		7.7 ± 3.2		12.7 ± 3.7		65.8 ± 3.9		41.7 ± 5.7		16.2 ± 4.4	
	fastest comfortable walking speed	30.3 ± 3.0	23.4 ± 2.6	6.8 ± 2.2	0.72 ± 1.9	6.5 ± 4.4	−7.2 ± 4.8	65.6 ± 4.3	1.4 ± 5.1	31.6 ± 4.8	9.4 ± 4.8	10.4 ± 3.6	3.7 ± 4.1
		54.0 ± 4.4		7.3 ± 2.4		14.9 ± 4.8		65.9 ± 4.5		41.5 ± 6.4		15.4 ± 5.0	
Schulte et al. [20]		30*	8*					65*	-5*	18*	10*		
Raghu et al. [21]	1 mph							48.14 ± 7.18			07.71 ± 3.32		
								47.89 ± 5.43		21.04 ± 5.17			
	2 mph							53.32 ± 5.65			06.81 ± 3.31		
								52.61 ± 4.93		25.51 ± 5.11			
	3 mph							55.16 ± 5.43	X	X	05.07 ± 3.23		
								53.53 ± 3.22		27.69 ± 6.38			
Mentiplay et al. [23]		33*	12*					62*	5*	20*	15*		

Table 4 describes the angular velocities for the lower limbs during different types of gait. Table 5, showing joint torques, contains only one row, as more information on this subject was not found in the literature. No studies were found on angular acceleration in the joints of the lower limb.

Table 3. Lower limb joints range of motion (deg) during an activity of daily living (ADL) (* values estimated based on the graphical representation, ** various possibilities of performing the activity presented -the largest angular values of these methods taken for the purposes of this study, flex.-flexion, ext.-extension, add.-adduction, abd.-abduction, int.rot.-internal rotation, ex.rot.-external rotation)

Author	Activity	Range of motion (deg)									
		HIP						Knee		Ankle	
		Flex	Ext	Add	Abd	Int. Rot	Ex. Rot	Flex	Ext	Flex	Ext
Das et al. [8]	Active range in the forward position							139.05 ± 0.88	138.23 ± 1.18		
Slade et al. [9]	Run (4m/s)	40*	15*	8*	8*	8*	10*	100*	0*	20*	20*
Cheatham et al. [13]	Passive knee flexion ROM (three groups)							135.24 ± 7.88			
								143.70 ± 8.12			
								134.12 ± 8.35			
Schulte et al. [20]	Stair Ascent	65*	–8*					100*	–20*	22*	15*
	Stair Descent	35*	–10*					100*	–15*	35*	22*
	Ramp Ascent	48*	3*					65*	–18*	25*	15*
	Ramp Descent	5*	–5*					70*	–5*	19*	10*
Gutierrez-Coronado et al. [21]	Knee extension with hip flexion to 90 deg								44.36 ± 8.51		
Kashitaro et al. [24]	putting on pants while sitting	86	–35	11	14	19	30	119	–42	37	11
	putting on pants while standing, putting on the first leg	86	4	12	13	24	29	115	–1	37	13

(*continued*)

Table 3. *(continued)*

Author	Activity	Range of motion (deg)									
		HIP						Knee		Ankle	
		Flex	Ext	Add	Abd	Int. Rot	Ex. Rot	Flex	Ext	Flex	Ext
	putting on pants while standing, putting on the second leg	18	3	4	12	1	34	24	–1	4	7
	put on the shoes while sitting, by crossing our legs and placing the knee of one leg on top of the other knee	85	–45	17	4	7	26	84	–55	19	5
	put on the shoes while sitting by crossing our legs and placing the ankle of one leg on top of the other knee	82	–43	0	18	6	61	103	–71	21	5
	tying shoelaces in the right shoe	89	–53	5	2	2	9	77	–73	2	3
	tying shoelaces in the left shoe	90	–54	–3	11	–4	18	78	–75	3	1
	wiping buttocks with the right hand	71	–54	3	4	–1	21	79	–71	1	4

(continued)

Table 3. (*continued*)

Author	Activity	Range of motion (deg)									
		HIP						Knee		Ankle	
		Flex	Ext	Add	Abd	Int. Rot	Ex. Rot	Flex	Ext	Flex	Ext
	wiping buttocks with the left hand	76	–55	–3	10	11	7	79	–75	4	–1
	getting in the bath**	87	8	12	31	31	37	128	–1	25	17
	getting out of the bath**	99	2	12	21	20	27	143	–1	35	28
	picking up object**	92	–53	10	13	2	18	79	–70	2	5
	trunk rotation during crouching	101	–82	–7	29	39	–10	149	–137	–18	28

Table 4. Lower limb angular velocity (deg/s) during different types of gait (* values estimated based on the graphical representation, flex.-flexion, ext.-extension, add.-adduction, abd.-abduction, int.rot.-internal rotation, ex.rot.-external rotation)

Author	Activity	Angular Velocity (deg/s)					
		HIP		Knee		Ankle	
		Flex	Ext	Flex	Ext	Flex	Ext
Martinez-Hernandez et al. [11]	Level-ground walking	160*	100*	390*	160*	410*	420*
	Ramp ascent	150*	95*	350*	180*	380*	415*
	Ramp descent	145*	90*	330*	175*	375*	410*
Mentiplay et al. [23]	Gait	210*	145*	320*	380*	300*	200*

4 Discussion

Rehabilitation aims to restore the patient to the highest possible level of activity and participation [25]. Therefore, when developing a rehabilitation robot, parameters such as range of motion and joint strength should be increased to enable the patient to perform daily activities [26]. Hence, they change within the recovery process, but the system has to be capable of assisting individuals of different anatomy and strength.

Walking is one of the most important activities performed using lower limbs. It is one of the primary forms of human locomotion. Studies show that joint motion during gait is multi-dimensional, but the most extensive ranges are activated in the parasagittal

Table 5. Joint moments [N] of hip and knee during gait (* values estimated based on the graphical representation, flex.-flexion, ext.-extension, add.-adduction, abd.-abduction, int.rot.-internal rotation, ex.rot.-external rotation, inv.-inversion, ev.-eversion)

Author	Activity	Joint moments [Nm]	
		HIP	Knee
		Flex./Ext	Flex./Ext
Song et al. [16]	Gait	60*	30*

plane. Knee flexion during gait can reach up to 70° with a maximum rotation of 8° [9, 14]. Gait rehabilitation robots typically operate with a maximum of 6 degrees of freedom. However, to achieve a level of gait that is optimal in terms of energy expenditure and as functional as possible, rehabilitation must strive to regain ranges of motion in all planes and avoid unanatomical compensations whenever possible [2, 28]. The same principle applies to running, where knee flexion reaches about 100° and hip adduction 8° [9]. Movements in which the largest ranges are activated mainly in one plane are climbing stairs or descending stairs [20]. Nevertheless, activating ranges in the other planes helps maintain balance or reduce energy expenditure during these activities [27].

Defining the use of rehabilitation robots in terms of supported activities is critical for building its kinematic chain and computing the required strength or driving joint torques. When patients reach a certain level of progress, available joint speeds, balance and endurance factors during walking become important. In advanced robotic systems, patient's gait in different environmental conditions (for example, underwater walking) can be simulated to strengthen their muscles. In such systems, control methods such as impedance and admittance are used.

The design problem becomes much more complex when it comes to daily activities that require much more range in different planes. For example, one way of putting on shoes, in addition to flexing the hip by 82°, requires external rotation of the hip by 61°. Stepping out of the bathtub may require a 31° hip visitation and a 39° internal rotation is needed in trunk rotation while squatting. The above examples show the complexity of movements in people's daily lives. In some of them, strongly dominant planes of mobility exist. However, even these often require joint activities from other planes [24]. Based on the literature review, lower limb rehabilitation robots must activate a wide range of hip joint motions in different planes to support daily life activities.

The degree of freedom of commercially available robots is typically lower than for the extremity movements of humans during daily life activities. The reasons for this include the difficulty of modelling these movements, as well as fitting high-power actuators with multiple sensors and the control hardware into a small volume. Despite this, the use of rehabilitation robots is increasing day by day. Thanks to their ability to perform repetitive movements with high accuracy and precision, these robots play an essential role in recovering muscles involved in most of the activities and, accordingly, the range of motion of the joints. In this way, they contribute to reducing the workload of physiotherapists or physicians and increasing treatment efficiency. They also allow remote rehabilitation [26]. With the developing technology and decreasing costs, it is estimated that more capable and cost-effective robots can be produced and will become widespread at a higher speed in the future.

5 Summary

The paper reviewed recent literature in terms of data on the kinematics and dynamics of the human lower limb. The collected numerical parameters can help in designing rehabilitation robots. However, the investigation shows a significant gap in experimental measurements of human lower extremity dynamics and kinematics parameters during task-oriented physiotherapy. Notably, data on angular accelerations in joints and involved joint torques (resulting from muscular forces) are missing from available sources.

As a continuation of the presented review, an experimental investigation of the kinematics and dynamics of the human lower extremity is planned. It is targeted to fill the current state-of-the-art gap.

It is also necessary to model the trajectories of the body segments accurately in activities of daily living. Thanks to this, robot-assisted rehabilitation will allow patients to gain more independence and increase their level of activity and participation. Moreover, the risk of physical injuries from joint misalignments will be minimised.

Acknowledgement. The paper is based on the results of the "Smart exoskeleton for remote rehabilitation with the digital twin technology" – *SmartEx-Twin* project, financed in 2021–2023 (265,800 EUR), in the scope of scientific research and development works by the National Center for Research and Development (funding programme 5th Call for Proposals Poland-Turkey).

References

1. Falkowski, P., Osiak, T., Pastor, A.: Analysis of needs and requirements of kinesiotherapy in Poland for robot design purposes. Prace Naukowe Politechniki Warszawskiej. Elektronika (2022)
2. Zhang, X., et al.: Robotics in lower-limb rehabilitation after stroke. Behav. Neurol. **2017**(1), 3731802 (2017). https://doi.org/10.1155/2017/3731802
3. Maggioni, S., et al.: Assessing walking ability using a robotic gait trainer: opportunities and limitations of assist-as-needed control in spinal cord injury. J. NeuroEng. Rehabil. **20**(1), 121 (2023). https://doi.org/10.1186/s12984-023-01226-4
4. Sumner, J., Lim, H.W., Chong, L.S., Bundele, A., Mukhopadhyay, A., Kayambu, G.: Artificial intelligence in physical rehabilitation: a systematic review. Artif. Intell. Med. **146**(6), 102693 (2023). https://doi.org/10.1016/j.artmed.2023.102693
5. Falkowski, P.: Predicting dynamics of a rehabilitation exoskeleton with free degrees of freedom. In: Conference on Automation. Springer, Cham (2022)
6. Wilk, J., Piotr, F.: A concept of detecting patient hazards during exoskeleton-aided remote home motor rehabilitation. Prace Naukowe Politechniki Warszawskiej. Elektronika, 101–112 (2022)
7. Falkowski, P.: Light exoskeleton design with topology optimisation and FEM simulations for FFF technology. J. Autom. Mob. Robot. Intell. Syst. **15**(2), 14–19 (2021). https://doi.org/10.14313/JAMRIS/2-2021/9

8. Das, R., Jhajharia, B., Ciocan, V.C., Majumdar, I., Sharma, A.: The relationship between latent myofascial trigger point and range of motion of knee flexor and extensor muscles. Phys. Educ. Theory Methodol. **23**(2), 192–198 (2023). https://doi.org/10.17309/tmfv.2023.2.06
9. Slade, P., Habib, A., Hicks, J.L., Delp, S.L.: An open-source and wearable system for measuring 3D human motion in real-time. IEEE Trans. Biomed. Eng. **69**(2), 678–688 (2021). https://doi.org/10.1109/TBME.2021.3103201
10. Filli, L., et al.: Characterizing cognitive-motor impairments in patients with myotonic dystrophy type 1. Neuromuscul. Disord. **30**(6), 510–520 (2020). https://doi.org/10.1016/j.nmd.2020.04.005
11. Martinez-Hernandez, U., Mahmood, I., Dehghani-Sanij, A.A.: Simultaneous Bayesian recognition of locomotion and gait phases with wearable sensors. IEEE Sens. J. **18**(3), 1282–1290 (2017). https://doi.org/10.1109/JSEN.2017.2782181
12. Langeard, A., Bigot, L., Loggia, G., Bherer, L., Chastan, N., Gauthier, A.: Ankle dorsiflexors and plantarflexors neuromuscular electrical stimulation training impacts gait kinematics in older adults: a pilot study. Gait Posture **84**, 335–339 (2021). https://doi.org/10.1016/j.gaitpost.2020.12.016
13. Cheatham, S.W., Stull, K.R.: Roller massage: comparison of three different surface type pattern foam rollers on passive knee range of motion and pain perception. J. Bodywork Mov. Ther. **23**(3), 555–560 (2019). https://doi.org/10.1016/j.jbmt.2019.05.002
14. Taborri, J., Santuz, A., Brüll, L., Arampatzis, A., Rossi, S.: Measuring kinematic response to perturbed locomotion in young adults. Sensors (Basel, Switzerland) **22**(2), 672 (2022). https://doi.org/10.3390/s22020672
15. Rebeyrat, G., et al.: Assessment of dynamic balance during walking in patients with adult spinal deformity. Eur. Spine J. **31**(7), 1736–1744 (2022). https://doi.org/10.1007/s00586-022-07199-7. official publication of the European Spine Society, the European Spinal Deformity Society, and the European Section of the Cervical Spine Research Society
16. Song, L., Wang, A., Zhong, J.: Inverse dynamics modeling and analysis of healthy human data for lower limb rehabilitation robots. Electronics (Switzerland) **11**(23), 3848 (2022). https://doi.org/10.3390/electronics11233848
17. Cash, J.J., van Werkhoven, H., Cole, K.J., Needle, A.R.: The effects of 72 h of dynamic ankle immobilization on neural excitability and lower extremity kinematics. Gait Posture **93**, 198–204 (2022). https://doi.org/10.1016/j.gaitpost.2022.02.005
18. Holcomb, A.E., Hunt, N.L., Ivy, A.K., Cormier, A.G., Brown, T.N., Fitzpatrick, C.K.: Musculoskeletal adaptation of young and older adults in response to challenging surface conditions. J. Biomech. **144**, 111270 (2022). https://doi.org/10.1016/j.jbiomech.2022.111270
19. Riazati, S., McGuirk, T.E., Perry, E.S., Sihanath, W.B., Patten, C.: Absolute reliability of gait parameters acquired with markerless motion capture in living domains. Front. Hum. Neurosci. **16**, 867474 (2022). https://doi.org/10.3389/fnhum.2022.867474
20. Schulte, R.V., et al.: Database of lower limb kinematics and electromyography during gait-related activities in able-bodied subjects. Sci. Data **10**(1), 461 (2023). https://doi.org/10.1038/s41597-023-02341-6
21. Raghu, S.L., Conners, R.T., Kang, C.K., Landrum, D.B., Whitehead, P.N.: Kinematic analysis of gait in an underwater treadmill using land-based Vicon T 40s motion capture cameras arranged externally. J. Biomech. **124**, 110553 (2021). https://doi.org/10.1016/j.jbiomech.2021.110553
22. Gutierrez-Coronado, J., et al.: The clinical benefits of a dynamic vs. static component as part of a comprehensive warm-up for recreational sports players with clinical histories of hamstring injuries: a randomized clinical trial. Int. J. Environ. Res. Public Health, **20**(1), 744 (2022). https://doi.org/10.3390/ijerph20010744

23. Mentiplay, B.F., Banky, M., Clark, R.A., Kahn, M.B., Williams, G.: Lower limb angular velocity during walking at various speeds. Gait Posture **65**, 190–196 (2018). https://doi.org/10.1016/j.gaitpost.2018.06.162
24. Hyodo, K., Masuda, T., Aizawa, J., Jinno, T., Morita, S.: Hip, knee, and ankle kinematics during activities of daily living: a cross-sectional study. Braz. J. Phys. Ther. **21**(3), 159–166 (2017). https://doi.org/10.1016/j.bjpt.2017.03.012
25. Habenicht, R., et al.: Machine learning based linking of patient reported outcome measures to WHO international classification of functioning, disability, and health activity/participation categories. J. Clin. Med. **12**(17), 5609 (2023). https://doi.org/10.3390/jcm12175609
26. Falkowski, P., et al.: Study on the applicability of digital twins for home remote motor rehabilitation. Sensors (Basel, Switzerland) **23**(2), 911 (2023). https://doi.org/10.3390/s23020911
27. Ma, Y., et al.: Gait characteristics of children with spastic cerebral palsy during inclined treadmill walking under a virtual reality environment. Appl. Bionics Biomech. **2019**(1), 8049156 (2019). https://doi.org/10.1155/2019/8049156
28. Falkowski, P.: An optimisation problem for exoskeleton-aided functional rehabilitation of an upper extremity. In: IOP Conference Series: Materials Science and Engineering. vol. 1239. no. 1. IOP Publishing (2022). https://doi.org/10.1088/1757-899X/1239/1/012012

Design and Construction of a 6-DOF Modular Robotic Arm-Z

Krzysztof Zawalski[1], Michał Jarek[2], Piotr Falkowski[1,2](✉), Ela Zawidzka[3], and Machi Zawidzki[3]

[1] Industrial Research Institute for Automation and Measurements PIAP, Warsaw, Poland
piotr.falkowski@piap.lukasiewicz.gov.pl
[2] Warsaw University of Technology, Warsaw, Poland
[3] Institute of Fundamental Technological Research, PAN, Warsaw, Poland

Abstract. This paper presents the design and construction process of the Arm-Z manipulator, which represents the concept of a hyper-redundant manipulator based on a connected sequence of identical modules. Each module provides a single degree of freedom (1-DOF), specifically a twist relative to the previous module. The uniformity of the modules gives Arm-Z the potential to be cost-effective and robust. Standardized modules allow for mass production and can be easily replaced in case of failure. This paper discusses each stage of the design process from mechanical perspective, including: module's geometry, simulation based on Denavit-Hartenberg parameters, and manufacturing process. The section covering work on the electrical system discusses issues such as selection of: motors, power supply, protections, and the network's topology connecting EPOS controllers.

Keywords: Robotic arm · Manipulator · Hyperredundant · Arm-Z · Extremely Modular System

1 Introduction

Diving into the world of animal biomechanics reveals a plethora of diverse solutions. Many of these solutions have "an edge" over what we currently know and employ in robotics. Presently, robots most commonly utilize movement on wheels or tracks, however, in unfavorable conditions, can limit the robot's capabilities. Biologically inspired robots hold tremendous potential and can achieve significantly better results than conventional robots [1,2].

The snake-like mechanism is an excellent example that, owing to its excessive degrees of freedom, can perform highly intricate tasks. It is well-suited for operating in challenging and irregular terrains [3]. Robots based on the snake-like mechanism possess excellent capabilities. This is especially true where exceptional precision, flexibility, and the ability to execute complex motion trajectories are required.

R. Szewczyk et al. (Eds.): AUTOMATION 2024, LNNS 1219, pp. 139–150, 2024.
https://doi.org/10.1007/978-3-031-78266-4_13

Research on snake robots commenced several decades ago, with the pioneering work of Shigeo Hirose, who developed the first snake robot in 1972 [4]. Other important researchers in this field include: Chirikjian [5,6], Ostrowski [7,8], Ma [9,10], Iwasaki [11,12], Ijspeert [13,14], Choset [15,16], Matsuno, [17,18]. For reviews of snake robotics research see: [2,4,19–21].

Systems with larger number of joints are called redundant robots, while term hyper-redundant refers to redundant manipulators with a very large, possibly infinite, number of DOFs [5]. Manipulators of this kind can perform highly responsible and complex tasks, such as medical surgeries [22].

Figure 1 shows examples of commercially available snake-like and trunk-like hyper-redundant manipulators.

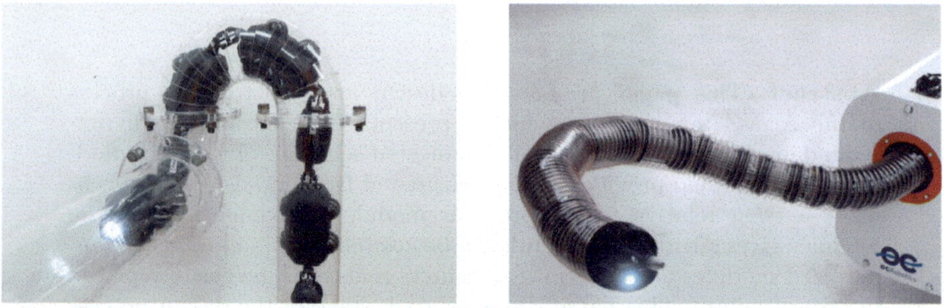

Fig. 1. On the left: A snake-like robot by HiBot (https://www.hibot.co.jp). On the right: A trunk-like robotic manipulator by OC Robotics (http://www.ocrobotics.com)

In typical manipulators with 6 degrees of freedom (6-DOF), it is sufficient to execute any movement in three-dimensional space by: displacement along three Cartesian axes: X, Y, Z, and three rotations: yaw, pitch, and roll. For such manipulators, the solution of inverse kinematics problem is relatively simple [23, 24]. However, control of hyper-redundant robotic systems is very challenging due to their high nonlinearity, and inverse kinematics for snake-like robots and trunk-like manipulators require application of sophisticated computational techniques, also involving artificial intelligence [25–27]. For an example of application of Particle Swarm Optimization for control of such HRM see [28].

2 Extremely Modular Systems

Extremely Modular System (EMS, for short) is a relatively new concept introduced a few years ago in [29]. It represents a new approach to the design of engineering structures and architectural objects where assembly of congruent units allows for the creation of free-form shapes. The main difference from the traditional modular systems used in engineering and building construction in particular, is the emphasis of the minimal diversity of types of modules, ideally - just one. This is why these system are *extremely* modular. These are five basic advantages of EMS:

1. **Economical** - as they are suitable for mass fabrication, thus lowering the cost so they can be broadly applied;
2. **Functional** - as they allow for reconfiguration, expansion, reduction;
3. **Robustness** - since every module which failed can be easily replaced with an identical but functional one;
4. **Discrete** - as they are suitable for intelligent mathematical modeling, and their configurations can be subjected to discrete (multi-objective) optimization using efficient search algorithms;
5. **Uniform** - this feature is advantageous for rapid deployment and automated assembly.

EMSs are part of discrete optimization approach to engineering and architecture introduced in [29] (for representative research see: [30–35]). Other areas are: fundamental aspects of architectural & urban layout [36] (for representative research see: [37–39]), and building envelope (including daylighting, shading and aesthetics) [40,41].

3 Arm-Z

Arm-Z is a concept of robotic manipulator introduced in [] composed of congruent modules each having one degree of freedom (1-DOF) - a relative twist. In principle, each module is defined by the following parameters: size r, offset d, and ζ, that is the angle between upper (**T**) and lower (**B**) faces of the module, as shown in Fig. 2.

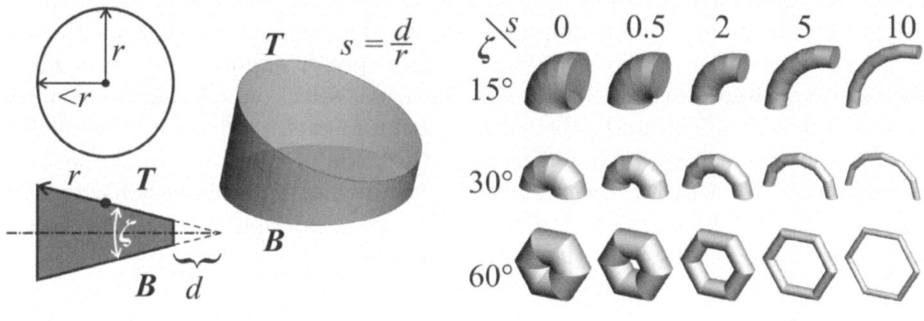

Fig. 2. On the left: visualization of the Arm-Z unit defined by three parameters: r, d and ζ. On the right: simple examples of assemblies of units for various values of ζ and s (slenderness), which is an additional parameter, a d to r ratio.

Figure 3 shows mock-up and early prototype of Arm-Z.

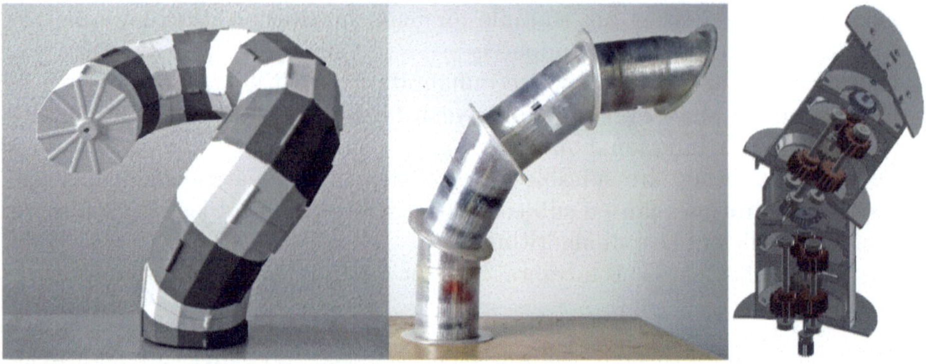

Fig. 3. Photograph of a mock-up illustrating the concept of the Arm-Z manipulator and an early prototype showing the internal gear system allowing for simple movements

For more information on the concept of Arm-Z including its folding study see [42], for early successful implementations of meta-heuristics to control of Arm-Z see [28], for documentation on early prototypes see [43,44].

4 New Construction of the Module

This paper documents the process of designing and constructing a modular 6-DOF robotic manipulator Arm-Z, exploring the its mechanical and electrical features. The mechanical design encompasses module construction, manufacturing technology, dynamics description using the Denavit-Hartenberg method, and MATLAB simulation of module moments. The electrical design includes a detailed examination of the electrical schemes, power supply, protection measures, drive system overview, motor and controller selection, communication via the CANopen protocol, and a list of used components. Additionally, we delve into the software aspect, highlighting the utilization of Epos Studio. As we traverse through the Arm-Z project's mechanical and electrical dimensions, we aim to provide a total understanding of the challenges, achievements, and outcomes, offering valuable insights into the construction of modular robotic systems.

4.1 Mechanical Design

Module Construction. The construction of this manipulator is based on parameterized modules of truncated cylinders. Each module is identical which results in maximum modularity of the robot which facilitates servicing and allows for its expansion (contraction). We used the following parameter values (see Fig. 2): $r = 95\,\text{mm}$, $s = 1/3$, $d = \frac{r}{3} = 31.6(6)$. The angle $\zeta = 25.4228°$. Due to the representative function of the prototype, these values were deemed most illustrative. Moreover, the overall shape of the manipulator is also determined by the number of modules. Six modules provide 6 degrees of freedom.

Fabrication. The modules were 3D-printed with Fused Deposition Modeling (FDM) method using PLA. It is important to note that the prototype is not intended for heavy-duty tasks. Due to the method's simplicity and the material's availability, we were able to conduct practical tests for several versions of the module. As a result, the final version stands out for its high quality and precise fitting of components.

Dynamics Description. The Denavit-Hartenberg method introduced in 1955 [45] is a mathematical tool widely used to analyze robotic manipulator kinematics. It provides a precise description of the kinematic structure of a manipulator by assigning coordinate systems to joints and links of the system. For each joint, four parameters are defined:

θ_i (rotation angle around the common axis)
d_i (distance along the previous joint)
a_i (length of the shared link)
α_i (rotation angle around the common link)

Setting these parameters allow for mathematical modeling of the manipulator's motion. The Denavit-Hartenberg method generates transformation matrices for each joint, enabling a comprehensive kinematic analysis. This method proves invaluable across various robotics applications, motion control, and solving inverse kinematics problems. The Denavit-Hartenberg parameters of the system are presented in Table 1.

Table 1. Denavit-Hartenberg parameters for each module of Arm-Z

Link number	θ_i	d_i mm	a_i mm	α_i
1	π	97.1429	0	0.1109
2	π	97.1429	0	0.1109
3	π	97.1429	0	0.1109
4	π	97.1429	0	0.1109
5	π	97.1429	0	0.1109
6	π	97.1429	0	0.1109
7	π	97.1429	0	0.1109

Using this method, the dynamics of six-module Arm-Z system has been described. By defining the above parameters, we easily obtain transformation matrices between consecutive modules, resulting in a comprehensive kinematic description of the manipulator.

Simulation. Before the final prototyping, an important step was to simulate a working system involving two basic steps. At first we processed the Denavit-Hartenberg parameters for randomly generated values for velocity and acceleration. In this way we produced matrix representing the manipulator's position considering the maximum angular velocity and acceleration conditions and preventing the manipulator from reaching a critical i.e. most extended position. Next, we use this matrix as input for the simulation which runs for specified time, determined by the number of steps. Random linear acceleration is selected in each iteration until the manipulator's height limit is reached. The acceleration is scaled to the maximum, ensuring that it does not exceed the set limits. Linear speed is calculated and checked against the maximum speed. The manipulator's position is updated, ensuring that it remains within the operational area. Similarly, angular acceleration is randomized until the calculated angular velocity meets the specified maximum. Finally, we estimated the maximum torques, and selected the suitable drive and gearhead.

4.2 Electrical Design

The analysis of components and the creation of electrical scheme was followed the simulations conducted in MATLAB environment. We needed to acquire precise information regarding the maximum torque that must be generated to select the motors appropriately. Regarding the drives, our selection was drawn from MAXON - a well-known motors manufacturer. Their strengths are high precision and advanced control capabilities. Another cause influencing our decision was the ease of selecting gears, a crucial aspect of this type of device due to the magnitude of the required torque. After analyzing various options, we opted for brushless EC flat motors. The determining factors were their compact size and high rotational speed. Due to the specific geometry of the module, the drive's compact size was crucial. Another significant element was the rotational speed of the motor, vital for the dynamic movement of the manipulator composed of multiple (in our case - six) modules. BLDC motors offer several positive attributes. They are known for their efficiency, as the absence of brushes reduces friction and energy loss. Additionally, BLDC motors tend to have a long lifespan due to decreased wear-and-tear compared to systems with physical brushes. Moreover, these motors typically exhibit a higher torque-to-weight ratio, making them suitable for applications where a high power-to-weight ratio is crucial. They offer precise speed control, making them well-suited for applications that demand accuracy and consistent speed. The final criterion in our considerations was the sensor element. We examined two types of encoders: incremental and absolute. Following a detailed analysis of needs and assumptions, our choice fell upon a high-resolution incremental encoder, facilitating precise tracking of the motor's movement and the entire module after the gear stage. The final configuration of the drive assemblies comprises of: the EC 45 flat motor, planetary gearhead GP 32, and MILE encoder with a resolution of 2048 counts per turn. After selecting the manipulator drive, the next step was to choose the remaining electrical parts. This involved selecting the appropriate wiring, protection against electric shock and over-current, as well as a power supply that meets the power requirements imposed by the rest of the system. Figure 4 shows the block diagram designed for 6-DOF Arm-Z.

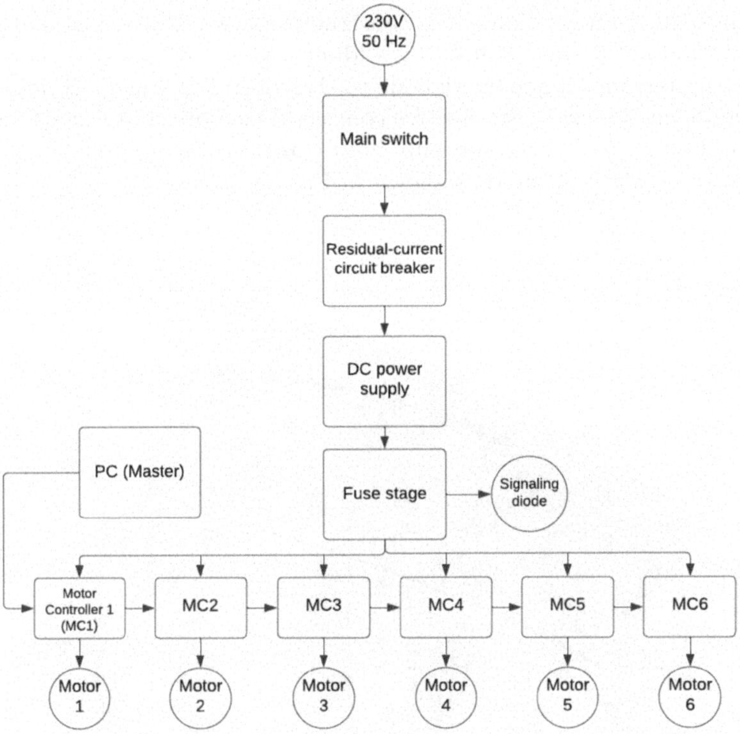

Fig. 4. Block diagram illustrating the Arm-Z electrical connections.

The first component in the system after the main switch is the residual-current circuit-breaker, an electrical protective device that disconnects the circuit when it detects that the electric current flowing out of it is not equal to the incoming current. Such a situation can occur in the case of an electric shock. We chose MSP-1000-24 Mean Well for power supply. It provides direct voltage of 24 V and power of 1000 W. Such power supplies are commonly used in robotic and automated applications. The power supply delivers energy to the controllers and ultimately to the motor set. Circuit-breakers are placed before each controller to prevent permanent damage in case of excessively high current values supplied to the controller. According to the theory, before the time elapsed, leading to potential damage, it should fail and interrupt the current in the circuit.

The next part of the system are the motor controllers, which, together with the master, form a serial network. In our case, a personal computer (PC) is utilized as the master device but can also be replaced with a programmable logic controller (PLC). Communication is executed by the CANopen protocol, a communication protocol for embedded systems commonly used in automation. Thanks to this type of network topology, we have the ability to control each of the motors, simultaneously reducing the need for connections. This is advantageous for the entire device, as it simplifies diagnostics in case of malfunction and expands the configuration and reconstruction possibilities. Installation of an additional module us done simply by adding a new controller to the last unit and assigning terminating network code to the controller. Such a serial topology can be significantly more cost-effective compared to more complex systems. The outcome of this project is the specially built control cabinet and a single module that could be easily replicated as shown in Fig. 5.

Fig. 5. The final result of our project.

After tuning the controllers in the control unit, it is possible to operate the motor using the EPOS Studio software. We developed the structure, conducted successful tests, and analyzed the kinematics and dynamics of Arm-Z.

Concurrently, we optimized the electrical system, ensuring precise control of the manipulator. The obtained results confirm the effectiveness of the design and construction process.

5 Discussion

The collaborative efforts put in designing, constructing, and testing the Arm-Z manipulator have delivered valuable insights and considerations within the field of robotics. This discussion addresses critical aspects of the project, examining its accomplishments, encountered challenges, and prospects for future development.

Simulations show that the highest torque is 24.14 Nm, indicating that the drive to be integrated into the module must match or exceed this value. The drive's nominal torque (maximal continuous torque) is 91.1 mNm, with a gear reduction of 66:1. Considering the internal gear ratio of 4:1 on the gear wheel, the maximum torque achievable is 24.05 Nm. The chosen drive and gearhead combination proves suitable for every module in our manipulator. Significantly, the torque generated surpasses the maximum torque specified in each module, ensuring the motor's capability to rotate the module and thereby ensuring the proper operation of the entire manipulator.

The presented electrical system is ready to work with software, opening up significant possibilities for future developments. The only necessary step is to connect the PC to the control cabinet using a USB cable and establish communication with the controllers. Potential directions encompass exploring advanced control algorithms, integrating machine learning for adaptive behavior, or adapting the modular design for specific applications such as collaborative robotics or autonomous systems.

6 Conclusions

The constructed prototype confirms the accuracy of the simulation and paves the way for further solutions related to Arm-Z. Thanks to this work, a method of designing and transferring concepts to reality using rapid prototyping methods has been documented. The electrical scheme can be expanded for larger and more complex manipulators with a similar structure. The serial network, communicating according to the CANopen protocol, does not impose serious constraints on the number of drive sets (individual modules). The limit of the quantity of devices in the network is sufficiently large that under normal circumstances, it is not a problem. A 6-axis manipulator is merely the beginning of the possibilities which this concept offers. Additional degrees of freedom in manipulators (beyond 6) provide the opportunity, for example, to avoid obstacles in the workspace and define complex trajectories. The distinctive geometry of a module and the void throughout the entire manipulator structure facilitate the delivery of substances to the end effector, allowing the utilization of the Arm-Z in various fields e.g. rescuing, delivering working substances, etc.

Acknowledgments. This research is a part of the project titled *Arm-Z: an extremely modular hyperredundant low-cost manipulator – development of control methods and efficiency analysis* and funded by *OPUS 17* research grant No. 2019/33/B/ST8/02791 supported by the National Science Centre, Poland.

References

1. Pettersen, K.Y.: Snake robots. Ann. Rev. Control **44**, 19–44 (2017)
2. Liljebäck, P., Pettersen, K.Y., Stavdahl, Ø., Gravdahl, J.T.: A review on modelling, implementation, and control of snake robots. Robot. Auton. Syst. **60**(1), 29–40 (2012)
3. Liljebäck, P., Pettersen, K.Y., Stavdahl, Ø., Gravdahl, J.T.: Snake Robots: Modelling, Mechatronics, and Control. Springer (2013)
4. Hirose, S., Yamada, H.: Snake-like robots [tutorial]. IEEE Robot. Autom. Mag. **16**(1), 88–98 (2009)
5. Chirikjian, G.S., Burdick, J.W.: An obstacle avoidance algorithm for hyper-redundant manipulators. In: Proceedings of the IEEE International Conference on Robotics and Automation, pp. 625–631. IEEE (1990)
6. Chirikjian, G.S., Burdick, J.W.: The kinematics of hyper-redundant robot locomotion. IEEE Trans. Robot. Autom. **11**(6), 781–793 (1995)
7. Ostrowski, J., Burdick, J.: Gait kinematics for a serpentine robot. In: Proceedings of IEEE International Conference on Robotics and Automation, vol. 2, pp. 1294–1299. IEEE (1996)
8. McIsaac, K.A., Ostrowski, J.P.: Motion planning for anguilliform locomotion. IEEE Trans. Robot. Autom. **19**(4), 637–652 (2003)
9. Ma, S.: Analysis of snake movement forms for realization of snake-like robots. In: Proceedings 1999 IEEE International Conference on Robotics and Automation (Cat. No. 99CH36288C), vol. 4, pp. 3007–3013. IEEE (1999)
10. Ma, S., Li, W.J., Wang, Y.: A simulator to analyze creeping locomotion of a snake-like robot. In: Proceedings 2001 ICRA. IEEE International Conference on Robotics and Automation (Cat. No. 01CH37164), vol. 4, pp. 3656–3661. IEEE (2001)
11. Prautsch, P., Mita, T., Iwasaki, T.: Analysis and control of a gait of snake robot. IEEJ Trans. Ind. Appl. **120**(3), 372–381 (2000)
12. Saito, M., Fukaya, M., Iwasaki, T., et al.: Modeling, analysis, and synthesis of serpentine locomotion with a multilink robotic snake. IEEE Control Syst. Mag. **22**(1), 64–81 (2002)
13. Crespi, A., Badertscher, A., Guignard, A., Ijspeert, A.J.: AmphiBot I: an amphibious snake-like robot. Robot. Auton. Syst. **50**(4), 163–175 (2005)
14. Crespi, A., Ijspeert, A.J.: Online optimization of swimming and crawling in an amphibious snake robot. IEEE Trans. Robot. **24**(1), 75–87 (2008)
15. Wright, C., et al.: Design of a modular snake robot. In: 2007 IEEE/RSJ International Conference on Intelligent Robots and Systems, pp. 2609–2614. IEEE (2007)
16. Tesch, M., et al.: Parameterized and scripted gaits for modular snake robots. Adv. Robot. **23**(9), 1131–1158 (2009)
17. Fukushima, H., Satomura, S., Kawai, T., Tanaka, M., Kamegawa, T., Matsuno, F.: Modeling and control of a snake-like robot using the screw-drive mechanism. IEEE Trans. Robot. **28**(3), 541–554 (2012)
18. Tanaka, M., Matsuno, F.: Control of snake robots with switching constraints: trajectory tracking with moving obstacle. Adv. Robot. **28**(6), 415–429 (2014)

19. Transeth, A.A., Pettersen, K.Y., Liljebäck, P.: A survey on snake robot modeling and locomotion. Robotica **27**(7), 999–1015 (2009)
20. Hopkins, J.K., Spranklin, B.W., Gupta, S.K.: A survey of snake-inspired robot designs. Bioinspiration Biomimetics **4**(2), 021001 (2009)
21. Sanfilippo, F., Azpiazu, J., Marafioti, G., Transeth, A.A., Stavdahl, Ø., Liljebäck, P.: Perception-driven obstacle-aided locomotion for snake robots: the state of the art, challenges and possibilities. Appl. Sci. **7**(4), 336 (2017)
22. Hu, Y., Zhang, L., Li, W., Yang, G.Z.: Design and fabrication of a 3-D printed metallic flexible joint for snake-like surgical robot. IEEE Robot. Autom. Lett. **4**(2), 1557–1563 (2019). https://doi.org/10.1109/LRA.2019.2896475. ISSN 23773766
23. Murray, R.M., Li, Z., Sastry, S.S.: A Mathematical Introduction to Robotic Manipulation. CRC Press (1994)
24. Bandyopadhyay, S.: Introduction to the inverse kinematics of serial manipulators. Department of Engineering Design, Indian Institute of Technology Madras, Chennai
25. Rolf, M., Steil, J.J.: Efficient exploratory learning of inverse kinematics on a bionic elephant trunk. IEEE Trans. Neural Netw. Learn. Syst. **25**(6), 1147–1160 (2014)
26. Melingui, A., Escande, C., Benoudjit, N., Merzouki, R., Mbede, J.B.: Qualitative approach for forward kinematic modeling of a compact bionic handling assistant trunk. IFAC Proc. Vol. **47**(3), 9353–9358 (2014)
27. Falkenhahn, V., Hildebrandt, A., Neumann, R., Sawodny, O.: Dynamic control of the bionic handling assistant. IEEE/ASME Trans. Mechatron. **22**(1), 6–17 (2016)
28. Zawidzki, M., Szklarski, J.: Transformations of Arm-Z modular manipulator with particle swarm optimization. Adv. Eng. Softw. **126**, 147–160 (2018)
29. Zawidzki, M.: Discrete Optimization in Architecture: Extremely Modular Systems. Springer (2017)
30. Zawidzki, M., Nishinari, K.: Modular Truss-Z system for self-supporting skeletal free-form pedestrian networks. Adv. Eng. Softw. **47**(1), 147–159 (2012)
31. Zawidzki, M., Nishinari, K.: Application of evolutionary algorithms for optimum layout of Truss-Z linkage in an environment with obstacles. Adv. Eng. Softw. **65**, 43–59 (2013)
32. Zawidzki, M.: Retrofitting of pedestrian overpass by Truss-Z modular systems using graph-theory approach. Adv. Eng. Softw. **81**, 41–49 (2015)
33. Zawidzki, M.: Optimization of multi-branch Truss-Z based on evolution strategy. Adv. Eng. Softw. **100**, 113–125 (2016)
34. Zawidzki, M., Szklarski, J.: Effective multi-objective discrete optimization of Truss-Z layouts using a GPU. Appl. Soft Comput. **70**, 501–512 (2018)
35. Zawidzki, M., Jankowski, Ł: Optimization of modular Truss-Z by minimum-mass design under equivalent stress constraint. Smart Struct. Syst. **21**(6), 715–725 (2018)
36. Zawidzki, M.: Discrete Optimization in Architecture: Architectural & Urban Layout. Springer (2016)
37. Zawidzki, M., Tateyama, K., Nishikawa, I.: The constraints satisfaction problem approach in the design of an architectural functional layout. Eng. Optim. **43**(9), 943–966 (2011)
38. Zawidzki, M., Chraibi, M., Nishinari, K.: Crowd-Z: the user-friendly framework for crowd simulation on an architectural floor plan. Pattern Recogn. Lett. **44**, 88–97 (2014)
39. Zawidzki, M.: Automated geometrical evaluation of a plaza (town square). Adv. Eng. Softw. **96**, 58–69 (2016)

40. Zawidzki, M.: Discrete Optimization in Architecture: Building Envelope. Springer (2017)
41. Zawidzki, M.: Dynamic shading of a building envelope based on rotating polarized film system controlled by one-dimensional cellular automata in regular tessellations (triangular, square and hexagonal). Adv. Eng. Inform. **29**(1), 87–100 (2015)
42. Zawidzki, M.: Deployable pipe-Z. Acta Astronaut. **127**, 20–30 (2016)
43. Zawidzka, E., Kiński, W., Zawidzki, M.: Preliminary prototype of a 4-unit Arm-Z hyper-redundant modular manipulator. In: Conference on Automation, pp. 285–294. Springer (2021)
44. Zawidzka, E., Szklarski, J., Kiński, W., Zawidzki, M.: Prototype of the Arm-Z modular solar tracker. In: Conference on Automation, pp. 273–282. Springer (2022)
45. Denavit, J., Hartenberg, R.S.: A kinematic notation for lower-pair mechanisms based on matrices (1955)

The Concept of the 5-DOF Passive Exoskeleton for Tracking the Motion of a Human Upper Extremity

Jan Oleksiuk[1,2], Mateusz Zadrożniak[2], and Piotr Falkowski[1,2](✉)

[1] Industrial Research Institute for Automation and Measurements PIAP, Warsaw, Poland
piotr.falkowski@piap.lukasiewicz.gov.pl
[2] Warsaw University of Technology, Warsaw, Poland

Abstract. With an increasing and ageing population and potential challenges in providing conventional, manual physiotherapy, this research paper proposes the use of exoskeleton robots for remote motor telerehabilitation. This approach, particularly task-oriented rehabilitation, is considered to be efficient and cost-effective. In order to create a reliable and safe rehabilitative exoskeleton system, precise data measurement during movements needs to be conducted. This study introduces a novel concept of an upper extremity trajectory recording system integrated with the construction of the passive exoskeleton with 5 degrees of freedom manufactured by the 3D-printing technology, granting both durability and light-weightness.

Keywords: robotized rehabilitation · telerehabilitation · exoskeleton · measurement system

1 Introduction

By the year 2050, 16% of global population is projected to be aged 65 or older and its total number will rise by 1.5 billion [1]. Due to growing and aging society the number of patients suffering from diseases such as stroke will increase [2]. Stroke survivors are commonly affected by paresis or hemiparesis which lead to movement system impairment. About 80% of patients deficits in upper extremity sensorimotor function [3]. As a result, their ability of performing activities of daily living (ADL), specified as tasks relevant for independent living [4], becomes restricted, significantly aggravating their quality of life and limiting participation in home and community activities which may additionally lead to several mental issues [5,6]. Motor rehabilitation with assistance of qualified physiotherapist is the primary treatment for restoring patients' mobility and independence [7,8].

Anticipated increase in the demand for rehabilitation, may result in more limited access to such services because of its time-consumption and restricted

R. Szewczyk et al. (Eds.): AUTOMATION 2024, LNNS 1219, pp. 151–161, 2024.
https://doi.org/10.1007/978-3-031-78266-4_14

number of specialized therapists. The pandemic reality is also remarkable impediment to the treatment process. The outbreak of another pandemic after SARS-CoV-2 is probable to occur due to various human activities such as mass deforestation or increased livestock grazing [9]. Social isolation is necessary to reduce disease transmission [10], but it prevents patients from obtaining service from physiotherapists in a consistent manner. Therefore, development of an alternative method (such as telerehabilitation) of conventional providing rehabilitation is a matter of priority.

Telerehabilitation is the provision of rehabilitation services to patients at a remote location using information and communication technologies [11]. Communication between the patient and the physiotherapist occurs through a variety of technologies such as audio-calls, videoconferencing and sensoring systems [12]. Teleconsultations encompass assessment, goal-setting, therapy and progress monitoring [13]. It may be additionally augmented through the use of Virtual Reality (VR) programs, based on which the patient completes therapy tasks within a computer generated virtual environment, and result data is sent to the rehabilitation specialist [14]. Growing interest in the use of varied telerehabilitation techniques has been observed over time as technologies have become more advanced [15], however remote rehabilitation still has its limitations and needs to be enhanced [16].

The implementation of exoskeleton robots for remote motor rehabilitation of stroke survivors and other patients suffering from mobility dysfunctions has became especially prospective [17] due to its potential of fulfilling the requirements of the effective telerehabilitation and has been shown to be (at the minimum) as effectual as high-intensity physical treatment [18]. Conducting robotized therapy can relieve physiotherapists of their workload, resulting in patients obtaining rehabilitation services consistently for longer duration of time irregardless of skill and fatigue level of therapist [19] or social isolation obligation which is the main goal of telerehabilitation. Additionally this approach has potential to be more cost-efficient [20].

Among motor rehabilitation approaches, task-oriented rehabilitation has demonstrated notable efficacy [21] and is considered to become a leading therapy strategy in the future [22] since it focuses not only on assessing factors such as muscle strength or range of motion but also on the evaluating the comprehensive functional condition of the patient [23]. This method incorporates performing motor tasks within a functional context [24]. In remote task-oriented treatment, patients conduct motions, such as ADL (e.g. picking up an object from a table), with the active assistance of an exoskeleton. It is essential for these movements to follow the required trajectory, as incorrect technique may lead to obtaining additional injuries and significantly hamper the convalescence process. Due to the absence of a therapist, collecting data on joint rotation angles during movements is necessary to evaluate their appropriateness.

Multiple measurement systems of an upper extremity joint angles have already been developed, such as e.g. accelerometry-based measurement [25] or motion capture systems [26]. However both systems involves using an additional equipment which results in generating higher costs of treatment. Additionally, their incorporation may complicate rehabilitation. Therefore, the most suitable option is to develop the trajectory-tracking system integrated with the exoskeleton, which would be intuitive to use by patients.

The primary purpose of this study is to design reliable and precise measurement system based on the passive exoskeleton construction, through which the physiotherapist, using collected data, can control the movements of the patient's active exoskeleton also grounded on the digital twin technology. The exoskeleton with which this model is ultimately intended to communicate is the ExoReha [17]. Measured angle values are input information to a control system of the device. It is an exoskeleton specifically designed to support the musculoskeletal system. Therefore, it can be utilized for conducting remote rehabilitation sessions.

2 Methodology

The entire construction was designed using Computer Aided Design (CAD) software, *Autodesk Inventor*, to offer a comprehensive visual representation of the final product and forestall extensive revisions. The components have been meticulously crafted from bioplastic polylactic acid (PLA) using 3D printing technology. For both strength and cost optimization, Fused Filament Fabrication (FFF) technology has been employed, which is a common choice for prototyping. To save time and filament, the interior of the components has a 20% infill which both granted sufficient durability and low weight of the output [17].

In the context of movement monitoring, data acquisition was executed via optical incremental encoders, specifically the *DFRobot 400 P/R* series. Subsequently, this data was channeled through a *STM32L476RG* microcontroller, and afterward, the data sets were systematically archived on a computer system for further filtering, analysis and evaluation.

3 Results and Discussion

Mechanical structure of the exoskeleton ExoReha [17] was adopted as the basis for the design. 5 degrees of freedom were introduced (3 degrees of freedom related with the shoulder joint adduction/abduction, flexion/extension, rotation and 2 related with the elbow joint flexion/extension and rotation), enabling the execution of most movements required during physiotherapy. The kinematic diagram of the adopted kinematic model of an upper extremity is presented in Fig. 1.

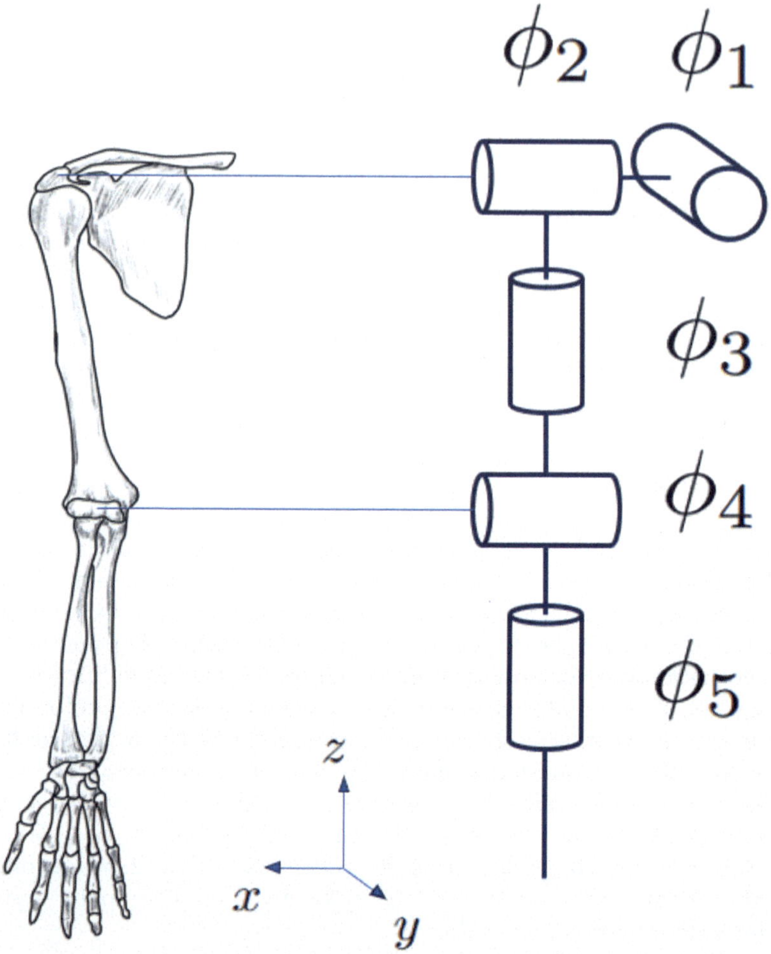

Fig. 1. The kinematic diagram of adopted model of an upper extremity

3.1 Mechanical Section

The exoskeleton was constructed from 13 different components (assembled to 6 main links), which were connected with kinematic pairs, either translational or rotational. The use of five rotational pairs allows for the execution of necessary movements, as these pairs contain the degrees of freedom, which are described by the Denavit-Hartenberg notation (DH) [27] in Fig. 2 and Table 1. Parameters d_i and a_i are adjustable before the trajectory tracking session, but they are constant during the record. Ranges of values of θ_i which are represented by ϕ_i angles of the adopted upper extremity model (Fig. 1) are displayed in Table 2. Encoders were mounted on the rotational axes of the rotational pairs to facilitate data collection essential for the proper conduct of physiotherapy sessions.

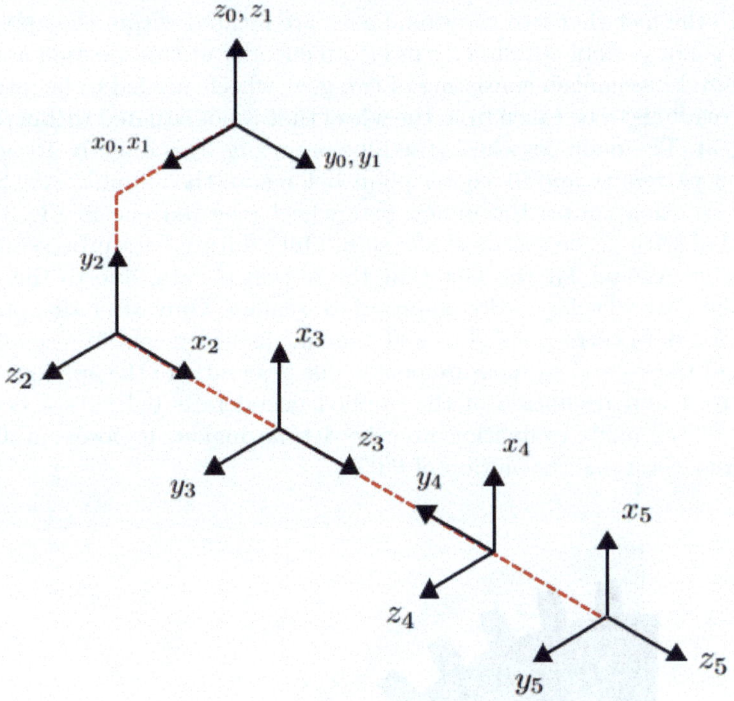

Fig. 2. Degrees of freedom of designed exoskeleton's kinematics in (DH)

Table 1. (DH) parameters of designed exoskeleton

i	θ_i [°]	d_i [mm]	a_i [mm]	ψ_i [°]
1	$\phi_1 = var.$	0	0	0
2	$\phi_2 = var.$	[−124; −100]	100	90
3	$\phi_3 = var.$	0	[160; 190]	90
4	$\phi_4 = var.$	90	0	−90
5	$\phi_5 = var.$	0	[160; 190]	90

Table 2. Ranges of ϕ angles

Angle	Value [°]
ϕ_1	[−135; 135]
ϕ_2	[−140; 140]
ϕ_3	[−90; 90]
ϕ_4	[−135; 135]
ϕ_5	[−90; 90]

Due to the fact that two rotational axes are located within the upper limb's axis, it was not possible to directly insert an encoder at this location and collect data. A novel mechanism consisting of two gear wheels needed to be introduced, enabling readings to be taken from the wheel that is not situated within the arm's axis (Fig. 3). The main cogwheel (the blue on in Fig. 3) is directly attached to a user upper extremity and its center point belongs to the a limb's axis. Shoulder or elbow rotation causes the minor gear wheel (the red one in Fig. 3), which is connected with an encoder's shaft, spin. Only during data processing, it was necessary to account for the fact that the obtained data, due to the different sizes of the gear wheels, needed appropriate scaling. Only after this procedure, the acquired data corresponded to real angular displacements. It also affects the precision of the ϕ_3 and ϕ_5 measurements. The gear ratio of the applied cogwheel is equal to 4 and resolution of the applied encoders is 0.2°. As a result, the accuracy of the angle evaluation may be 4 time higher, it poses an ability to spot inferior rotations (resolution of 0.05°).

Fig. 3. Introduced cogwheel mechanism to measure shoulder and elbow rotation

The translational pairs, on the other hand, were introduced to allow for the adjustment of distances between the kinematic rotational pairs. This enables the customization of the exoskeleton's dimensions for individuals of varying heights and weights. Once specific distances are determined, the members of these pairs are then locked into position, preventing them from moving relative to each other during the execution of movements. The component located on the scapula and the gear wheels passing through the axis of the arm feature specially designed openings. These openings facilitated the easy attachment of easily-resizable straps with buckles required for securing the exoskeleton to the upper limb and chest. Moreover, this element is adapted to be fixed to a frame, which grants stability and static position of the exoskeleton during the tracking sessions. Additionally, specially prepared pads were mounted on the inner side of these wheels to refine the positioning of the arm, ensuring alignment with the axis of rotation of the gear wheel. The use of pads and straps also enhances the comfort of using this exoskeleton. As a result, construction of the exoskeleton is intented to suit to human upper extremity size belonging between 5-percentile polish female height and 95-percentile polish male height. It can be easily attached both on left and right upper limb. Render of the designed exoskeleton is presented in Fig. 4 The final result of the work, after assembling all the components and attaching the straps required to stabilize the upper limb in the exoskeleton, can be seen in Fig. 5.

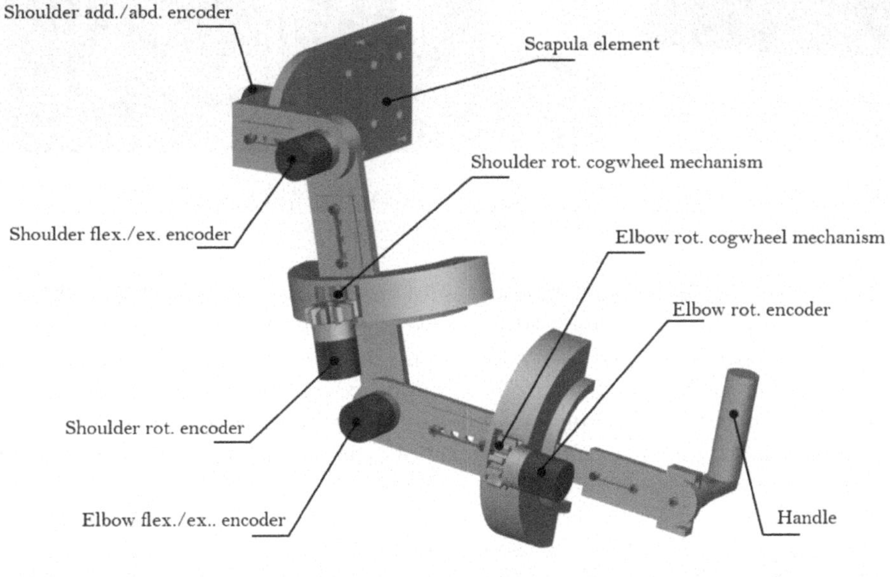

Fig. 4. Render of the final outcome from Autodesk Inventor

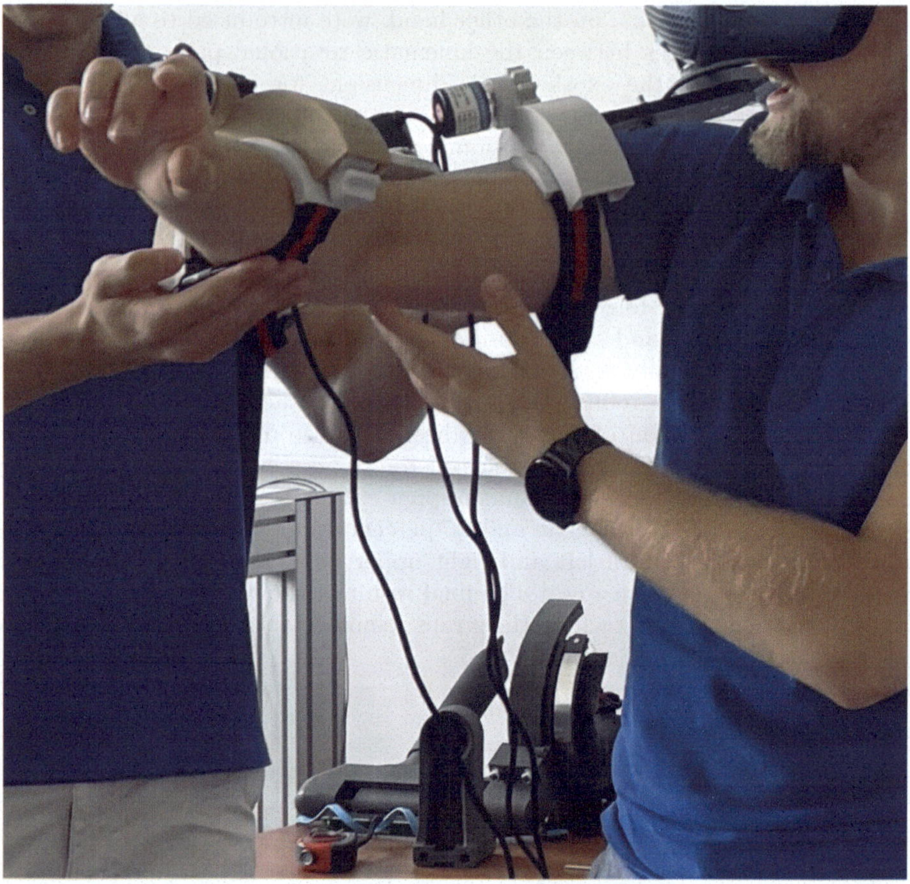

Fig. 5. The final outcome attached to a volunteer

During the initial tests of functioning of the designed exoskeleton, outcomes were satisfying, but some limitations were observed. While performing movements, kinematic singularities could be encountered, such as during shoulder joint extension by 90° placing extended upper extremity parallel to the ground. Additionally, events of exoskeleton structure constraining user's range of motion were spotted such as preventing the lifting of arms above the head. The example measured angles scopes are presented in Fig. 6.

Moreover, the precision of 3D-printing technology may affect the functioning of the structure, especially concerning cogwheel mechanisms within the device. A matter of a priority is minimizing friction forces between movable elements through the use of meticulously selected bearing applications to ensure measurements quality.

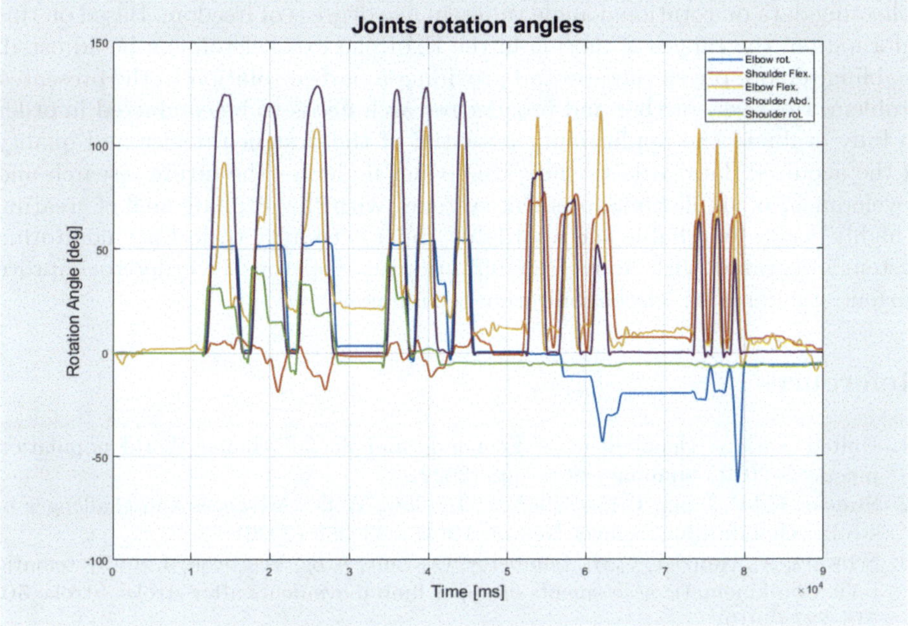

Fig. 6. An example measured angle data

3.2 Electronical Section

Each revolut joint of the device was related to an optic, incremental encoder *DFRobot 400 P/R*, with 5V electric power input, which recorded ϕ angles of the adopted upper extremity model related to the initial, standard anatomic position, at a sampling period of 12 ms and resolution of 0,2°.

An encoder system was connected to a *STM32L476RG* microcontroller on *NUCLEO-L476RG* board, which was transmitting in real-time recorded data through the use of USART protocol to the connected personal computer in a stable manner, where output was saved in a .txt file and then filtered and analyzed in *MATLAB* environment. The measuring application could be also adjusted to save data on the SD-Card, with no necessity of the connected PC in order to improve its usability. Cables connecting the encoders with the board were managed to maximize the workspace.

The encoder calibration process was setting the initial value of each encoder as 0 in the standard anatomical position of a user.

4 Conclusion

The implementation of exoskeleton robots for task-oriented telerehabilitation has the potential to satisfy the demands of the projected reality. The concept of this passive exoskeleton presented in this paper fulfills its designated purpose of collecting data on rotational angle values in five degrees of freedom. Based on this information, the ranges of motion in the ExoReha exoskeleton can be adjusted, enabling remote physiotherapy and providing a partial solution to the presented problem. However, further and broader research needs to be conducted in order to fully evaluate the applicability potential of the designed device and quality of the acquired data. Additionally, this paper underlies the future research and development of similar measurement systems, with the ultimate goal of creating a highly accurate, reliable and affordable upper extremity trajectory monitoring system integrated with active rehabilitation exoskeletons in order to improve current standards of telerehabilitation techniques.

References

1. United Nations Department of Economic and Social Affairs: World population prospects 2022 - summary of results (2022)
2. Stinear, C.M., Lang, C.E., Zeiler, S., Byblow, W.D.: Advances and challenges in stroke rehabilitation. Lancet Neurol. **19**(4), 348–360 (2020)
3. Schwarz, A., Kanzler, C.M., Lambercy, O., Luft, A.R., Veerbeek, J.M.: Systematic review on kinematic assessments of upper limb movements after stroke. Stroke **50**, 718–727 (2019)
4. Kaptain, R.J., Helle, T., Patomella, A.-H., Weinreich, U.M., Kottorp, A.: New insights into activities of daily living performance in chronic obstructive pulmonary disease. Int. J. Chronic Obstructive Pulm. Disease **16**, 1–12 (2021)
5. Yeon, K.S., Hong, P.J., Seok, K.W., Kyungdo, H., Yookyung, L., Jong, P.N.: Health-related quality of life and related factors in stroke survivors: data from Korea National Health and Nutrition Examination Survey (KNHANES) 2008 to 2014. PLoS ONE **13**(4), 2018 (2008)
6. Probst, M.: Physiotherapy and mental health. Clin. Phys. Ther. **230**, 59–68 (2017). https://doi.org/10.5772/67595
7. Maureen, L.D.: Stroke rehabilitation. Crit. Care Nurs. Clin. North Am. **32**(1), 97–108 (2020)
8. Nielsen, G., et al.: Physiotherapy for functional motor disorders: a consensus recommendation **86**(10), 1113–1119 (2015). https://doi.org/10.1136/jnnp-2014-309255
9. Paolo, C., Nicola, D., Alessio, L.: SARS-CoV-2 pandemic: not the first, not the last. Microorganisms **9**(2), 433 (2021)
10. Faruque, A., Nicole, Z., Amra, U.: Effectiveness of workplace social distancing measures in reducing influenza transmission: a systematic review. BMC Public Health **18**(1), 1–13 (2018)
11. Brennan, D., Mawson, S., Brownsell, S.: Telerehabilitation: enabling the remote delivery of healthcare, rehabilitation and self management. In: Advanced Technologies in Rehabilitation, pp. 231–248. IOS Press, Amsterdam (2009)
12. Laver, K.E., Adey-Wakeling, Z., Crotty, M., Lannin, N.A., George, S., Sherrington, C.: Telerehabilitation services for stroke. Cochrane Database Syst. Rev. (2020)
13. Russel, T.: Telerehabilitation: a coming of age. Aust. J. Physiotherapy **55**, 5–6 (2009)
14. Rogante, M., Grigioni, M., Cordella, D., Giacomozzi, C.: Ten years of telerehabilitation: a literature overview of technologies and clinical applications. NeuroRehabilitation **27**, 287–304 (2010)

15. Galea, M.: Telemedicine in rehabilitation. Phys. Med. Rehabil. Clin. N. Am. **30**(2), 473–483 (2019)
16. Standing, C., Standing, S., McDermott, M.L., Gururajan, R., Kiani, M.R.: The paradoxes of telehealth: a review of the literature 2000–2015. Syst. Res. Behav. Sci. **35**(1), 90–101 (2018)
17. Falkowski, P.: Light exoskeleton design with topology optimisation and FEM simulations for FFF technology. J. Autom. Mob. Robot. Intell. Syst. 14–19 (2022). https://doi.org/10.14313/jamris/2-2021/9
18. Andrea, D., Carpino, G., Zollo, L., Johnson, M.J.: Affordable robotics for upper limb stroke rehabilitation in developing countries: a systematic review. IEEE Trans. Med. Robot. Bionics **3**(1), 11–20 (2021)
19. Di, S., Wuxiang, Z., Wei, Z., Xilun, D.: A review on lower limb rehabilitation exoskeleton robots. Chin. J. Mech. Eng. (Engl. Ed.) **32**(1), 1–11 (2019)
20. Clark, W.E., Sivan, M., O'Connor, R.J.: Evaluating the use of robotic and virtual reality rehabilitation technologies to improve function in stroke survivors: a narrative review. J. Rehabil. Assistive Technol. Eng. **6**, 1–7 (2019)
21. Marryam, M., Umar, M.: Effectiveness of task oriented training in improving upper limb function after stroke. Rawal Med. J. **42**(3), 341–343 (2017)
22. Falkowski, P.: An optimisation problem for exoskeleton-aided functional rehabilitation of an upper extremity. In: IOP Conference Series: Materials Science and Engineering, vol. 1239, p. 012012. IOP Publishing (2022). https://doi.org/10.1088/1757-899X/1239/1/012012
23. Falkowski, P., et al.: Study on the applicability of digital twins for home remote motor rehabilitation. Sensors **23**(2), 911 (2023)
24. da Silva, E.S.M., et al.: The effect of priming on outcomes of task-oriented training for the upper extremity in chronic stroke: a systematic review and meta-analysis. Neurorehabil. Neural Repair **34**(6), 479–504 (2020)
25. Lum, P.S., et al.: Improving accelerometry-based measurement of functional use of the upper extremity after stroke: Machine learning versus counts threshold method. Neurorehabil. Neural Repair **34**(12), 1078–1087 (2020)
26. Leczkowski, B., Falkowski, P.: Analiza kinematyczna ruchów kończyny górnej w celu opracowania testowego treningu funkcjonalnego do zastosowań w zrobotyzowanej rehabilitacji ruchowej. Aktualne Problemy Biomechaniki **21**, 15–23 (2021)
27. Hartenberg, R.S., Denavit, J.: A kinematic notation for lower pair mechanisms based on matrices. J. Appl. Mech. **77**(2), 215–221 (1955)

Positioning of Mobile Robots on the Ground

Shatokhin Oleksii, Vytautas Bučinskas(✉), and Andrius Dzedzickis

Vilnius Gediminas Technical University, Saulėtekio al. 11, 10223 Vilnius, Lithuania
oleksii.shatokhin@vilniustech.lt

Abstract. Mobile robots are becoming increasingly popular due to their relevance and widespread use in modern society. Their ability to navigate their environment without the need for physical or electromechanical guidance devices makes them promising and useful. Autonomous mobile robots are applied in various fields, including enterprises, medicine, education, agriculture, and consumer services, with the aim of improving services and daily life. With the advancement of technology, there is a growing demand for mobile robots that can successfully perform tasks such as transporting heavy objects, monitoring, search and rescue operations, and others. Various studies conducted by scientists highlight the importance of mobile robots, their varied applications, and the challenges they face. The review covers the current literature and examines the key issues faced by mobile robots. The text presents a comprehensive study of devices and sensors, along with common methods for combining them, aimed at solving problems such as localization, estimation, and navigation in mobile robots. The aspects are organized based on their relevance, strengths, and weaknesses, providing a solid foundation for future research and development of methods to overcome the challenges faced by autonomous mobile robots.

Keywords: Mobile robot · Motion · Positioning · Sensors

1 Introduction

In modern times, with the acceleration of production, many processes are being optimized and automated. Initially, machines were introduced to supplement manual labor, which expedited and simplified production. Subsequently, machines were automated, leading to increased manufacturing accuracy and reduced human involvement in production. The decreasing cost of components and the improving computing performance of computers have made it possible to create increasingly intelligent machines capable of performing complex tasks. However, to achieve full automation, it is still necessary to transport materials, workpieces, or products between machines or from a warehouse. While a pipeline can partially solve this problem, it cannot completely eliminate it. Mobile robots can be programmed for new tasks and used to move loads from place to place. Fully automated production is possible with the use of mobile robots. They can be a simple platform on wheels, moving products between machines, production lines, or warehouses. Alternatively, they can be a special automatic loader that can load and

R. Szewczyk et al. (Eds.): AUTOMATION 2024, LNNS 1219, pp. 162–175, 2024.
https://doi.org/10.1007/978-3-031-78266-4_15

unload trucks. For mobile robots to function properly, they must accurately determine their location and detect and respond to unexpected obstacles, such as people or other robots. Various sensors are built into the robot's design to perform these tasks [1]. Robotic forklifts, for example, can use special Time-of-Flight (ToF) cameras to determine the location of loads and place them rationally in the truck.

Mobile robots have a wide range of applications, including in industry, such as robotic vacuum cleaners, automatic lawn mowers, robot consultants, and automated harvesters. Even a car with an autopilot can be considered a mobile robot [2, 3].

2 Kinematics of the Robot

The method of movement is a crucial aspect of a mobile robot. The appropriate drives are selected depending on the task and the area in which the robot will operate. This is necessary to ensure stability, maneuverability, and, in some cases, high speed. The design of the robot is directly dependent on the tasks it performs. A mobile robot can move on the ground, underground through tunnels, swim on and under water, and fly in the air [1, 4]. For instance, a robot designed to move on the ground can utilize various methods [5–8]. The most popular form of locomotion for mobile robots is the use of wheels. Wheels are a simple and reliable method of locomotion, making it easier to design, build, and program robots that need to move on a flat surface. Common types of wheels are showed on Fig. 1. Compared to legged robots, wheeled robots are cheaper, more accurate, and easier to program. Additionally, there are no balancing difficulties since the robot always maintains contact with the surface.

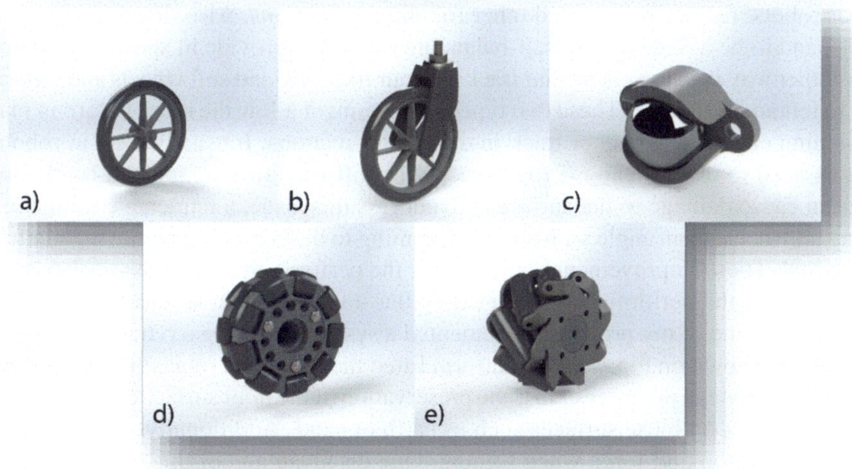

Fig. 1. Type of wheels in mobile robots. a. regular fixed wheel, b. castor wheel, c. ball roller, d. omni-wheel, e. mecanum wheel

One disadvantage is that they are not well-suited for off-roading or traveling over rocky terrain or low-friction areas.

Common types of wheels:

- Fixed rotating wheel – have one degree of freedom, on the axle.
- Castor wheel – have two degree of freedom, on the wheel axle and vertical axis to rotate.
- Ball roller – omnidirectional way of movement, usually used as support roller.
- Omnidirectional wheels, such as mecanum or omni wheels, are wheels with rollers that can be driven in any direction. These wheels are useful for their ability to move in multiple directions, making them ideal for certain applications.

In a wheeled method of locomotion, it is crucial that all mounted wheels maintain good contact with the surface. This affects walkability, stability, and maneuverability. Wheeled robots are designed to ensure that all wheels touch the ground, making it pretty stable. Also, mobile robots can be classified by drive system:

- Differential drive – most popular drive system, usually have two wheels and free roller.
- Car-type drive – traditional four-wheel platform with two or four steering wheels.
- Omnidirectional drive – three- or four-wheel drive with wheels with integrated rollers.
- Synchro drive – three or more-wheel drive where all wheels are turning and always directed in the same direction.

Differential drive is most popular in robots, it uses two motorized wheels and an omnidirectional roller or free castor wheel. This design is simple and enables the robot to turn around easily in place and it very stable, because have three points of contact, that means that all wheels will be in contact with surface. It is commonly used in vacuum cleaner robots, factory robots, and other robotic applications. Also, it can use just two wheels, and to be a Segway like self-balancing robot that provide hi speed movement.

Another way is to use caterpillar tracks or four fixed big and soft wheels and separate drive to left and right side. These two types of movement allow the robot to turn in place by spinning each side wheel or track in different directions. It usually used in robotics when its need to drive on an off-road space. This off-road turning method is effective, but it can cause damage to flat surfaces. To address this issue, a retractable turntable or turning all wheels to an angle with all axles pointing to the center of the moving platform can be used. These improvements can enhance the performance of these travel systems.

To enhance the performance of the caterpillar tracks on a flat surface and minimize wear and tear, one of my projects implemented a system featuring a retractable rotating platform that shown on Fig. 2. This platform lifted the robot and rotated it to the desired angle. The advantage of this system is the preservation of the floor surface and the absence of burying effect on loose surfaces such as earth or sand. Additionally, it ensures even weight distribution and force when turning. One disadvantage of this system is that it requires lifting the robot at each turn, which affects its speed. Additionally, this system can be used for wheeled robots, adding maneuverability and passability. For example, a robot with a car-like system of movement can turn on the spot, or if the robot gets stuck on a loose surface, it can rise on the platform, move the wheels to a more passable place, and continue moving.

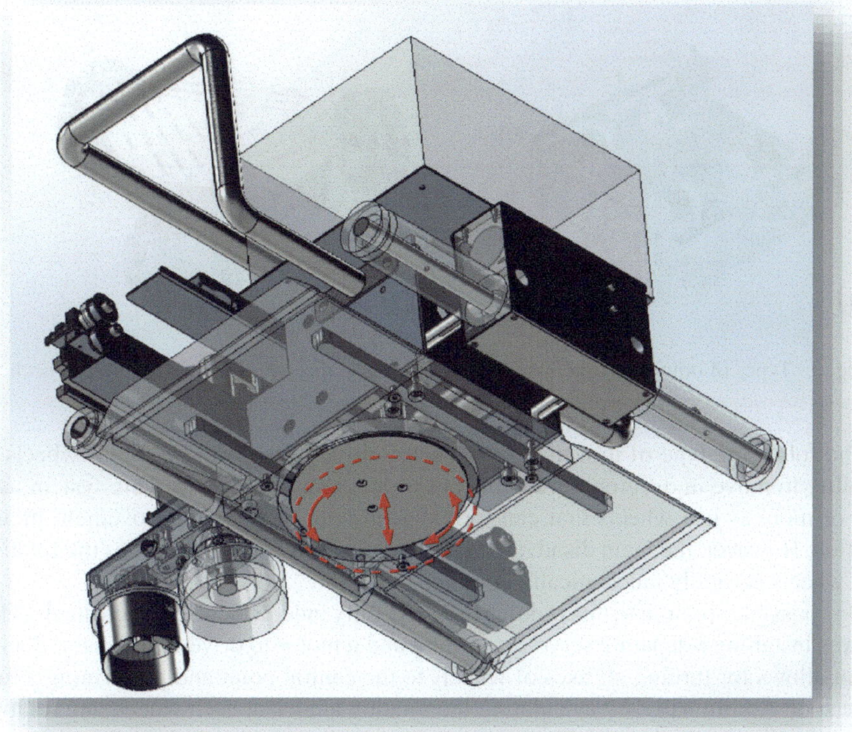

Fig. 2. Retractable rotating platform of the developed caterpillar track robot

A classic system with four wheels, such as a car-like scheme with one rotary axle and steering, is also stable. This is the best choice for fast-moving mobile robots, but the control algorithms become more complicated [2, 3]. If a suspension and all-wheel drive are added to that platform, it will have excellent off-road capability. But for off-road mobile robots, it is recommended to use caterpillar tracks or four large wheels without steering. If good maneuverability and movement in any direction are required, three or four omni wheels with rollers can be installed on the robot. However, they are more suitable for small robots as they should be installed at an equal distance from the center along the perimeter of the circle. So, this wheel is perfectly fit to the square or circle shaped robot.

Mecanum wheels are a suitable choice for larger robots due to their ability to move in any direction. Comparison of Omni and Mecanum wheels are showed on Fig. 3. They consist of four pieces and are installed similarly to car wheels. It is important to note that the ratio of the distances between the wheels does not affect their performance [9].

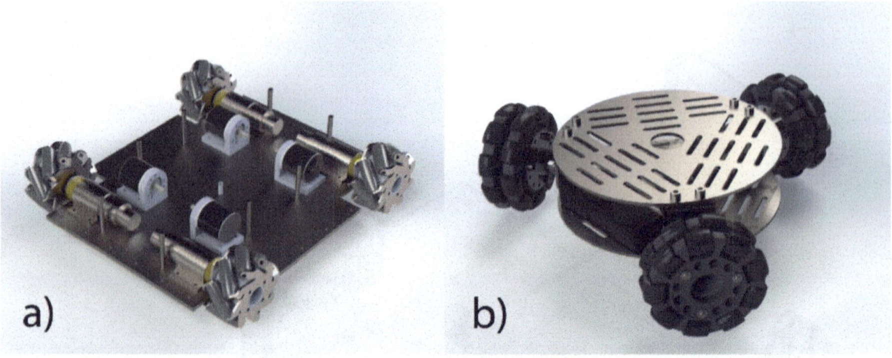

Fig. 3. Types of omnidirectional type of movement. a. mecanum wheels, b. omni-wheel

The following type of movement is Synchro drive. It is similar to omni-wheels in its ability to move in different directions but is structurally different. This system uses three or more swivel wheels that can turn synchronously, allowing movement in any direction. However, the main disadvantage is that the robot cannot rotate, as the rotation of all wheels is rigidly mechanically coupled.

One possible solution to this problem is to use fully independent turning wheels. This involves installing a separate servo for turning and a motor to drive each wheel. Such a system allows for turning all axes of wheels to the central point and performing a turn on the spot. Additionally, by calculating the rotation angles of each wheel, any rotation of the robot can be performed. This system also enables the installation of wheels with larger treads, which enhances off-road performance.

However, it is worth noting that these wheels are typically designed for use on flat, level floors and may not be suitable for off-road use. In such cases, it is recommended to use regular wheels with larger treads, tracks, or walking robots. For instance, a robot designed to move on the ground can utilize various methods [5]. Hybrid or transformable transportation methods are also feasible, such as wheels that can transform their shape to be more traversable [7], wheels with tracks, wheels and legs, wheels and propellers, or wheels and propellers combined.

On Table 1 showed caparison of different types of movement by its off-road capacity, maneuverability, price, stability and turn issues. By this table we can see that differential drive with two wheels have good characteristics for use on a flat surface at a low price. Also, after modifying, previously described, synchro drive, caterpillar tracks and drive with fixed four wheels can improve their specs.

Table 1. Comparison table of wheel-based methods of move.

Type of movement	Off-road capability	Maneuverability	Price	Stability	Speed	Turn issues
2 Fixed wheels, with roller	Bad	Best	Cheap	Good	Good	No
4 wheels with steering	Medium	Medium	Medium	Good	Best	Have turn radius
4 wheels with suspension and steering	Good	Medium	Expensive	Best	Best	Have turn radius
4 Fixed big wheels	Good	Good	Cheap	Good	Best	Have high friction with surface
Omni wheels	Bad	Best	Medium	Good	Medium	No
Mecanum wheels	Bad	Best	Medium	Good	Medium	No
Caterpillar tracks	Good	Good	Medium	Best	Medium	Have high friction with surface
Synchro drive	Good	Good	Expensive	Good	Best	Can't turn if made by classic scheme

3 Types of Positioning – Sensors

Accurately navigating and orienting mobile robots in space requires the use of positioning systems. Mobile robots use two main types of positioning and navigation: global positioning and local positioning [12].

3.1 Global Positioning

An essential aspect of designing mobile robots is their capability to navigate through open terrain, particularly for those that operate or travel outdoors. Typically, satellite navigation systems are utilized for positioning, relying on a network of satellites in space that transmit signals containing information about their location and transmission time. The mobile robot's receiver obtains signals from multiple satellites to determine its absolute location in world coordinates. The sensors measure the distance between the robot and the satellites, which is used to determine the robot's coordinates in a global coordinate system, such as GPS, GALILEO, or GLONASS. This enables the robot to navigate large areas, such as cities or warehouses. GPS systems offer precise location information, which is crucial for tasks that require global orientation, such as autonomous driving in open spaces.

Mobile robots can utilize an electronic compass to determine their direction of travel relative to magnetic north. This feature is especially useful for autonomous transportation systems or robots, as it allows them to follow a given direction or course. Additionally, while traveling, mobile robots can use data from the electronic compass to correct their orientation. For instance, if they stray from their intended path due to external factors, the electronic compass can assist them in returning to the correct direction of travel.

Another effective navigation method for mobile robots, especially in environments where other methods such as GPS or magnetic compass are unavailable or unreliable, is determining direction using polarization filters to locate the sun [10, 11]. Sunlight reflected from the sky or other surfaces becomes polarized depending on the angle at which the sunbeam is incident. A polarizing filter permits the passage of light with a specific polarization while blocking light with other polarizations. Mobile robots can be equipped with multiple cameras or light sensors that have polarization filters at different angles, enabling them to analyze the polarization of light falling on them. By analyzing the angle of polarization of the light, they can determine the direction from which the sun's light is coming. By knowing the angle of the sun's direction and the exact time, the mobile robot can determine its direction of travel. This method is effective in areas with strong electromagnetic interference, but it is limited to daytime use only.

3.2 Local Positioning

Local positioning enables the robot to determine its coordinates in a local coordinate system, such as a building or warehouse map. This allows the robot to navigate small spaces, identify obstacles, and move around them. Additionally, optimizing movement planning can increase task completion speed [12, 13].

Various sensors, including distance sensors, cameras, gyroscopes, accelerometers, and encoders, can be used for local positioning.

The most basic mobile robots are carts that follow a guide rail to reach a specific location. These systems are primarily utilized in manufacturing, such as in chip production. The following positioning option involved using a line drawn on the floor. Illumination sensors are utilized in mobile robots for line following based on the principle of detecting differences in illumination between the line surface and its surroundings. This is achieved through sensors that measure the intensity of light reflected from the surface. The robot is equipped with multiple sensors located along the chassis or under the bottom. As the robot moves along the line, its sensors detect changes in brightness. If the robot deviates from the line, one or more sensors will detect a difference in luminance compared to the others. This information is then used to adjust the robot's trajectory and guide it back onto the line.

The use of distance sensors has become more sophisticated, allowing the robot to avoid collisions with obstacles and prevent falls down stairs. With a specific location, the robot's movement can be corrected, aligned with the walls, and navigated in space. There are several types of distance sensors, including ultrasonic, infrared, and laser.

The most commonly used sensors for measuring distance in robotics:

- Ultrasonic sensors,
- Infrared sensors,
- Laser rangefinders.

Ultrasonic sensors emit ultrasonic waves and measure the time it takes for the reflected signal to return to the sensor, providing accurate measurements with a 10mm error margin. These sensors have a sufficient operating range and can detect objects made of different materials. This type of sensor typically has a measuring range of 20 to 4000 mm, although this may vary depending on the model.

Infrared sensors, which use infrared radiation to measure distance, are also commonly used. While these sensors are usually small and inexpensive, they may be less reliable in varying lighting conditions and have a limited range of 20 to 500 mm. The readings may be distorted based on the surface and material being measured, and at times, the sensor may not detect the obstacle. As a result, robots equipped with this type of sensor often have a bumper with touch sensors to accurately detect obstacles.

Laser rangefinders are more advanced and accurate sensors that use laser light to measure the distance to objects with high precision, down to 1 mm. They have a wide measurement range from 20 to 40000 mm, making them particularly useful for navigation and precise positioning. However, they are also the most expensive type of sensor on the list.

The most advanced sensor used in robotics is lidar. It is a laser rangefinder attached to a rotating platform that enables the robot to scan the environment quickly and regularly, allowing it to position itself accurately in space. Lidars can create precise two-dimensional maps of the environment, which the robot can use for navigation and route planning. Additionally, lidars can detect objects around the robot and measure their distance and size. This helps them avoid collisions and interact with surrounding objects. With high-speed scanning and data processing, lidars provide robots with real-time navigation capabilities, reacting to changes in the environment instantly and efficiently [16–18].

Solid-state lidars, which do not contain moving parts such as rotating mirrors or prisms, also exist. Instead, laser light is generated and detected using semiconductor components such as photodiodes and laser diodes. A microelectromechanical mirror system (MEMS) is used for positioning, which allows the laser beam to scan the surrounding area. This results in faster scanning speeds and a smaller size [18].

The utilization of cameras in mobile robots offers a multitude of opportunities for perceiving the environment and executing various tasks. Cameras enable navigation by visible landmarks, detection and recognition of objects in the surroundings, and reading data from QR codes, barcodes, or text signs. This functionality can be beneficial for avoiding obstacles, recognizing signs and signals, or even identifying individuals. QR codes can be scanned on various surfaces, including floors, ceilings, walls, and objects. They enable the adjustment of the robot's position, such as its angle and location, as well as the verification of whether it has reached the desired destination [17]. Advanced algorithms and artificial intelligence make it possible to recognize objects or create 3D maps using only one camera [19]. The utilization of cameras in mobile robots allows them to gather extensive information about their surroundings and make intelligent decisions based on this data. Furthermore, with the advancement of computer vision and deep learning technologies, cameras have become an even more potent tool for robotics, enabling robots to analyze and comprehend their environment in a progressively sophisticated manner.

The use of Time-of-Flight (ToF) cameras in mobile robots is an advanced technology that enables accurate measurement of the distance to objects and creation of a three-dimensional representation of the environment. ToF cameras utilize the principle of measuring the time it takes for a laser signal to travel to and from an object to determine its distance. ToF cameras provide the robot with the ability to accurately measure the distance to objects in the environment, which is crucial for navigation and collision avoidance. They can scan the environment and create three-dimensional maps, enabling the robot to gain a better understanding of its surroundings. This facilitates navigation in complex environments and allows the robot to plan its route efficiently. ToF cameras enable robots to detect objects and obstacles in their environment, accurately measuring distance and ensuring safe movement to avoid collisions. This allows robots to adapt to dynamic conditions and respond to changes in their surroundings. However, data processing requires high computational performance. Fortunately, the use of neural modules and network systems is simplifying this task [14, 15].

In addition to measuring distances to objects, sensors are also used to determine the position of the actuators, specifically encoders. Encoders come in several types, including optical, magnetic, and sliding contact.

An optical encoder is composed of a disk or ribbon with slots and an optical system that includes an LED and a photodiode. When the disk moves, light passes through the slots and reaches the photodiode, which sends electrical impulses to the controller. The impulses are then processed to determine the movement.

A magnetic encoder, which uses a magnet and a hall-effect sensor, is used to detect the position of the actuator. Magnetic encoders typically have a high degree of protection against external influences and can operate in high temperatures and humidity.

A sliding contact encoder is a simple disk with contact pads arranged in a specific sequence and a sliding contact that is connected to an actuator. As the encoder rotates, the sliding contact moves and connects and disconnects the contacts in a specific sequence.

Encoders are crucial components for precise motion control and positioning of mobile robots.

They can measure the speed of wheels or other moving parts of the robot. The robot can precisely control its motion by managing the speed and direction of movement. It tracks the number of encoder revolutions to determine the distance traveled. This feature is especially useful for autonomous movement and navigation, as well as for tasks requiring precise movement over a specific distance. Some systems rely solely on encoders for positioning, such as inkjet printers. Encoders enable the robot to correct its position and compensate for any errors that may occur during movement. They help the robot maintain a straight line of motion and avoid drifting or deviations from a predetermined path. The robot receives feedback on the current position and speed of the motors from the encoders, allowing it to react to changing conditions and adjust its motion in real-time.

One more option for positioning indoors is to use a light-based indoor positioning system. This method uses light sources to determine the location of mobile robots indoors. The technology is based on detecting and analyzing light signals, which allows the robot to determine its location relative to known points or beacons in the room.

Special light beacons or LEDs are installed in the room, which emit light signals with a unique identification. Beacons can be placed on walls, ceilings, or other visible locations for the robot to detect at various points in the room. The mobile robot is equipped with sensors or cameras capable of detecting light signals from the beacons. These detectors can be conventional photodiodes, phototransistors, or cameras with the function of detecting light signals. The robot determines its location in the room by using data about the location and identification of beacons, as well as information about the strength and time of arrival of light signals after detecting them [20].

Sensors are crucial for the operation of mobile robots as they provide information about the environment and enable them to perceive and interact with it. Table 2 combines the general information discussed earlier. A variety of sensor types, including ultrasonic sensors, infrared sensors, cameras, lidars, ToF cameras, and encoders, provide robots with diverse sensing capabilities in various environments and scenarios.

Table 2. Comparison table sensors for mobile robots.

Sensor	Type of measurement	Price	Precision	Function	Measurements
Touch, button	Contact with object, obstacle	Very cheap	Logical sensor, yes/no	Allows to detect obstacles by touching it, for example bumper of robotic vacuum cleaner	Logical sensor, yes/no
Pressure, tension sensor	Measures pressure, tension	Medium	Medium	Possible to measure loads on robot mechanics, force of grip	

(continued)

Table 2. (*continued*)

Sensor	Type of measurement	Price	Precision	Function	Measurements
Light sensor	Detecting surface color change	Very cheap	Depends from model	With its robot can move by line on the floor, for example some factory robots orientate by these marks to drive precise in special parking zones	
Ir distance meter	Measurement of distance	Very cheap	Medium 10mm	It has close range of detection, so Robot can detect nearest obstacles	20–500 mm
Ultrasonic sensor	Measurement of distance	Cheap	High 1mm	Could be used for robot navigation for cheap	20–4000 mm
Laser distance meter	Measurement of distance	Expensive	High 1mm	It uses if need precise measurements	20–100000 mm
Lidar	Measurement of distance in 360	Expensive	Medium 10mm	Very popular precise solution for navigation, 1 sensor enough for robot	20–100000 mm
Camera	Detects visible objects	Medium	Depends from model	Optical object recondition, could be used for navigation	100–5000 mm
ToF Camera	Detects objects and distance to it	Expensive	High 1mm	Object recondition and distance to it, could be used for navigation	3000 mm

(*continued*)

Table 2. *(continued)*

Sensor	Type of measurement	Price	Precision	Function	Measurements
Contact encoder	Drive motion control rotary and linear	Medium	High	Have sliding contacts, have hermetic enclosure	
Optical encoder	Drive motion control rotary and linear	Cheap	High	Contactless, can be used in high speed measurements, but need space to install encoder disc	
Magnetic encoder	Drive motion control rotary and linear	Medium	Very high	Contactless, can detect position trough nonmagnetic materials	
Potentiometer	Drive motion control rotary and linear	Very cheap	Medium	Have sliding contacts, need ADC	

3.3 Combination of Sensors

Mobile robots typically use a combination of sensors, such as lidar and GPS or lidar and encoders on the robot's wheels, to increase positioning accuracy by complementing each other's capabilities. The use of different types of sensors is crucial in the development of mobile robots as it enables them to collect more complete and accurate information about the environment. Multiple sensors work together to enhance navigation, safety, and task efficiency. By combining GPS with local sensors like laser rangefinders and cameras, mobile robots can accurately determine their location both indoors and in open spaces. Robots can detect obstacles in their path and take appropriate actions to avoid collisions using ultrasonic sensors, infrared sensors, and cameras. Additionally, gyroscopes and compasses provide robots with information about their orientation and direction. The robot's movement is controlled by comparing the distance traveled using a gyroscope, accelerometer, encoder, and GPS. This is crucial for motion control and stabilization in various conditions. Combining different sensors makes the robot more adaptable to a variety of environmental conditions, such as changes in light and temperature. Visual sensors, such as cameras and laser rangefinders, enable robots to build maps of their surroundings. This feature is particularly beneficial for self-navigation in unknown or dynamic surroundings. In open areas, a mobile robot can use GPS to determine its location roughly. To ensure obstacle avoidance, laser rangefinders and a camera can be used for more precise localization within the building. Additionally, gyroscopes and compasses can provide information about the robot's current orientation.

The combined approach enables the robot to optimize its resources, resulting in more dependable and adaptable operation in diverse conditions.

For example, the mobile robot consultant Sanbot S1 Elf, shown in Fig. 4, has multiple sensors installed. However, these sensors are not fully utilized for robot positioning.

Fig. 4. Sanbot S1 elf sensors types. 1. 8MP color camera, 2. 3D ToF camera, 3. LED light, 4. Interactive touch screen display, 5. Motion detector, 6. Speaker, 7. IR reciver for charger search, 8. IR distance meter, 9. Omni-wheels

The robot is equipped with 9 IR rangefinders positioned along its front circumference to detect obstacles. It also has optical encoders on all three wheels, a gyroscope, and an accelerometer to monitor its path and turns. Although the robot has a color and ToF camera, they are only utilized for detecting people. The addition of cameras to the IR sensors can significantly enhance the robot's positioning accuracy and speed. The robot's IR sensors are located only at the bottom, which can cause problems with orientation in a room with furniture such as a table. As a result, the robot may not notice obstacles and continue to move, potentially hitting the upper part and rolling over. To improve the robot's positioning and avoid collisions, it would be beneficial to use both types of sensors simultaneously.

4 Conclusions

This work describes modern methods for movement and positioning of mobile robots. It presents visualizations and descriptions of different types of movement, and proposes options for modernizing designs to improve the characteristics of tracked and wheeled robots based on our own development experience. This text describes methods for positioning mobile robots using standard and original techniques. It includes determining direction using polarizing filters and local navigation using modulated light. The text also provides comparative tables of movement methods and sensors. This information can be useful when designing mobile robots for specific tasks, such as selecting a method of movement and considering the characteristics and operating features of popular sensors. A solid comprehension of various systems will aid future developers in enhancing mobile robot designs.

References

1. Rubio, F., Valero, F., Llopis-Albert, C.: A review of mobile robots: concepts, methods, theoretical framework, and applications, April 2019
2. Ni, J., Chen, Y., Chen, Y.: A Survey on Theories and Applications for Self-Driving Cars Based on Deep Learning Methods, April 2020
3. Self-Driving RC Car. https://www.youtube.com/watch?v=zuyOdaQ2xuw&ab_channel=StevenGong
4. Biswal, P., Mohanty, K.: Development of quadruped walking robots: a review, June 2021
5. Niloy, A., Shama, A., Chakrabortty, R.: Critical Design and Control Issues of Indoor Autonomous Mobile Robots: A Review, February 2021
6. Bruzzone, L., Edin Nodehi, S., Fanghella, P.: Tracked Locomotion Systems for Ground Mobile Robots: A Review, August 2022
7. Mertyüz, İ., Tanyıldızı, A.K., Taşar, B.: FUHAR: a transformable wheel-legged hybrid mobile robot, November 2020
8. Trojnacki, M., Dąbek, P.: Mechanical properties of modern wheeled mobile robots. J. Autom. Mob. Robot. Intell. Syst. **13**(3), 3–13 (2019)
9. Tagliavini, L., Colucci, G., Botta, A.: Wheeled Mobile Robots: State of the Art Overview and Kinematic Comparison Among Three Omnidirectional Locomotion Strategies, August 2022
10. Du, T., Zeng, Y.H., Yang, J.: Multi-sensor fusion SLAM approach for the mobile robot with a bio-inspired polarised skylight sensor, October 2020
11. Dupeyroux, J., Viollet, S., Serres, J.R.: An ant-inspired celestial compass applied to autonomous outdoor robot navigation, July 2019
12. Gia Luan, P., Truong Thinh, N.: Real-Time Hybrid Navigation System-Based Path Planning and Obstacle Avoidance for Mobile Robots, May 2020
13. Hichri, B., Fauroux, J.-C., Adouane, L.: Design of cooperative mobile robots for co-manipulation and transportation tasks, June 2019
14. Chen, C., Wang, B., Lu, C.X.: A Survey on Deep Learning for Localization and Mapping: Towards the Age of Spatial Machine Intelligence, June 2020
15. Al-dabbagh, A.H.A., Ronsse, R.: A review of terrain detection systems for applications in locomotion assistance, November 2020
16. Iqbal, J., Xu, R., Sun, S.: Simulation of an Autonomous Mobile Robot for LiDAR-Based In-Field Phenotyping and Navigation, June 2020
17. Takamatsu, J., Ruchanurucks, M.: ROS based Autonomous Mobile Robot Navigation using 2D LiDAR and RGB-D Camera, January 2019
18. Velodyne Lidar, comparison of sensors. https://velodynelidar.com/product-comparison/
19. Luan, P.G., Thinh, N.T.: Real-Time Hybrid Navigation System-Based Path Planning and Obstacle Avoidance for Mobile Robots, May 2020
20. Maheepala, M., Kouzani, A., Joordens, M.: Light-Based Indoor Positioning Systems: A Review, January 2020

Simulation of 6-DOF Modular Arm-Z Manipulator in MATLAB Simulink

Michał Jarek[2], Krzysztof Zawalski[1], Piotr Falkowski[1,2(✉)], Ela Zawidzka[3], and Machi Zawidzki[3]

[1] Industrial Research Institute for Automation and Measurements PIAP, Warsaw, Poland
piotr.falkowski@piap.lukasiewicz.gov.pl
[2] Warsaw University of Technology, Warsaw, Poland
[3] Institute of Fundamental Technological Research PAN, Warsaw, Poland

Abstract. Arm-Z is a concept of a hyper-redundant manipulator based on linearly joined sequence of congruent units. Each unit has only one degree of freedom (1-DOF) - the twist relative to the previous unit. This paper presents the approach to design of modular robotic manipulator based on computer simulations. The approach is rather standard nowadays, however, it is applied to rather unique robotic system. The CAD model of the module structure for a manipulator with 6 degrees of freedom and respective Denavit-Hartenberg notation are presented. Simulink - a MATLAB-based graphical programming environment was used to determine the torques of individual modules. The results of three simulations are presented and discussed. Finally, based on critical observations, the drive and gear for the module of Arm-Z system were selected.

Keywords: Simulation · Robotic arm · Manipulator · Arm-Z · MATLAB Simulink

1 Introduction

Robotic manipulator is an electronically controlled mechanical device with several links, and it is used to perform any specific task under given circumstances [1]. The goal of manipulator kinematics is to build mathematical relationship between the movement of each connected link joint and the space position and orientation of the end-effector. Some constraints can be also given to the middle part of manipulator e.g. in order to avoid collisions with obstacles [2,3]. Bio-inspired trunk-like manipulators have very specific movement characteristics which makes them especially useful in situations where classic robotic arms cannot be used [4–6]. Just like elephant trunk, they can enter into a rugged and narrow environment while actively avoiding obstacles [7]. The first trunk-like locomotors and manipulators have been proposed in [8]. Trunk-like manipulators can transport working heads in very complex, narrow, confined, cluttered spaces which are inaccessible by any other means. Reaching range can be of order

R. Szewczyk et al. (Eds.): AUTOMATION 2024, LNNS 1219, pp. 176–191, 2024.
https://doi.org/10.1007/978-3-031-78266-4_16

of tens of meters and actions can be accomplished in hazardous environments. They can perform tasks like welding, cleaning, visual inspection and any other which can be realized by the head mounted on the manipulator tips. Manipulators of this kind can also perform highly responsible and complex tasks, such as medical surgeries [9,10]. However, in general it has proved very difficult to design a real arm mechanism. As a result only a few companies offer such type of manipulators, and these are rather experimental [11]. For reviews of this kind of robotic systems see: [5,12–15].

Trunk-like manipulators have relatively large number of degrees of freedom (DOFs) compared to classic robotic arms with low number of DOFs (usually 6, sometimes 7) which are widely used in the industry. This paper presents design of modular robotic manipulator Arm-Z based on computer simulations. The concept of Arm-Z is highly scalable, both in size and functionality. It is worth mentioning that it belongs to Extremely Modular Systems (EMS, for short) introduced in [16], which present a novel approach to modular design, where the number of types of modules is just one. The main advantages of EMS are:

1. **Economy** - as the mass production is possible, which reduces costs and allows for wide application;
2. **Functionality** - as it allows for reconfiguration, expansion, reduction;
3. **Reliability** - since every damaged module can be easily replaced with an identical but functional one;
4. **Discreteness** - as they are suitable for intelligent mathematical modeling, and their configurations can be subjected to discrete (multi-objective) optimization using efficient search algorithms;
5. **Uniformity** - this feature is beneficial for quick deployment and automatic assembly.

These systems are part of discrete optimization approach to design introduced in [16] (for representative research see: [17–22]). Other areas are: fundamental aspects of architectural & urban layout [23] (for representative research see: [24–26]), and building envelope (including daylighting, shading and aesthetics) [27,28].

This paper is organized as follows: Sect. 2 presents the construction of Arm-Z, its Denavit-Hartenberg model and the structure of the program in MATLAB Simulink. The results are presented in Sect. 3. Discussion is carried out in Sect. 3.1, followed by Conclusions in Sect. 4.

2 Arm-Z Modular Manipulator

In Arm-Z, the number of DOFs is equal to the total number of the connected modules. Besides the possibility of performing complex actions, large number of DOF introduces the possibility of high fault tolerance and robustness. As Figs. 1.1&2 indicate, its operational space depends on the number of modules.

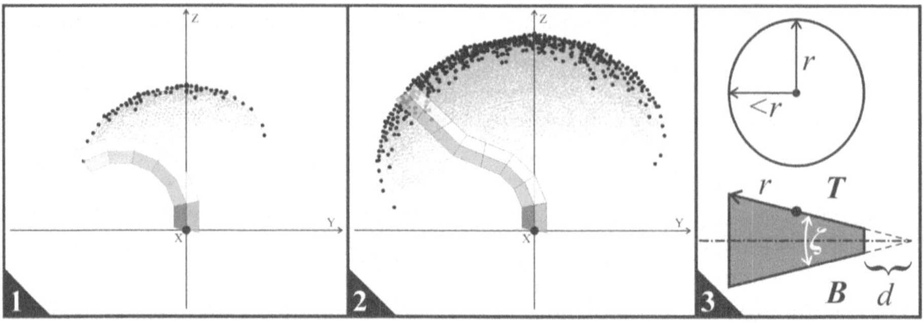

Fig. 1. 1. The operational space for a 6-unit Arm-Z. 2. The operational space for 8-unit Arm-Z. 3. Three parameters defining Arm-Z module: r, d and ζ.

The purpose of this project is to establish through computer simulations the design of the Arm-Z physical demonstrator. It is based on computer-controlled servo drives. The kinetic system of Arm-Z mechanism is scalable in such a way, that at the beginning of the kinematic chain the servos are supplied with greater power which is reduced towards the end-effector (tip). Nevertheless, the geometry of Arm-Z modular elements (Fig. 1.3) is constant throughout the entire length of the mechanism. Thus, it is be possible to simultaneously examine the behavior of the Arm-Z mechanism and optimize the power supply for its movement. Constantly improving computer technology allows to design and interactively manipulate virtual models with high precision incomparably less expensively and more quickly than by physical means. In virtual robotics toolbox we conveniently established the spatial model of the manipulator with its kinematics analysis, simulation, and trajectory planning. Next, by the computer simulation of the virtual model, we obtained the motion curve of each joint of the manipulator. Using simulation, it is possible to study various cases and thus select appropriate elements and reject incorrect ones. In this project there are six modules of Arm-Z and six rotating joints. Each module is defined by the following parameters: size r, offset d, and angle ζ between bottom (**B**) and top (**T**) faces of the module, as illustrated in Fig. 1.3. Slenderness $s = \frac{d}{r}$ is an additional convenient parameter. The following values of geometric parameters have been assumed: $r = 95\,\mathrm{mm}$, $s = \frac{1}{3}$, thus $d = \frac{r}{3} = 31.667\,\mathrm{mm}$; $\zeta = 25.4228°$ (Fig. 3).

Figure 2 shows section of virtual model of the module, and Fig. 4 shows the entire 6-DOF virtual manipulator.

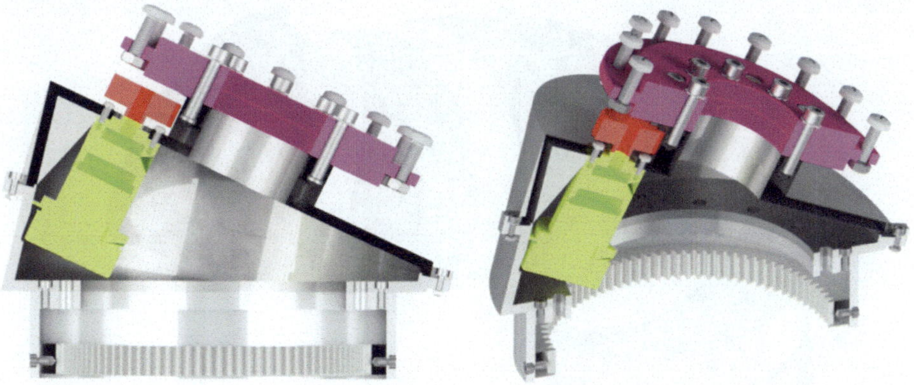

Fig. 2. Parallel and isometric views of the section through the virtual model of Arm-Z module created in Autodesk Inventor.

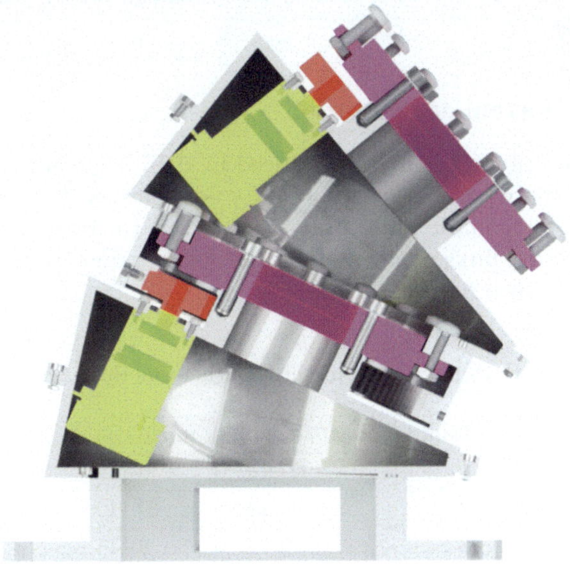

Fig. 3. Parallel view of the section through an assembly of two modules of the Arm-Z created in Autodesk Inventor.

Fig. 4. Isomeric view of the virtual model of modular 6-DOF Arm-Z created in Autodesk Inventor.

2.1 Denavit-Hartenberg Notation

The Denavit-Hartenberg method introduced in 1955 [29] is a mathematical tool widely used to analyze robotic manipulator kinematics. It provides a precise description of the kinematic structure of a manipulator by assigning coordinate systems to joints and links of the system [30–35]. Parameters of each link of the manipulator are shown in Table 1.

Table 1. Denavit-Hartenberg parameters for the 6-module Arm-Z manipulator.

Link number	θ_i	d_i mm	a_i mm	α_i
1	π	97.14	0	0.11
2	π	97.14	0	0.11
3	π	97.14	0	0.11
4	π	97.14	0	0.11
5	π	97.14	0	0.11
6	π	97.14	0	0.11
7	π	97.14	0	0.11

To align all modules in a straight vertical line, θ_i parameters in the Denavit-Hartenberg method presented in Table 1 are equal to π. Since the first unit of the chain of modules is fixed, there is formally one more element included in the simulation than the number of DOFs. See for example the table above and Figs. 5 and 6.

After the parameters of all the links have been determined, the relationships between adjacent link $i-1$ and i are established and collected in the transformation matrix:

$$^{i-1}T_i = \begin{bmatrix} \cos\theta_i & -\sin\theta_i \cdot \cos\alpha_i & -\sin\theta_i \cdot \sin\alpha_i & a_i \cdot \cos\theta_i \\ \sin\theta_i & \cos\theta_i \cdot \cos\alpha_i & -\cos\theta_i \cdot \sin\alpha_i & a_i \cdot \sin\theta_i \\ 0 & \sin\alpha_i & \cos\alpha_i & d_i \\ 0 & 0 & 0 & 1 \end{bmatrix}$$

$$^{0}T_6 = ^{0} T_1^1 T_2^2 T_3^3 T_4^4 T_5^5 T_6 = \begin{bmatrix} n_x & o_x & a_x & p_x \\ n_y & o_y & a_y & p_y \\ n_z & o_z & a_z & p_z \\ 0 & 0 & 0 & 1 \end{bmatrix}$$

This matrix shows the transposition relations between the location, the posture of the end-effector (tip) and the base coordinate system origin.

2.2 Simulation of Arm-Z in MATLAB Simulink

In this part of the project we used free *Robotics Toolbox* - a MATLAB toolbox software which supports research [36] and teaching [37] into arm-type and mobile robotics. Particularly relevant are Refs. [38] and [39] where simulation researches of a 6-DOF manipulator kinematics based on this toolbox are presented.

The Robotics Toolbox is a user-friendly simulation platform. It was easy to simulate random movement within the available ranges. Nest we examined the maximum torques and finally we selected the drive and gearbox with higher torque. The tree structure of Arm-Z manipulator is similar to the methodology described in [40]. First, we created a rigid body tree object of the manipulator. This tree was initialized with a base coordinate frame with subsequent bodies to be attached to it. Next, we created a rigid body object and joint assigned to it. We used *setFixedTransform* function to specify the body-to-body transformation using Denavit-Hartenberg parameters. The last parameter *theta* was ignored since this angle depends on the joint position. The first body was added to the base of the tree with *addBody* function. Figure 5 shows the parameters of rigid tree body of Arm-Z.

Idx	Body Name	Joint Name	Joint Type	Parent Name(Idx)	Children Name(s)
1	b0	bj0	fixed	base(0)	m1(2)
2	m1	mj1	revolute	b0(1)	m2(3)
3	m2	mj2	revolute	m1(2)	m3(4)
4	m3	mj3	revolute	m2(3)	m4(5)
5	m4	mj4	revolute	m3(4)	m5(6)
6	m5	mj5	revolute	m4(5)	m6(7)
7	m6	mj6	revolute	m5(6)	

Fig. 5. Rigid tree body of Arm-Z

The fixed transform defined previously is from the base (parent) to the first body. The category of the first body is "fixed". All subsequent bodies are of the revolut type. We specified the dynamic properties for the bodies: mass, center of mass and inertia of the body. The next step was to create the second body. We defined its properties and attached it to the first rigid body. Then we defined the transformation relative to the previous body frame, and so forth. In this type of modular manipulator, the subsequent bodies are identical to the second body. However, the mass, center of mass and inertia of the body need to be determined. We repeated the sequence for the remaining four bodies. Finally, *showdetails* function displayed our robot. At this point we verified that the joint types were correct by using *config*. Function *show* displays a stick-model of the manipulator, as shown in Fig. 6.

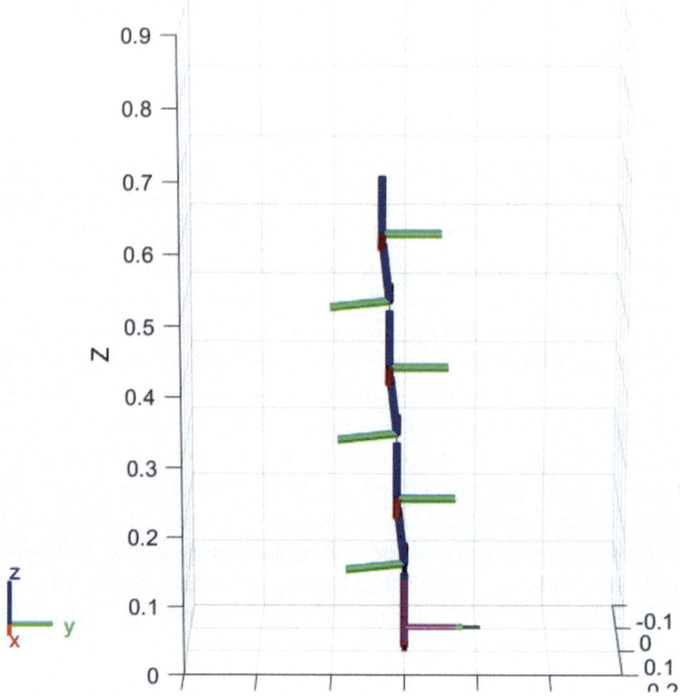

Fig. 6. Configuration of the stick-model of the Arm-Z

Figure 7 illustrates the procedure of building virtual model of Arm-Z.

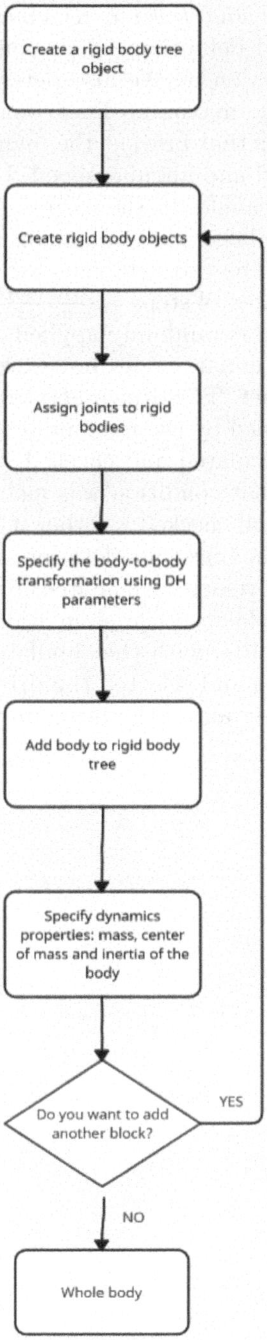

Fig. 7. Diagram for the procedure in MATLAB Simulink

In order to show the trajectory of the manipulator, we developed code in MATLAB Simulink [41–43] using *Inverse Kinematics* and *Inverse Dynamics* blocks. These blocks required definitions of the initial position of the manipulator and the force that acts on it. We also determined the maximum linear and angular velocities and the maximum linear and angular accelerations. We also established the conditions that prevent the manipulator from exceeding the maximum length, linear speed and angular speed. Using the *eul2rotm* function, we converted the set of Euler angles to the corresponding rotation matrix. As a result, we received the manipulator position matrix so the program in MATLAB Simulink could be executed. The program runs for a set time which is equal to the number of steps. The number of steps $= \frac{simulationtime}{simulationstep+1}$. At each subsequent iteration, a linear acceleration is randomly applied and continues until the condition ensuring that the manipulator does not exceed its physical length (here the height of 6 modules) is met. The drawn acceleration is then divided by the total acceleration and multiplied by the maximum acceleration. Next, the linear speed in a given step was calculated and checked whether it stayed within the limit of maximum speed. If that condition was met, the current position of the manipulator was calculated and checked whether it remained within the robot's operating area. Similarly to the linear acceleration, the angular acceleration was randomized until the angular velocity calculated from this acceleration was less than the preset maximum velocity. The simulation time in Simulink was set to 200 s and in MATLAB to 10 s, with the simulation step of 0.1 s. Finally, we checked the maximum torques and selected the drive and gearhead with higher torque. Figure 8 shows the diagram of the program in MATLAB for Simulink

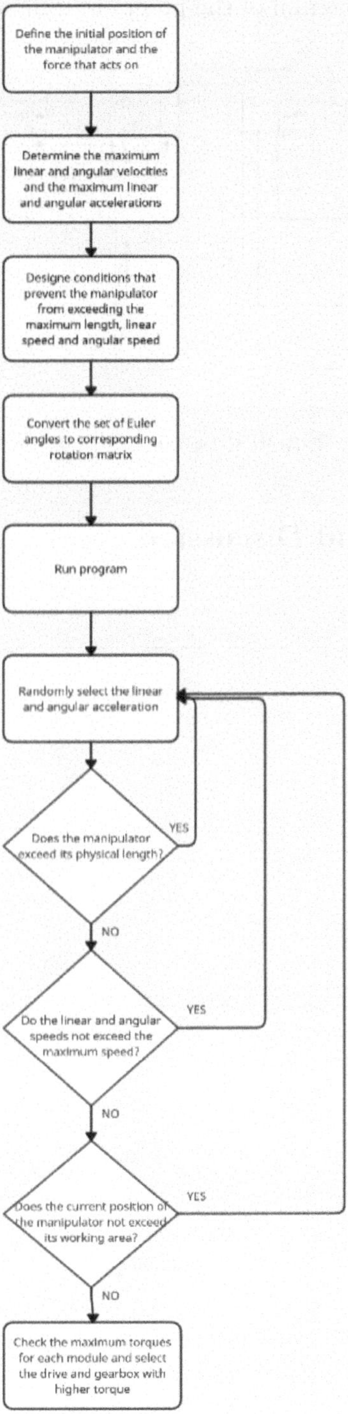

Fig. 8. Diagram for program in MATLAB for Simulink.

Figure 9 shows the diagram of the program in Simulink.

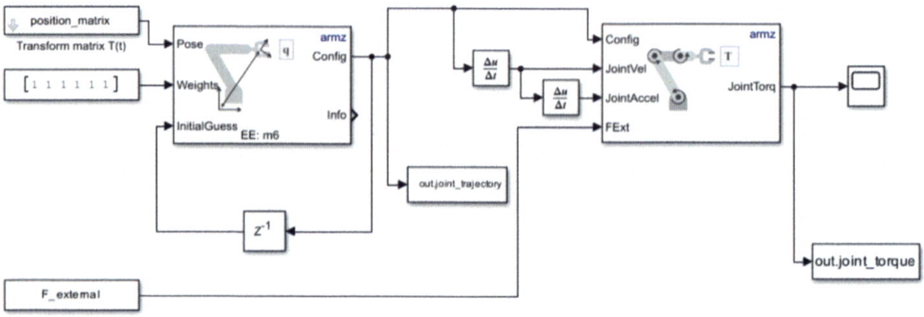

Fig. 9. Program in Simulink.

3 Three Cases and Discussion

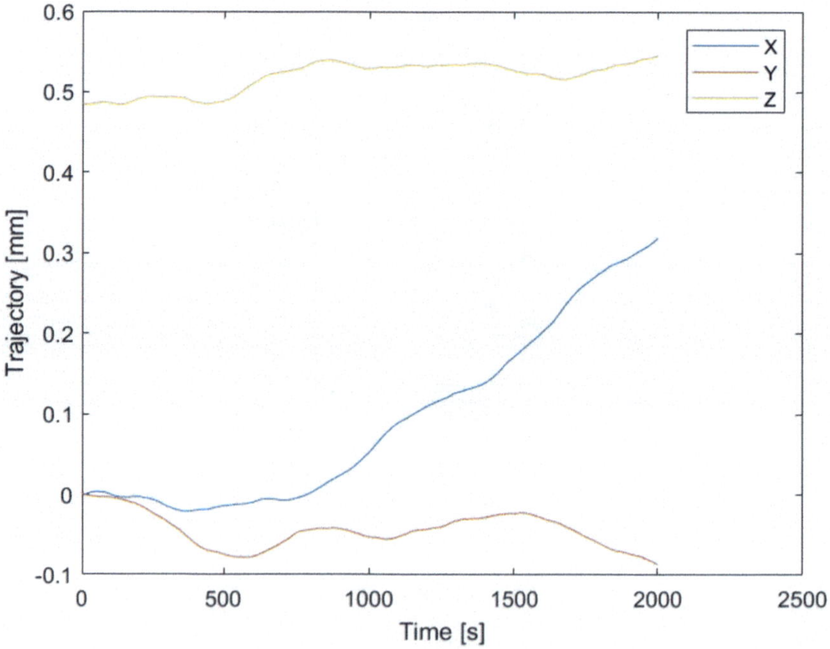

Fig. 10. Arm-Z movement trajectory in Case 1

Table 2. Maximum rigid joint torque value for case 1

Module 2	Module 3	Module 4	Module 5	Module 6	Module 7
15.66	15.51	13.93	12.25	11.12	9.75

3.1 Discussion

Three cases simulated using simulation are presented. The obtained graphs in Figs. 10, 11 and 12 show the trajectory of the manipulator tip movement over time. Tables 2, 3 and 4 provide the maximum torques of subsequent modules. The values of the torques at the first two modules are close to each other. The largest torque occur between modules 2–3 and 4–5. The torque values of the two farthest modules 5 and 6 are close to each other and they are the smallest. Smaller difference between modules 3–4 is noticeable. However, for such a small number of modules, the selected drive torques do not differ much. Thanks to the observed torque and knowledge of the limitations related to the price and structure of the module, the appropriate drive for each module can be determined. Depending on the manipulator's path, the power for motors of each module

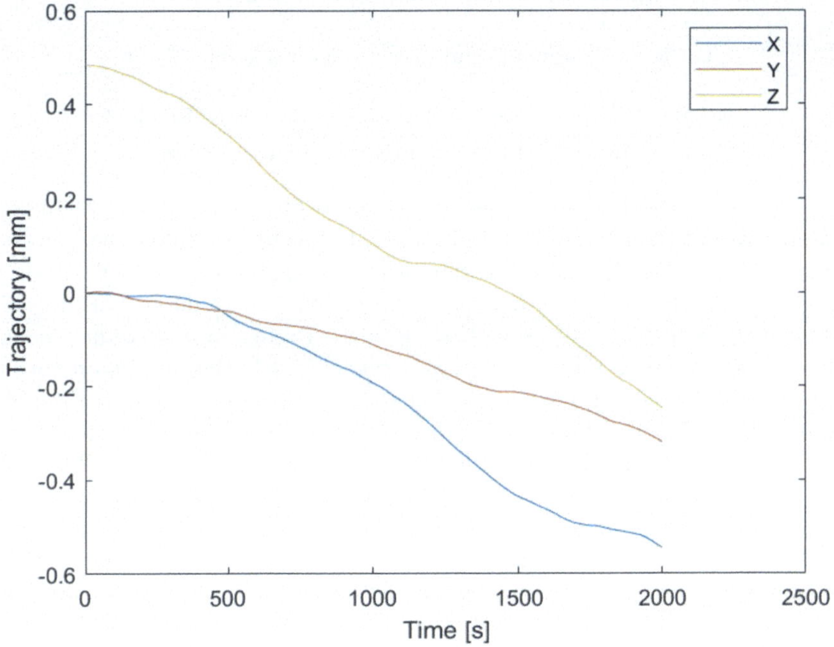

Fig. 11. Arm-Z movement trajectory in Case 2

Table 3. Maximum rigid joint torque value for Case 2

Module 2	Module 3	Module 4	Module 5	Module 6	Module 7
14.88	14.86	13.96	12.92	11.78	9.82

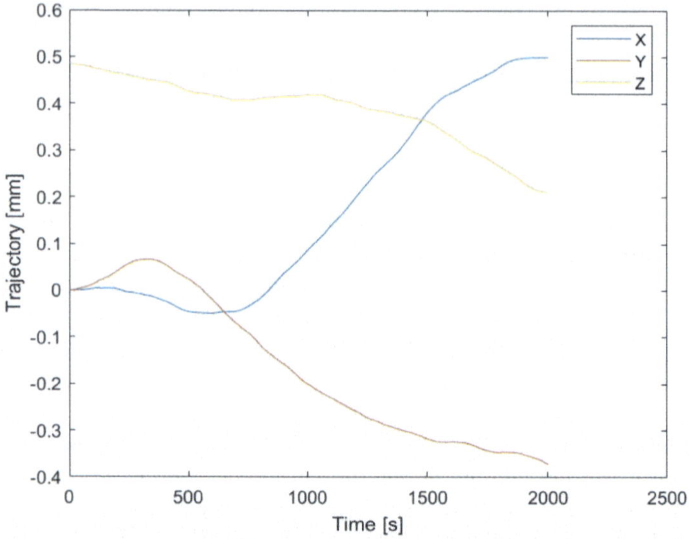

Fig. 12. Arm-Z movement trajectory in Case 3

Table 4. Maximum rigid joint torque value for Case 3

Module 2	Module 3	Module 4	Module 5	Module 6	Module 7
15.97	15.82	14.15	13.32	12.37	10.1

varies. Moreover, the highest torque shown in Tables 2, 3 and 4 is 24.1428 Nm, which means that the drive that will be mounted in the module must have equal or greater torque. Finally, we chose flat brushless engine EC 45 50 W, and planetary gearhead GP 32 C. The nominal torque (max. continuous torque) of the drive is 91.1 mNm, the gear reduction is 66:1. Taking into account the internal gear ratio of the member on the gear wheel is 4:1, the maximum torque is 24.0504 Nm. The selected drive and gearhead can be used in each module of our manipulator. The torque produced by our drive and gearhead is greater than the maximum torque specified in each module in each shown simulation. This means that the motor is able to rotate any module and therefore the entire manipulator works properly. If the mass of the module increased, thus the mass of the entire manipulator, the torques do not show a significant change compared to the initially assumed mass. The reason why the difference in torque is so small may be due to a very small change in rotational speed, because the speed at which we move the module has a large impact on the torque value.

It is also worth mentioning that presented here although presented here concept of hyper-redundant manipulator [44] has several advantages, they are difficult to control. The inverse kinematics problem for a typical industrial robotic arm can be solved relatively easily [45], which is definitely not the case of hyper-redundant manipulators [46–48]. However, an effective geometrical method for inverse kinematics of *planar* hyper-redundant manipulators has been presented in [49]. For further details on hyper-redundant robotic manipulators see [50].

4 Conclusions

In this project, we presented a virtual manipulator module, calculation of the Denavit-Hartenberg parameters and three simulations. Based on them, we select the appropriate drives and gearheads that provide the appropriate torque for the 6-DOF Arm-Z manipulator. Our code created in the MATLAB environment not only efficiently simulates modular Arm-Z manipulator with 6-DOF, but is capable of simulating much larger virtual modular manipulators, in particular, hyper-redundant manipulators.

Computer simulations naturally reduce errors related to incorrect selection of drives and gears. Our program allows you to perform multiple simulations in succession and increase the accuracy of the results. Thanks to this, the simulation reduces the time spent selecting parts and produces accurate results. Presently, the fully operational physical prototype is presently being developed. In the future, it will be possible to build a larger manipulator and select parts for it using the presented simulation.

Acknowledgments. This research is a part of the project titled *Arm-Z: an extremely modular hyperredundant low-cost manipulator – development of control methods and efficiency analysis* and funded by *OPUS 17* research grant No. 2019/33/B/ST8/02791 supported by the National Science Centre, Poland.

References

1. Sahu, V.S.D.M., Samal, P., Panigrahi, C.K.: Modelling, and control techniques of robotic manipulators: a review. Mater. Today Proc. **56**, 2758–2766 (2022)
2. Rubrecht, S., Singla, E., Padois, V., Bidaud, P., de Broissia, M.: Evolutionary design of a robotic manipulator for a highly constrained environment. In: New Horizons in Evolutionary Robotics: Extended Contributions from the 2009 EvoDeRob Workshop, pp. 109–121. Springer (2011)
3. Singh, S., Singla, A., Singla, E.: Modular manipulators for cluttered environments: a task-based configuration design approach. J. Mech. Robot. **10**(5), 051010 (2018)
4. Pettersen, K.Y.: Snake robots. Ann. Rev. Control **44**, 19–44 (2017)
5. Liljebäck, P., Pettersen, K.Y., Stavdahl, Ø., Gravdahl, J.T.: A review on modelling, implementation, and control of snake robots. Robot. Auton. Syst. **60**(1), 29–40 (2012)
6. Liljebäck, P., Pettersen, K.Y., Stavdahl, Ø., Gravdahl, J.T.: Snake Robots: Modelling, Mechatronics, and Control. Springer (2013)
7. Hirose, S.: Design of hyper-redundant arm. In: Robotics Research: The Seventh International Symposium, pp. 548–557. Springer (1996)
8. Hirose, S.: Biologically Inspired Robots: Snake-Like Locomotors and Manipulators. Oxford University Press (1993)
9. Su, H., Li, S., Manivannan, J., Bascetta, L., Ferrigno, G., De Momi, E.: Manipulability optimization control of a serial redundant robot for robot-assisted minimally invasive surgery. In: 2019 International Conference on Robotics and Automation (ICRA), pp. 1323–1328. IEEE (2019)

10. Hu, Y., Su, H., Chen, G., Ferrigno, G., De Momi, E., Knoll, A.: Hierarchical optimization control of redundant manipulator for robot-assisted minimally invasive surgery. In: 2020 IEEE/RSJ International Conference on Intelligent Robots and Systems (IROS), pp. 2929–2934. IEEE (2020)
11. Zawidzki, M., Szklarski, J.: Transformations of Arm-Z modular manipulator with particle swarm optimization. Adv. Eng. Softw. **126**, 147–160 (2018)
12. Transeth, A.A., Pettersen, K.Y., Liljebäck, P.: A survey on snake robot modeling and locomotion. Robotica **27**(7), 999–1015 (2009)
13. Hirose, S., Yamada, H.: Snake-like robots [tutorial]. IEEE Robot. Autom. Mag. **16**(1), 88–98 (2009)
14. Hopkins, J.K., Spranklin, B.W., Gupta, S.K.: A survey of snake-inspired robot designs. Bioinspiration Biomimetics **4**(2), 021001 (2009)
15. Sanfilippo, F., Azpiazu, J., Marafioti, G., Transeth, A.A., Stavdahl, Ø., Liljebäck, P.: Perception-driven obstacle-aided locomotion for snake robots: the state of the art, challenges and possibilities. Appl. Sci. **7**(4), 336 (2017)
16. Zawidzki, M.: Discrete Optimization in Architecture: Extremely Modular Systems. Springer (2017)
17. Zawidzki, M., Nishinari, K.: Modular Truss-Z system for self-supporting skeletal free-form pedestrian networks. Adv. Eng. Softw. **47**(1), 147–159 (2012)
18. Zawidzki, M., Nishinari, K.: Application of evolutionary algorithms for optimum layout of Truss-Z linkage in an environment with obstacles. Adv. Eng. Softw. **65**, 43–59 (2013)
19. Zawidzki, M.: Retrofitting of pedestrian overpass by Truss-Z modular systems using graph-theory approach. Adv. Eng. Softw. **81**, 41–49 (2015)
20. Zawidzki, M.: Optimization of multi-branch Truss-Z based on evolution strategy. Adv. Eng. Softw. **100**, 113–125 (2016)
21. Zawidzki, M., Szklarski, J.: Effective multi-objective discrete optimization of Truss-Z layouts using a GPU. Appl. Soft Comput. **70**, 501–512 (2018)
22. Zawidzki, M., Jankowski, Ł: Optimization of modular Truss-Z by minimum-mass design under equivalent stress constraint. Smart Struct. Syst. **21**(6), 715–725 (2018)
23. Zawidzki, M.: Discrete Optimization in Architecture: Architectural & Urban Layout. Springer (2016)
24. Zawidzki, M., Tateyama, K., Nishikawa, I.: The constraints satisfaction problem approach in the design of an architectural functional layout. Eng. Optim. **43**(9), 943–966 (2011)
25. Zawidzki, M., Chraibi, M., Nishinari, K.: Crowd-Z: the user-friendly framework for crowd simulation on an architectural floor plan. Pattern Recogn. Lett. **44**, 88–97 (2014)
26. Zawidzki, M.: Automated geometrical evaluation of a plaza (town square). Adv. Eng. Softw. **96**, 58–69 (2016)
27. Zawidzki, M.: Discrete Optimization in Architecture: Building Envelope. Springer (2017)
28. Zawidzki, M.: Dynamic shading of a building envelope based on rotating polarized film system controlled by one-dimensional cellular automata in regular tessellations (triangular, square and hexagonal). Adv. Eng. Inform. **29**(1), 87–100 (2015)
29. Denavit, J., Hartenberg, R.S.: A kinematic notation for lower-pair mechanisms based on matrices (1955)
30. Lewis, F.L., Dawson, D.M., Abdallah, C.T.: Robot Manipulator Control: Theory and Practice. CRC Press (2003)

31. Pratheep, V.G., Chinnathambi, M., Priyanka, E.B., Ponmurugan, P., Thiagarajan, P.: Design and analysis of six DOF robotic manipulator. In: IOP Conference Series: Materials Science and Engineering, vol. 1055, p. 012005. IOP Publishing (2021)
32. Iqbal, J., Islam, R.U., Khan, H., et al.: Modeling and analysis of a 6 DOF robotic arm manipulator. Can. J. Electr. Electron. Eng. **3**(6), 300–306 (2012)
33. Afloare, A.I., Apostolescu, N., Chira, A.I., Munteanu, C.E.: Inverse kinematic solution of a 6 DOF serial manipulator. In: DAAAM 2019, vol. 30, pp. 628–631. Danube Adria Association for Automation and Manufacturing (2019). https://doi.org/10.2507/30th.daaam.proceedings.086
34. Denavit, J., Hartenberg, R.S.: A kinemati nation for lower-pair mechanisms based on matirics. J. Appl. Mech. **21**(5), 212–215 (1995)
35. Corke, P.I.: A simple and systematic approach to assigning Denavit-Hartenberg parameters. IEEE Trans. Robot. **23**(3), 590–594 (2007)
36. Staranowicz, A., Mariottini, G.L.: A survey and comparison of commercial and open-source robotic simulator software. In: Proceedings of the 4th International Conference on PErvasive Technologies Related to Assistive Environments, pp. 1–8 (2011)
37. Aliane, N.: Teaching fundamentals of robotics to computer scientists. Comput. Appl. Eng. Educ. **19**(3), 615–620 (2011)
38. Xiao, J., Han, W., Wang, A.: Simulation research of a six degrees of freedom manipulator kinematics based on MATLAB toolbox. In: 2017 International Conference on Advanced Mechatronic Systems (ICAMechS), pp. 376–380. IEEE (2017)
39. Yilmaz, Z., Yilmaz, O., Bingul, Z.: Design, analysis and simulation of a 6-DOF serial manipulator. Kocaeli J. Sci. Eng. **3**(1), 9–15 (2020)
40. Velarde-Sanchez, J.A., Rodriguez-Gutierrez, S.A., Garcia-Valdovinos, L.G., Pedraza-Ortega, J.C.: 5-DOF manipulator simulation based on MATLAB-Simulink methodology. In: 2010 20th International Conference on Electronics Communications and Computers (CONIELECOMP), pp. 295–300. IEEE (2010)
41. Klee, H., Allen, R.: Simulation of Dynamic Systems with MATLAB® and Simulink®. CRC Press (2018)
42. Xue, D., Chen, Y.: System Simulation Techniques with MATLAB and Simulink. Wiley (2013)
43. Yang, B., Abramova, I.: Dynamic Systems: Modeling, Simulation, and Analysis. Cambridge University Press (2022)
44. Ning, K., Wörgötter, F.: A novel concept for building a hyper-redundant chain robot. IEEE Trans. Robot. **25**(6), 1237–1248 (2009)
45. Murray, R.M., Li, Z., Sastry, S.S.: A Mathematical Introduction to Robotic Manipulation. CRC Press (1994)
46. Rolf, M., Steil, J.J.: Efficient exploratory learning of inverse kinematics on a bionic elephant trunk. IEEE Trans. Neural Netw. Learn. Syst. **25**(6), 1147–1160 (2014)
47. Melingui, A., et al.: Qualitative approach for forward kinematic modeling of a compact bionic handling assistant trunk. IFAC Proc. Vol. **47**(3), 9353–9358 (2014)
48. Falkenhahn, V., Hildebrandt, A., Neumann, R., Sawodny, O.: Dynamic control of the bionic handling assistant. IEEE/ASME Trans. Mechatron. **22**(1), 6–17 (2017)
49. Yahya, S., Moghavvemi, M., Mohamed, H.A.F.: Geometrical approach of planar hyper-redundant manipulators: Inverse kinematics, path planning and workspace. Simul. Model. Pract. Theory **19**(1), 406–422 (2011)
50. Chirikjian, G.S., Burdick, J.W.: A hyper-redundant manipulator. IEEE Robot. Autom. Mag. **1**(4), 22–29 (1994)

Conception of the Channel Robot Navigation

Vygantas Ušinskis(✉), Nikolaj Šešok, Tadas Rasimavičius, Dominykas Čičiurėnas, Muhammad Zulkifal, and Vytautas Bučinskas

Vilnius Gediminas Technical University, Vilnius Plytines str. 25, Vilnius, Lithuania
vygantas.usinskis@vilniustech.lt

Abstract. Mobile robot navigation has gained increasing interest in academical and industrial fields. Navigation task mainly consists of three main parts including object detection and mapping, path planning and motion execution [1]. Various sensors exist which separately or in fusion can acquire different kind of information from surrounding environment recognizing distances, shapes, colors etc. Nevertheless, navigation tasks are becoming more difficult requiring to adjust to varying static and dynamic obstacles and unknown environments. To efficiently deal with adaptive navigation tasks it is important not only to acquire correct data from environment but also to process this data using algorithms for path planning. Adaptivity challenges classical path planning methods like potential field histogram (PFH), visual graph (VG) and Dijkstra [2] and require to look into heuristic methods and in some cases the use of artificial intelligence. This paper contains a review of classical and heuristic methods of path planning and possible combinations analyzing main advantages and disadvantages. Moreover, taking into account presented analysis specific obstacle detection and path planning method is presented for autonomous mobile robot used to navigate in under soil tunnels. Finally, conclusions and research situations are presented.

Keywords: channel robot · mobile robot · path planning · autonomous navigation · sensor fusion

1 Introduction

Path tracking robots are evolving rapidly but tasks are becoming increasingly difficult challenging old methods and demanding to find new autonomous solutions. New cognitive tasks in household, industry, traffic fields require not only to detect objects but to evaluate them without direct recognition. To deal with this challenge sensor fusion can be used [3]. But sensors of different physical nature cannot always directly provide required information thus demand of AI approach is needed to interpret data.

Challenges also arises in path planning stage then interpreted data must be converted to path. Navigation task for mobile robot in general can be divided in four stages as shown in Image 1.

First the workspace has to be established. Then environment has to be interpreted by localizing and mapping obstacles and the target. Acquired information has to be

R. Szewczyk et al. (Eds.): AUTOMATION 2024, LNNS 1219, pp. 192–205, 2024.
https://doi.org/10.1007/978-3-031-78266-4_17

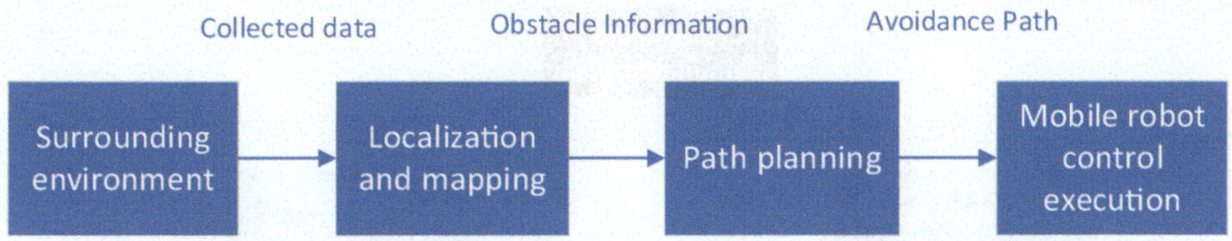

Fig. 1. Implementation of obstacle avoidance for mobile robot

processed so that path could be generated in path planning stage and eventually this path must be executed by mobile robot [1, 2]. As obstacles are becoming more complex and workspaces can contain both static dynamic objects path planning methods need to be adaptable solving both local and global problems.

One of the fields which is in need of autonomous navigation solution is channel inspection which are difficult to reach physically by human. This raises several challenges. First of all, environment is enclosed which hinder ability of outside communication. Moreover, it is completely dark which requires corresponding sensing devices. Because these tunnels are vacated there is a possibility of random obstacles.

In this article navigation method will be proposed for this specific field. Article will consist of three main parts. Evaluation of existing path planning methods. Then method of localization and mapping for channel inspection will be introduced. In addition, path planning method will be selected according to evaluation results. Lastly conclusion will be made.

2 Methodology of Path Planning

Path planning for mobile robot generally can be divided into two main parts including local and global methods [4]. Main difference is that global methods require prior knowledge about environment including static obstacles determined in advance while local methods can work in unfamiliar environment and enable to adapt to new static and dynamic obstacles [5]. There are many methods developed in both of these groups which alternatively can be divided according to completeness and admissibility as shown in Fig. 2.

The main criteria for path planning are to find most efficient, accurate and safe path. Because of this many heuristic methods are being researched which are able to find sub-optimal solution with high efficiency. This is advantageous comparing to classical methods which can be much more time consuming, nevertheless depending on the method it possible to ensure most optimal solution [6, 7].

2.1 Global Methods

As mentioned before global methods require prior knowledge about environment. Algorithms dedicated for this method use the available information to find the best or optimal solution to reach destination. These algorithms most commonly cannot adapt efficiently and if something changes in the environment all calculation need to be repeated. Before

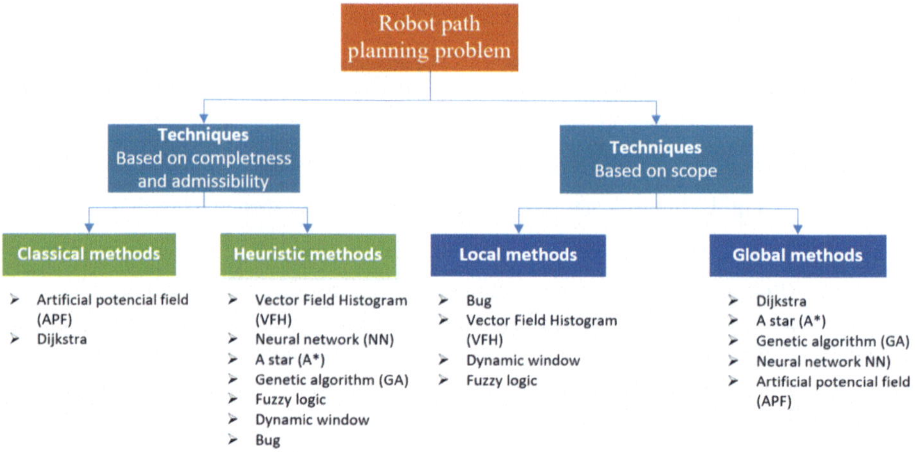

Fig. 2. Navigation techniques for mobile robot

applying these methods map has to be generated. There many ways to do this. For example article suggest [8, 9] to contract topological map with interconnected nodes and vertices from obtain visual data, images which represents environment. Another way to establish global map is to use GPS system as suggested in articles [10, 11] which allows to use global information received from satellites to establish nodes which are later used for path planning using global methods.

Dijkstra Method

Dijkstra method is path planning algorithm which searches for the shortest path from one node to another in interconnected node map which is fully defined. It searches for the shortest path by recursively checking all nodes starting from the starting node until goal node is found excluding any path which is longer. It is an effective method for finding best possible solution in fully defined map as researched in articles [12–14].

Dijkstra method is commonly known as global path planning method which need well defined nodes. Nevertheless, this method is also improved by some researchers to be able to avoid obstacles and adjust the path for mobile robots. One example presented in article [15] is to use ultrasonics sensor as obstacle localization device and to procced with localization on each node. If obstacle is detected map is updated by excluding corresponding node and robot procced with updated path using Dijkstra method.

A* Star Method

Another method which is commonly used as alternative to Dijsktra is A* star method. It is a search algorithm mainly used in topological configuration workspaces. It is slightly different from Dijkstra method because it uses heuristic mathematical functions to graphically find optimal path from start to destination node. It continuously searches for the new location in the graph until destination is reached and if it is not the case all neighbor nodes are set for exploration [2, 16]. Nevertheless, as Dijkstra method it is computationally demanding especially but it can find suboptimal path faster.

Genetic Algorithm

Genetic algorithm is a more advanced heuristic method which uses biological principals of genetics to find optimal solution from initial population of input information. As explained in articles [17, 18] it uses fitness function to evaluate all possible solution in initial population and select acceptable candidates. Then by applying reproduction and convergence steps new solutions are generated and also best solutions can be transferred to a new generation. With each generation cycle continues to produce more accurate solution according to the fitness function until optimal solution is found.

It is possible to apply this method in various fields including navigation. For example, in article [19] mobile robot workspace consists of grid in 2D coordinate system. Each path is defined by points consisting of x and y coordinate. Map has predefined obstacles which occupies specific cells. Using genetic algorithm initial population consists of several different paths cycle continues until optimal path is found which does not coincide with obstacles and is relatively short.

Neural Network

Like previous method neural network algorithm applies biological principles for finding optimal solution. Neural network introduced a term machine learning. By providing algorithm with different variants of defined input information about the workspace machine is able to learn and detect patterns which allow for more efficient and accurate path planning [20]. This method can help to improve path smoothness and slow convergence of more straightforward methods including Dijkstra and A*. As example, input data can consist of sampling distribution of the map where each pixel is encoded with value – empty, obstacle, goal, start as presented in articles [21–23]. Output consists of pixels probability of the same map which dictates if it will be chosen during the sampling process. Training stage is executed using A* star method and then optimal path is found, path is encoded in pixels which is used for training. In such a way algorithm is trained which can converge optimal path faster.

2.2 Local Methods

Local path planning for mobile robot requires sensory devices to get input about workspace in real time then map is not necessarily defined in advance and also contains static and dynamic obstacles as stated in articles [24, 25]. Adaptivity in real time is the main advantage comparing to global path planners, nevertheless because these methods are commonly based on fastest descend optimization they tend to get trapped in local minimum [26]. Some of the widely used and researched methods are presented below.

Bug Method

Bug algorithm is a straightforward but effective method for local path planning with low computational demand. The idea is that mobile robot has distance sensors and moves in straight line to defined destination until obstacle is detected. Then obstacle is detected robot moves around the obstacle and measures the distance to the nearest obstacle in direction of the destination and jumps to the next obstacle or final destination [3]. Many variants of bug method exist as stated in articles [27, 28] like Bug 1, Bug2, VistBug,

DistBug, PointBug, etc. While these methods are not very demanding computationally, they are not effective in complex environments because they can ether take a very long time before finding destination or have a high risk of getting trapped in local minimum.

Artificial Potential Field

The basic idea of APF method is that workspace is transformed into virtual force field motion like artificial gravity or magnetic field. Obstacles inside environment have repulsive forces while destination has attractive force. The resultant force is the direction of motion which correspond to the obstacle free path leading to the goal [29]. This method is widely implemented because of simplicity and minimal computational resources for determining the path from start to finish [30]. Nevertheless, this method has a risk of getting stuck in local minimum but handful of research exists which succeeded in countering this problem to some degree [31–34]. As example in article [34] APF in combination with A* method is used for global direction.

Vector Field Histogram

Vector field histogram is an improved APF method which uses similar principal of repulsive and attractive force but additionally considers robot geometry and dynamics as stated in [35, 36]. It avoids obstacle by simultaneously steering the robot towards goal. VFH algorithm use data from range and vision sensors to generate certainty grid. Certainty grid values change in real time according to the approached obstacles thus avoiding them. For attractive force goal can be defined globally or the wall can be used as a reference by adjusting VFH algorithm like presented in article [37]. This method is efficient and computationally not demanding. Also, even if it has probability to encounter local minimum as stated in article [38] this problem can be countered with intermediate data representation.

Dynamic Window

Dynamic windows approach is based on sampling velocities which determine curvatures and by evaluating these curvatures optimal path is determined as explained in article [39]. Curvatures are defined by velocity vector (v, ω) which represents rotational and transitional velocities which are restricted by obstacles thus providing distance to the closest obstacle [40]. Dynamic window itself is reduces searched field containing only velocities which can be reached in the next time interval defined in the algorithm. Dynamic window is purely reactive method for obstacle detection and thus it is usually combined with other path planning methods as A* [41].

Fuzzy Logic

Fuzzy logic is a very versatile local path planning method which can work with imprecise information interpreted from sensors for finding an optimal path [42]. Fuzzy algorithm works on rule-based principle. Then input data acquired from the sensors is interpreted in fuzzification process transforming them to fuzzy numbers which are then mapped with membership function according to defined rules. The resulting values then are transformed to precise values using defuzzification which yield required direction or position of robot movement according to the defined algorithm. Fuzzy based algorithms are quite effective especially then working with uncertainty and imprecision from input data. Nevertheless, then systems require high accuracy fuzzy methods might not be suitable.

All local path planning methods mentioned can be used effectively depending on the workspace conditions and mobile robot configuration. To choose the most suitable method it is important to evaluate main properties. One of these properties is adaptivity to complex obstacles including complex geometry and dynamic obstacles. It is important to be able to generate smooth and short path which takes less time but optimization can also decrease reliability. Moreover, computational resources are becoming increasingly important as systems can be configured of various input devices. Furthermore, uncertainty is significant as optimizing path planning tends to become uncertain at some degree which can lead to local minimum. Also, algorithms which work with uncertain data tend to be less predictable and reliable. Lastly, implementation and adaptivity is important. Implementation means that some straightforward methods like Bug method require simple programming while Fuzzy based methods require more complex mathematics and interference rules which require expertise. Also, even though specific path planning algorithm can deal with complex environment it might not be able to work in another kind of workspace because it is tuned to specific conditions which corresponds to adaptivity.

Table 1. Evaluation of local path planning methods

Method	Criteria				
	Complex obstacles	Computational resources	Local minima	Implementation/adaptivity	Resources
Bug	**Low**	*Low*	*Low*	*High*	[3, 27, 28, 43, 44]
APF	Moderate-low	Moderate	**High**	**Low**	[29–34, 45]
VFH	Moderate	Moderate	Moderate	Moderate high	[35–38]
Dynamic window	Moderate-high	Moderate	Moderate	**Low**	[39–41, 46]
Fuzzy logic	*High*	**High**	Moderate-low	Moderate	[42, 47–49]

As seen in Table 1 starting from the top Bug method it has several advantages. It is simple and straightforward method which require low computational resources. It is easy to implement. Robot only needs direction defined by position of the goal defined in the global map and range sensor for detecting obstacles. It also deals efficiently with local minima problem. Bug 1 method has high reliability to reach destination.

On the other hand, even though it is possible to reliably reach destination using Bug1 method it might take a long time because trajectory will not be optimized. And if alternatives like Bug2 are used which work according to the fastest descend principle local minima becomes a major problem.

APF method improves upon dealing with complex shapes comparing to bug method with more optimized trajectory. It also requires relatively low computational resources.

Nevertheless, it suffers from local minimum problems especially with U-formed shapes. Also, it is mostly designed for static environment and traditional methods have to be fused to accommodate for this shortcoming. For example in article [50] local minimum problem is improved and mobile robot is able to navigate in dynamic workspace by using a hybrid method of APF, genetic algorithm and membrane computing. Similarly in article [51] hybrid fuzzy logic and APF method is proposed where fuzzy logic expert system provides with most suitable heading to the destination thus optimizing trajectory and minimizing local minima.

Vector field histogram is improved version of APF path planning algorithm. It can deal with complex and dynamic obstacles better because local minimum is minimized. There are no repulsive or attractive forces in polar histogram which points the robot away from obstacle and, in some cases, away from target searching for best valleys to reach the goal. Also, it takes into account robot geometry and dynamics making is easier to implement and polar histogram generation does not require high computational resources.

Nevertheless, it still has a risk of local minima and path is not optimized as the robot is sometimes pointed away from the target. Moreover, it is challenging to deal with dynamic obstacles resulting in insufficient trajectory. But various methods are suggested by researcher to deal with these problems to some extent. For example in article [52] hybrid method of VFH and Fuzzy logic algorithm which generates the most appropriate heading toward the goal thus optimizing the path and reducing local minimum risk.

Dynamic window is a good alternative to VFH which excels in avoiding complex obstacles by generating smooth curvatures around them. In addition, it is not computationally demanding and deals with local minimum quite effectively.

On the other hand, it is purely reactive method used for obstacle avoidance with no functionality for direction thus integration for path planning requires combination of additional techniques. To provide this method with direction and path towards the goal global method like A* is frequently used [53–55]. Also combination Theta* (alternative to A*) dynamic window and fuzzy logic approach is suggested in article [56], where Theta* is responsible for global path ant position of the target and fuzzy logic fused with Dynamic windows methodic allow for smooth avoidance of dynamic and static obstacles towards the goal.

Lastly, Fuzzy logic approach allows to work in vague rules for interpreting complex environments with controlled uncertainty allowing to significantly optimize path planning. In addition, local minimum problem can be avoided quite easily and rule-based approach makes it not so difficult to implement. Also, it can be effectively combined with other previously mentioned methods to optimize the path and deal with local minimum. For example, in articles [57, 58] APF effectiveness of avoiding obstacles is combined with Fuzzy capability to deal with noise creating more reliable and optimized path planning.

Nevertheless, this method is computationally more demanding comparing to previous methods because it requires more complex math. Also, this method may lack adaptivity if designed system is applied in different workspace because established rules are usually oriented for certain geometry of obstacles and terrain of the map. Besides this, Fuzzy path planning methods can be improved by combining with other methods like mentioned previously. Also, as presented in articles [59–61] fuzzy logic is commonly combined with AI neural network approach to train a specific model of NN neural network path planner thus creating improved method for specific environment.

3 Obstacle Localization and Path Planning for Channel Robot

Navigating in closed dark environments like tunnels which are difficult to access directly by human propose several challenges. Firstly, mobile robot can be difficult to reach remotely because of isolation inside the tunnels. That means that robot has to be able to perform tasks with minimum surveillance. Secondly, uncared for environments which are difficult to inspect can propose a challenge of random unexpected obstacles. Taking these challenges into account requires for reliable and adaptive solution for obstacle detection and path planning.

We propose Pseudo Lidar and laser triangulation-based method for navigation in uncared for difficult to access tunnels. As shown in Fig. 3 hardware of this system consists of Robot tank-type chassis, pseudo-Lidar with fixed angle ω_w, one red light laser pointer rotated by angle ω_g, three green light laser pointers rotated by angles ω_r, ω_l, ω_u and RGB camera rotated by angle ω_c.

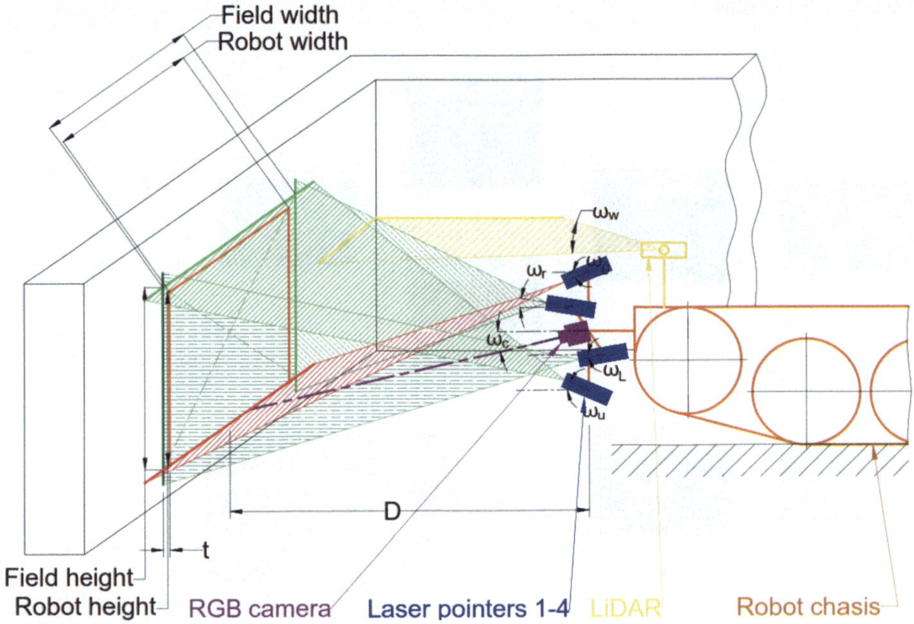

Fig. 3. Pseudo-Lidar and laser triangulation-based channel mobile robot navigation method

This method is composed of two perception techniques. First technique is based on laser triangulation. As presented in articles [62] linear laser pointer is used to project line on the surface. Then RGB camera is pointed in an angle to filter and inspect projected line. The line is distorted then obstacle is approached or surface terrain changes and it is possible to detect this change with camera and evaluate geometry, distance and position of detected object. This is the basis of laser triangulation principle. It is also possible to perform scanning of object geometry using combination of two laser pointers as researched in article [63].

In our method red light laser pointer is used to observe changes on the ground and it is tuned angularly with the camera for more precise measurement of distance and dimension of an encountered obstacle. Some obstacles like small holes or bums can be ignored thus more accurate evaluation from the ground is required. Additionally, three green laser pointers are mounted for detecting obstacles from right, left and upper directions which are not connected to the ground in monitored space. Green light is used to distinguish surveillance of obstacles which are not on the ground during RGB filtering stage. These lasers also are not tuned with the camera in an angle because it is not so important to evaluate the distance and dimensions of that kind of obstacles as it is more important to declare a fact that obstacle is present in the field and must be avoided. Monitored field is constructed from four laser projected lines on the plane which is positioned in distance D from the robot chassis. Field dimensions equal to the mobile robot height and width with additional clearance – t. This method allows for reliable obstacle detection in wide range using simple hardware with relatively small computational resources comparing to 3D Lidar. Also, laser pointers allow to detect obstacles in the dark.

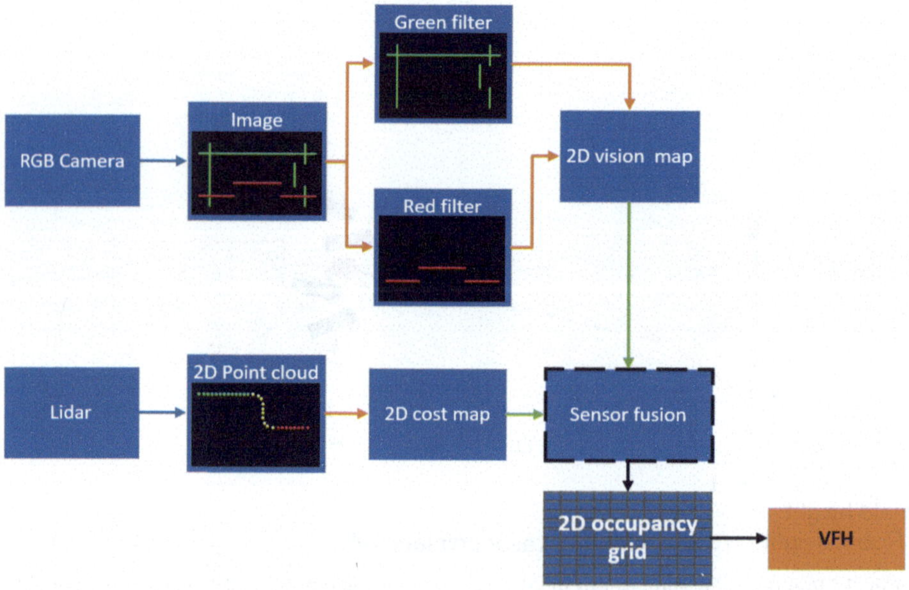

Fig. 4. Sensor fusion scheme of proposed channel robot navigation method

Proposed laser triangulation technique works as purely reactive method for obstacle detection. Additionally, pseudo 2D pseudo-Lidar with fixed angle ω_w is integrated to enable wall detection which will be used inside the channel as the guide. Lidar data fused with camera data can also double check obstacles detected by camera to reduce noise and uncertainty of the measurements. General sensor fusion scheme is presented in Fig. 4.

Sensors by abstraction level can be fused in low-level meaning direct raw data is used and connected which reduced uncertainty and high resolution as all data is taken into account [64]. On the other hand, fusing raw data is complex and require high computational resources. In our case mid-level fusion will be used. Raw data first will be processed for each sensor so that it could be fused and integrated in chosen path planning algorithm – VFH.

Vector field histogram method was chosen for development of this system because of adaptivity and capability to deal with complex obstacles, reasonable computational resources at low local minimum risk, which can be avoided by tunning the algorithm. On the other hand, trajectory optimization is lacking comparing to such heuristic methods as fuzzy logic. Depending on obstacles robot can be pointed backwards before finding free valley to pass. But for this project purposes it is more important to reliably reach the target by inspecting required channels even if optimization has to be slightly sacrificed.

VFH requires certainty grid which represents varying density on detected obstacles thus allowing to maneuverer. In our case two separate certainty grids will be generated for each sensory devices as show in Image 4. After that data of both certainty grids would be fused to the main polar histogram which will be used to avoid obstacles and follow the wall.

4 Conclusions

Local path planning methods are necessary edition for navigating in complex workspaces with static and dynamic obstacles of various geometry. On the other hand, Local methods mainly work reactively to avoid obstacles and lack optimization to reach the target. In many cases combination with global methods significantly improve reliability and optimization of the path by providing additional direction.

Advancement of path planning methods is growing increasingly by combining different methods and also incorporating artificial intelligence. This allows for more efficient and accurate trajectory in complex workspace allowing to reach the target. Nevertheless, development stage can be challenging. For example, neural networks enable machine learning to develop efficient path planning algorithm but it requires specific teaching methods for specific conditions of the workspace. Also, if conditions change and mobile robot is placed in a different environment it might be difficult to adapt developed algorithm. In comparison, less complex methods like Bug, VFH, Dynamic windows are not most efficient in many cases but they are easier to implement and can be more reliable depending on the system.

Method for mobile robot navigation inside channels was suggested which incorporates fusion of laser triangulation and pseudo-Lidar sensory devices. Laser triangulation combined of four laser pointers and RGB camera allow to fully observe the field in front of the robot covering robot width and height with reasonable computational. Pseudo-lidar with fixed angle double checks approached obstacles and locates the wall which is a reference for direction.

Sensor data is first processed and converted 2D grid maps which are then fused to generate field histogram for path planning using VFH method. VFH method is computationally not expensive, deals with complex obstacles relatedly good, it is reliable and adaptable. Nevertheless, it is important to know that VHF does not deal well with dynamic obstacles, but for channel inspection this is not an issue.

References

1. Jain, S., Malhotra, I.: A review on obstacle avoidance techniques for self-driving vehicle. Int. J. Adv. Sci. Technol. **29**(6), 5159–5167 (2020)
2. Gul, F., Rahiman, W., Alhady, S.S.N.: A comprehensive study for robot navigation techniques. Cogent Eng. **6**(1), 1632046 (2019). https://doi.org/10.1080/23311916.2019.1632046
3. Kolar, P., Benavidez, P., Jamshidi, M.: Survey of datafusion techniques for laser and vision based sensor integration for autonomous navigation. Sensors (Switzerland) **20**(8), 1–49 (2020). https://doi.org/10.3390/s20082180
4. Tsukiyama, T.: RFID based navigation system for indoor mobile robots. **44**(1) PART 1. IFAC (2011)
5. Li, J., Qin, H., Wang, J., Li, J.: OpenStreetMap-based autonomous navigation for the four wheel-legged robot via 3D-lidar and CCD camera. IEEE Trans. Ind. Electron. **69**(3), 2708–2717 (2022). https://doi.org/10.1109/TIE.2021.3070508
6. Abdulsaheb, J.A., Kadhim, D.J.: Classical and heuristic approaches for mobile robot path planning: a survey. Robotics **12**(4), 93 (2023). https://doi.org/10.3390/robotics12040093
7. Atyabi, A., Powers, D.M.W.: Review of classical and heuristic-based navigation and path planning approaches. Int. J. Adv. Comput. Technol. **5**(14), 1–14 (2013)
8. Wang, F., et al.: Object-based reliable visual navigation for mobile robot. Sensors **22**(6), 1–24 (2022). https://doi.org/10.3390/s22062387
9. Kwon, O., Kim, N., Choi, Y., Yoo, H., Park, J., Oh, S.: Visual graph memory with unsupervised representation for visual navigation. In: Proceedings of the IEEE International Conference on Computer Vision, pp. 15870–15879 (2021). https://doi.org/10.1109/ICCV48922.2021.01559
10. Akhshirsh, G.S., Al-Salihi, N.K., Hamid, O.H.: A cost-effective GPS-aided autonomous guided vehicle for global path planning. Bull. Electr. Eng. Inform. **10**(2), 650–657 (2021). https://doi.org/10.11591/eei.v10i2.2734
11. Lee, H., Seo, J., Kassas, Z.M.: Integrity-based path planning strategy for Urban autonomous vehicular navigation using GPS and cellular signals. In: Proceedings of the 33rd International Technical Meeting of the Satellite Division of the Institute of Navigation, ION GNSS+ 2020, pp. 2347–2357 (2020). https://doi.org/10.33012/2020.17589
12. Arafat, M.Y., Alam, M.M., Moh, S.: Vision-based navigation techniques for unmanned aerial vehicles: review and challenges. Drones **7**(2), 89 (2023). https://doi.org/10.3390/drones7020089
13. Zhou, K., Lingli, Yu., Long, Z., Mo, S.: Local path planning of driverless car navigation based on jump point search method under urban environment. Future Internet **9**(3), 51 (2017). https://doi.org/10.3390/fi9030051
14. Ivanov, M., et al.: Influence of data clouds fusion from 3D real-time vision system on robotic group dead reckoning in unknown terrain. IEEE/CAA J. Autom. Sin. **7**(2), 368–385 (2020). https://doi.org/10.1109/JAS.2020.1003027
15. Alshammrei, S., Boubaker, S., Kolsi, L.: Improved Dijkstra algorithm for mobile robot path planning and obstacle avoidance. Comput. Mater. Contin. **72**(3), 5939–5954 (2022). https://doi.org/10.32604/cmc.2022.028165

16. Martins, O.O., Adekunle, A.A., Olaniyan, O.M., Bolaji, B.O.: An improved multi-objective a-star algorithm for path planning in a large workspace: design, implementation, and evaluation. Sci. Afr. **15**, e01068 (2022). https://doi.org/10.1016/j.sciaf.2021.e01068
17. Rahmaniar, W., Rakhmania, A.E.: Mobile robot path planning in a trajectory with multiple obstacles using genetic algorithms. J. Robot. Control **3**(1), 1–7 (2022). https://doi.org/10.18196/jrc.v3i1.11024
18. Liang, L., et al.: RLaGA: a reinforcement learning augmented genetic algorithm for searching real and diverse marker-based landing violations. **1**(1), 1–19 (2023). http://arxiv.org/abs/2310.07378
19. Manikas, T.W., Ashenayi, K., Wainwright, R.L.: Genetic algorithms for autonomous robot navigation. IEEE Instrum. Meas. Mag. **10**(6), 26–31 (2007). https://doi.org/10.1109/MIM.2007.4428579
20. Yu, J., Su, Y., Liao, Y.: The path planning of mobile robot by neural networks and hierarchical reinforcement learning. Front. Neurorobot. **14**, 1–12 (2020). https://doi.org/10.3389/fnbot.2020.00063
21. Janji, S., Kliks, A.: Neural networks for path planning, pp. 1–6 (2022). http://arxiv.org/abs/2207.00874
22. Yonetani, R., Taniai, T., Barekatain, M., Nishimura, M., Kanezaki, A.: Path planning using neural A* search. Proc. Mach. Learn. Res. **139**, 12029–12039 (2021)
23. Bhatt, S., Jain, A., Maheshwari, P., Jha, A., Chakravarty, D.: Reproducibility Report Path Planning using Neural A* Search (2022). http://arxiv.org/abs/2208.04153
24. Dong, H., Weng, C.Y., Guo, C., Yu, H., Chen, I.M.: Real-time avoidance strategy of dynamic obstacles via half model-free detection and tracking with 2D lidar for mobile robots. IEEE/ASME Trans. Mechatronics **26**(4), 2215–2225 (2021). https://doi.org/10.1109/TMECH.2020.3034982
25. Wang, B., Liu, Z., Li, Q., Prorok, A.: Mobile robot path planning in dynamic environments through globally guided reinforcement learning. IEEE Robot. Autom. Lett. **5**(4), 6932–6939 (2020). https://doi.org/10.1109/LRA.2020.3026638
26. Wang, L.C., Yong, L.S., Ang, M.H.: Hybrid of global path planning and local navigation implemented on a mobile robot in indoor environment. In: IEEE International Symposium on Intelligent Control - Proceedings, no. February 2002, pp. 821–826 (2002). https://doi.org/10.1109/isic.2002.1157868
27. Ali, M.A.H., Mailah, M.: Path planning and control of mobile robot in road environments using sensor fusion and active force control. IEEE Trans. Veh. Technol. **68**(3), 2176–2195 (2019). https://doi.org/10.1109/TVT.2019.2893878
28. Kim, P., Chen, J., Kim, J., Cho, Y.K.: SLAM-driven intelligent autonomous mobile robot navigation for construction applications. In: Smith, I.F.C., Domer, B. (eds.) Advanced Computing Strategies for Engineering, pp. 254–269. Springer, Cham (2018). https://doi.org/10.1007/978-3-319-91635-4_14
29. Fan, X., Guo, Y., Liu, H., Wei, B., Lyu, W.: Improved artificial potential field method applied for AUV path planning. Math. Probl. Eng. **2020**, 1–21 (2020). https://doi.org/10.1155/2020/6523158
30. Ravankar, A.A., Ravankar, A., Emaru, T., Kobayashi, Y.: HPPRM: hybrid potential based probabilistic roadmap algorithm for improved dynamic path planning of mobile robots. IEEE Access **8**, 221743–221766 (2020). https://doi.org/10.1109/ACCESS.2020.3043333
31. Sfeir, J., Saad, M., Saliah-Hassane, H.: An improved Artificial Potential Field approach to real-time mobile robot path planning in an unknown environment (2011). https://doi.org/10.1109/ROSE.2011.6058518
32. Shi, P., Zhao, Y.: An Efficient Path Planning Algorithm for Mobile Robot Using Improved Potential Field (2010)

33. Raja, R., Dutta, A., Venkatesh, K.S.: New potential field method for rough terrain path planning using genetic algorithm for a 6-wheel rover. Robot. Auton. Syst. **72**, 295–306 (2015). https://doi.org/10.1016/j.robot.2015.06.002
34. Zhang, E.: Path planning algorithm based on improved artificial potential field method. Appl. Comput. Eng. **10**(1), 167–174 (2023). https://doi.org/10.54254/2755-2721/10/20230170
35. Oroko, J., Nyakoe, G.N.: Obstacle avoidance and path planning schemes for autonomous navigation of a mobile robot: a review. In: Proceedings of the Sustainable Research and Innovation Conference, vol. 4, no. May, pp. 314–318 (2014). http://jkuat-sri.com/ojs/index.php/proceedings/article/view/237
36. Baltzakis, H., Argyros, A., Trahanias, P.: Fusion of laser and visual data for robot motion planning and collision avoidance. Mach. Vis. Appl. **15**(2), 92–100 (2003). https://doi.org/10.1007/s00138-003-0133-2
37. Wang, J., Luo, C.: Automatic wall defect detection using an autonomous robot: a focus on data collection. In: ASCE International Conference on Computing in Civil Engineering 2019, no. January, pp. 312–319 (2019). https://doi.org/10.1061/9780784482438.040
38. Luo, C., Gao, J., Li, X., Mo, H., Jiang, Q.: Sensor-based Autonomous Robot Navigation Under Unknown Environments with Grid Map Representation (2014)
39. Cheng, Y., Wang, G.Y.: Mobile robot navigation based on lidar. In: Proceedings of the 30th Chinese Control Decision Conference, CCDC 2018, pp. 1243–1246 (2018). https://doi.org/10.1109/CCDC.2018.8407319
40. Fox, D., Burgard, W., Thrun, S.: The Dynamic Window Approach to Collision Avoidance, pp. 137–146 (2016). https://www.ri.cmu.edu/pub_files/pub1/fox_dieter_1997_1/fox_dieter_1997_1.pdf
41. Molinos, E.J., Llamazares, Á., Ocaña, M.: Dynamic window based approaches for avoiding obstacles in moving. Rob. Auton. Syst. **118**, 112–130 (2019). https://doi.org/10.1016/j.robot.2019.05.003
42. Davis, D., Supriya, P.: Implementation of fuzzy-based robotic path planning. In: Satapathy, S.C., Srujan Raju, K., Mandal, J.K., Bhateja, V. (eds.) IC3T 2015, pp. 375–383. Springer, New Delhi (2016). https://doi.org/10.1007/978-81-322-2523-2_36
43. Zohaib, M., Pasha, S.M., Javaid, N., Iqbal, J.: IBA: intelligent bug algorithm – a novel strategy to navigate mobile robots autonomously. Commun. Comput. Inf. Sci. **414**, 291–299 (2013). https://doi.org/10.1007/978-3-319-10987-9_27
44. Sivaranjani, S., Nandesh, D.A., Gayathri, K., Ramanathan, R.: An investigation of bug algorithms for mobile robot navigation and obstacle avoidance in two-dimensional unknown static environments. In: 2021 International Conference on Communication information and Computing Technology (ICCICT), pp. 1–6 (2021). https://doi.org/10.1109/ICCICT50803.2021.9510118
45. Gyenes, Z., Szádeczky-Kardoss, E.G.: Uncertain estimation-based motion-planning algorithms for mobile robots. Acta IMEKO **10**(3), 51–60 (2021). https://doi.org/10.21014/ACTA_IMEKO.V10I3.1035
46. Kobayashi, M., Motoi, N.: Local path planning: dynamic window approach with virtual manipulators considering dynamic obstacles. IEEE Access **10**, 17018–17029 (2022). https://doi.org/10.1109/ACCESS.2022.3150036
47. Ali, M.A.H., Mailah, M., Jabbar, W.A., Moiduddin, K., Ameen, W., Alkhalefah, H.: Autonomous road roundabout detection and navigation system for smart vehicles and cities using laser simulator–fuzzy logic algorithms and sensor fusion. Sensors (Switzerland) **20**(13), 1–28 (2020). https://doi.org/10.3390/s20133694
48. Aouf, A., Boussaid, L., Sakly, A.: Same fuzzy logic controller for two-wheeled mobile robot navigation in strange environments. J. Robot. **2019**, 1–11 (2019). https://doi.org/10.1155/2019/2465219

49. Teli, T.A., Arif Wani, M.: A fuzzy based local minima avoidance path planning in autonomous robots. Int. J. Inf. Technol. **13**(1), 33–40 (2020). https://doi.org/10.1007/s41870-020-00547-0
50. Orozco-Rosas, U., Montiel, O., Sepúlveda, R.: Mobile robot path planning using membrane evolutionary artificial potential field. Appl. Soft Comput. J. **77**, 236–251 (2019). https://doi.org/10.1016/j.asoc.2019.01.036
51. Jaradat, M.A.K., Garibeh, M.H., Feilat, E.A.: Autonomous mobile robot dynamic motion planning using hybrid fuzzy potential field. Soft. Comput. **16**(1), 153–164 (2012). https://doi.org/10.1007/s00500-011-0742-z
52. Babinec, A., Duchoň, F., Dekan, M., Pásztó, P., Kelemen, M.: V FH * TDT (V FH * with Time Dependent Tree): A new laser rangefinder based obstacle avoidance method designed for environment with non-static obstacles. Rob. Auton. Syst. **62**(8), 1098–1115 (2014). https://doi.org/10.1016/j.robot.2014.05.003
53. Guan, W., Wang, K.: Autonomous collision avoidance of unmanned surface vehicles based on improved a-star and dynamic window approach algorithms. IEEE Intell. Transp. Syst. Mag. **15**(3), 36–50 (2023). https://doi.org/10.1109/MITS.2022.3229109
54. Fu, Q., Ning, Y., Ji, Y., Xiang, H.: Global dynamic path planning based on fusion improved A-star and dynamic window method. In: 2021 IEEE 5th Information Technology, Networking, Electronic and Automation Control Conference (ITNEC), vol. 5, pp. 243–249 (2021). https://doi.org/10.1109/ITNEC52019.2021.9587060
55. Li, K., Gong, X., Tahir, M., Wang, T., Kumar, R.: Towards path planning algorithm combining with A-star algorithm and dynamic window approach algorithm. Int. J. Adv. Comput. Sci. Appl. **14**(6), 511–519 (2023). https://doi.org/10.14569/IJACSA.2023.0140655
56. Wang, N., Xu, H.: Dynamics-constrained global-local hybrid path planning of an autonomous surface vehicle. IEEE Trans. Veh. Technol. **69**(7), 6928–6942 (2020). https://doi.org/10.1109/TVT.2020.2991220
57. Fakoor, M., Kosari, A., Jafarzadeh, M.: Revision on fuzzy artificial potential field for humanoid robot path planning in unknown environment. Int. J. Adv. Mechatron. Syst. **6**(4), 174–183 (2015). https://doi.org/10.1504/ijamechs.2015.072707
58. Wang, D., Chen, S., Zhang, Y., Liu, L.: Path planning of mobile robot in dynamic environment: fuzzy artificial potential field and extensible neural network. Artif. Life Robot. **26**(1), 129–139 (2021). https://doi.org/10.1007/s10015-020-00630-6
59. Guo, Na., Li, C., Wang, Di., Song, Y., Liu, G., Gao, T.: Local path planning of mobile robot based on long short-term memory neural network. Autom. Control. Comput. Sci. **55**(1), 53–65 (2021). https://doi.org/10.3103/S014641162101003X
60. Wang, G., Zhou, J.: Dynamic robot path planning system using neural network. J. Intell. Fuzzy Syst. **40**, 3055–3063 (2021). https://doi.org/10.3233/JIFS-189344
61. Mishra, D.K., Thomas, A., Kuruvilla, J., Kalyanasundaram, P., Prasad, K.R., Haldorai, A.: Design of mobile robot navigation controller using neuro-fuzzy logic system. Comput. Electr. Eng. **101**, 108044 (2022). https://doi.org/10.1016/j.compeleceng.2022.108044
62. Utaminingrum, F., et al.: A laser-vision based obstacle detection and distance estimation for smart wheelchair navigation. In: 2016 IEEE International Conference on Signal Image Process, ICSIP 2016, pp. 123–127 (2017). https://doi.org/10.1109/SIPROCESS.2016.7888236
63. So, E.W.Y., Munaro, M., Michieletto, S., Antonello, M., Menegatti, E.: Real-time 3D model reconstruction with a dual-laser triangulation system for assembly line completeness inspection. In: Lee, S., Cho, H., Yoon, K.-J., Lee, J. (eds.) Intelligent Autonomous Systems, pp. 707–716. Springer, Heidelberg (2013). https://doi.org/10.1007/978-3-642-33932-5_66
64. Vasquez, B.P.E.A., Gonzalez, R., Matia, F., De La Puente, P.: Sensor fusion for tour-guide robot localization. IEEE Access **6**, 78947–78964 (2018). https://doi.org/10.1109/ACCESS.2018.2885648

Experimental Analysis for the Enhancement of Industrial Robot Tool Path Planning

Mantas Makulavičius(✉), Tadas Rasimavičius, Sigitas Petkevičius, Justas Makutėnas, and Andrius Dzedzickis

Vilnius Gediminas Technical University, Saulėtekio al. 11, 10223 Vilnius, Lithuania
m.makulavicius@vilniustech.lt

Abstract. Industrial robots play a key role in modern manufacturing processes, contributing to increased efficiency, precision and productivity. However, the dynamic nature of robotic movements can cause vibrations that may impact performance and product quality. In this study, experimental analysis for the enhancement of the industrial robot tool path planning is implemented to get the initial knowledge of less robust robot configuration. To demonstrate the effect of different robotic configurations of the RoboDK path generation software, a Yaskawa Motoman HC10DTP industrial robot with an acceleration measurement setup was used to analyse the smoothness of the robot path. The information obtained is presented in the time and frequency domains and highlights key points such as mean square error and main frequencies in specific configurations.

Keywords: industrial robot · path smoothness · vibration analysis · RoboDK

1 Introduction

Industrial robots play a key role in modern manufacturing processes, contributing to increased efficiency, precision and productivity. However, the dynamic nature of robotic movements can cause vibrations that may impact performance and product quality [1]. Vibration analysis is essential for various reasons, not only including the need to enhance accuracy, ensure product quality, but at the same time to extend the lifespan of robotic setup. Vibrations in industrial robots can come from a variety of sources, such as sharp changes in motion, mechanical and structural resonances or external disturbances [2].

There are several options to prevent the influence of the vibrations to make robotic movements smoother optimizing its tool path. Firstly, main parameters identification is essential to better understand and make further actions for the robust robot path generation [3]. Further, structural robot pose design, like the enhancement of the rigidity of the robot [4, 5] implementation of additional damping means or tools [6]. The use of advance control algorithms with real life tracking [7–9] can be another example to achieve smoother performance of generated robot path tracking. Additional means like machine learning methods for the vibration prediction and suspension can be implemented as well, requiring additional computing recourses [10].

R. Szewczyk et al. (Eds.): AUTOMATION 2024, LNNS 1219, pp. 206–212, 2024.
https://doi.org/10.1007/978-3-031-78266-4_18

The robot tool path vibration and optimization analysis are often case in robotic machining like milling [4], grinding [5], polishing [8], chamfering [9] and other manufacturing operations like forming [11], welding [12], or 3D printing [13] to achieve better overall performance. To simulate various industrial robotic manufacturing fields with their tools and generate corresponding tool paths, RoboDK software can be utilized for its versatility, which is consisting of wide range of industrial robots in its library [14]. Other monitoring tools like laser tracking [15], motion capture [16] systems can be implemented as well.

In this study, an experimental analysis was performed to improve the tool path planning of an industrial robot to obtain preliminary knowledge about the less robust robot configuration for further research. The offline simulation tool RoboDK for trajectory generation and Yaskawa Motoman HC10DTP robot with acceleration measurement setup have been utilized to analyse robot path following smoothness. Collected data is represented in time series domain and frequency domain, and key points were underlined. This data can be beneficial for the further optimization of industrial robot path planning implementing together computer vision systems.

2 Methodology for Path Following Analysis

During the experiment 4 different shape paths were programmed, which are shown in the figure below (Fig. 1). Each path is named and has two variable parameters: relative height to the robot's base and velocity. Relative height is defined by two values 100 mm and 300 mm from the robot's base, velocity also is defined by two different values: 100 mm/s and 200 mm/s. These parameters are combined and results to 4 different sets of values for one trajectory, and in total 16 measurements are carried out.

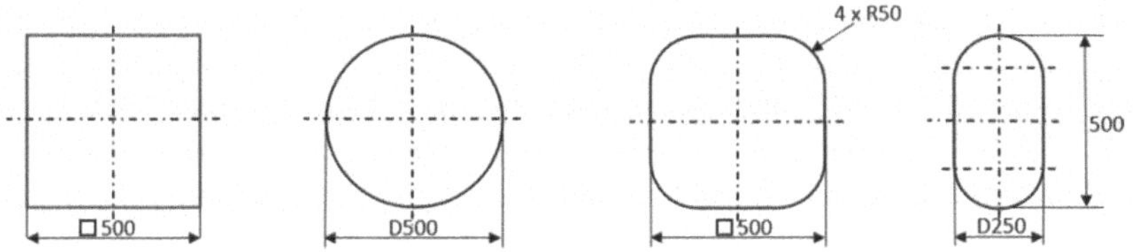

Fig. 1. Performed paths (in mm) during the analysis (names of the trajectories from left to right: square, circle, fillet, wedge)

The main object of the research is the industrial 6 degrees of freedom Yaskawa Motoman HC10DTP robot with the 10 kg of payload, 0,05 mm repeatability, and 1379 mm maximum reach and is controlled with the Yaskawa Motoman YRC1000 controller. The trajectories for the industrial robot were generated using RoboDK v5.6.7.23825 release for 64-bit systems (Fig. 2(a)). RoboDK is simulation and offline programming software for industrial robots, offering users a range of possibilities to enhance their workflow. The software's support for offline programming is another notable feature, allowing users to create and test robot programs without the need for physical robot hardware, thereby saving time and increasing flexibility. Compatibility is a key advantage, with RoboDK supporting various industrial robots and facilitating integration with CAD/CAM software. Therefore, the industrial robot is chosen from RoboDK online library, then all the path trajectories provided before are generated. To achieve the tool path – the characteristic position points must be selected, e.g., if a square must be drawn, then it means that it has 4 characteristic positions – one at each corner, this path as an example it given in the figure below (Fig. 2(a)).

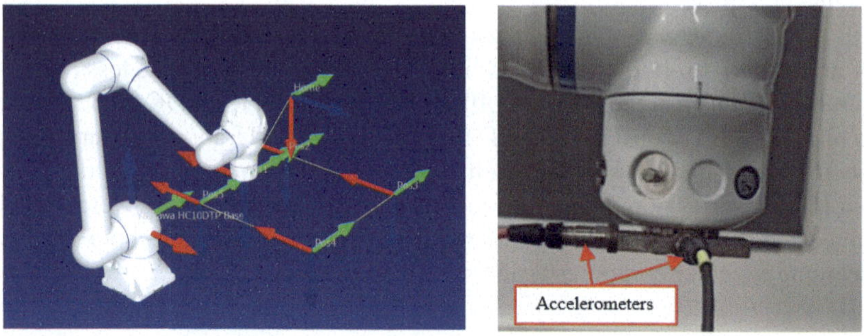

(a) (b)

Fig. 2. Analysis setup: (a) simulation implementation in RoboDK; (b) sensors placement for the analysis.

Piezoelectric accelerometers (Fig. 2(b)) are sensors designed to measure acceleration by exploiting the piezoelectric effect, where certain materials generate an electrical charge in response to mechanical deformation. Their advantages include high sensitivity, wide frequency response, and suitability for both dynamic and static acceleration measurements. For sensors data acquisition, the National Instruments USB-4432 Analogue-to-Digital Converter with sampling rate of 102,4 kS/s and 24 bits resolution with connected piezo-accelerometers and the National Instruments SignalExpress 2014 software was used. Accelerometers are magnetically attached to the robot. However, the shell of the robot is paramagnetic, therefore ferromagnetic steel was attached to the flange of the robot in X and Y axis directions.

3 Results and Discussion

Several trajectories following tasks were performed, and measurements were collected for further analysis. The example representation of collected data from the square shape trajectory following performed in the height of the 300 mm from the robot base coordinate system with the 200 mm/s following velocity is shown in Fig. 3. From here it can be visible the standardised acceleration over the trajectory in X axis (Fig. 3, top-left) and Y axis (Fig. 3, bottom-left) in time domain. From these and other trajectory graphs respectively, peak values of acceleration have been collected and root mean square (RMS) values of the standardised data have been calculated and saved. The graphs in frequency domain are also plotted using Discrete Fourier transform (DFT) algorithm for X axis (Fig. 3, top-right) and for Y axis (Fig. 3, bottom-right).

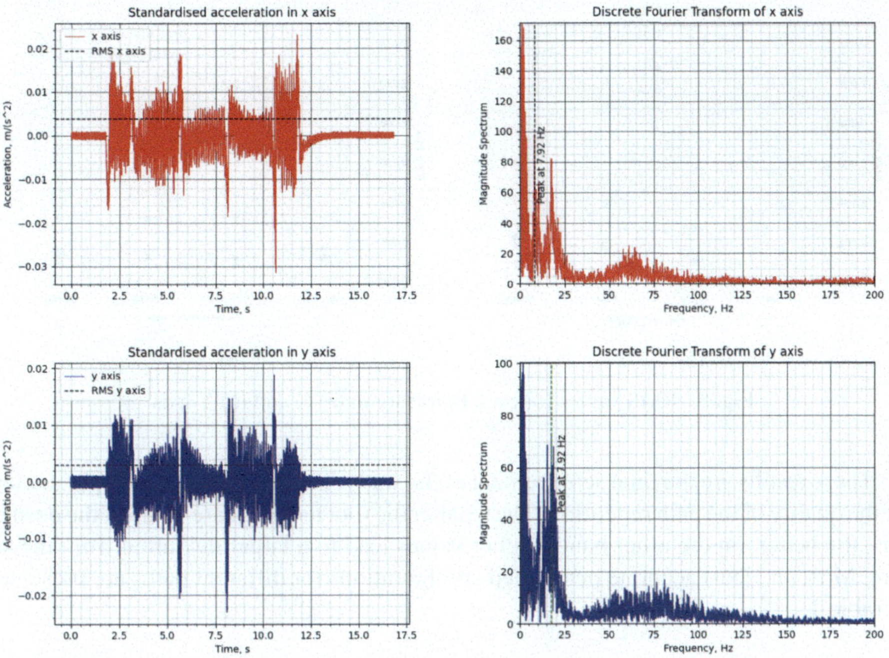

Fig. 3. Example representation of collected data performing square shape trajectory when velocity = 200 mm/s, height = 300 mm.

From DFT, the ranges from 0 Hz to 5 Hz and from 50 Hz are excluded for the analysis because the lower frequencies spectrum can be interpeted as they are coming from the mechanical part of the industrial robot (joint rotation) while higher frequencies spectrum is being affected by the electrical system of the industrial robot control [17].

All the retrieved values of acceleration peaks and RMS are presented in Fig. 4 and Fig. 5, respectively.

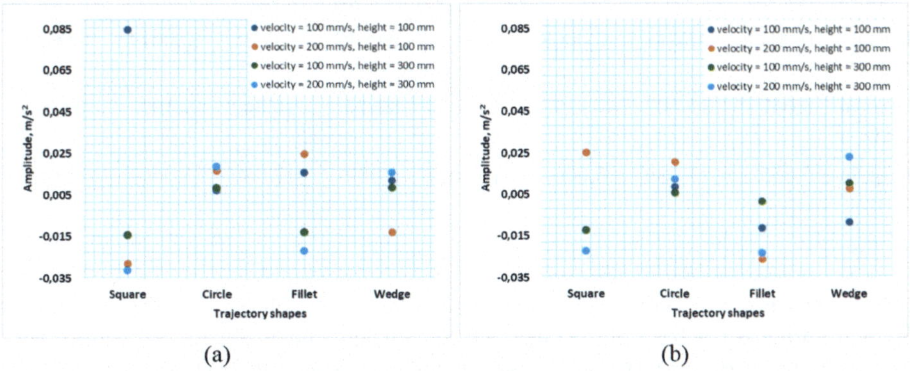

Fig. 4. Acceleration peak values for each trajectory in (a) X and (b) Y axis.

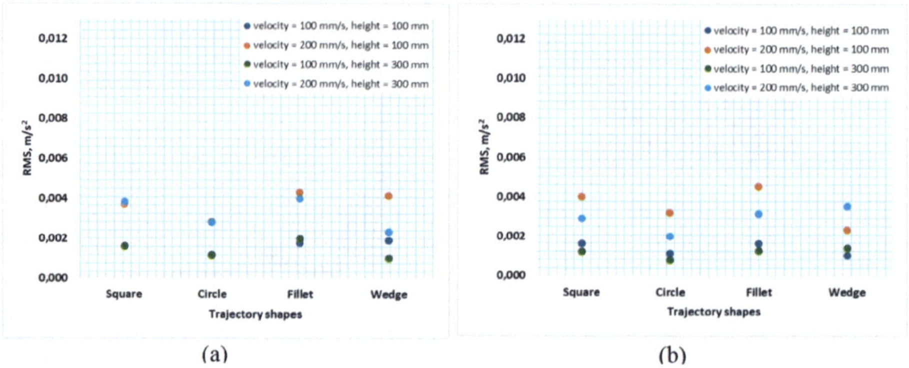

Fig. 5. RMS values for each trajectory in (a) X and (b) Y axis.

These results are the final achievement of the study. They indicate the best parameters for the quality of the path repetition. The smaller the root mean square value of the desired axis, the better, on the other hand, higher values lead to a larger deviation from the set path. Most and least accurate performed configurations for different paths are presented in Table 1.

Table 1. Most and least accurate performed configurations for different paths.

Shape	Square		Circle		Fillet		Wedge	
	Most accurate	Least accurate	Most accurate	Least accurate	Most accurate	Least accurate	Most accurate	Least accurate
Velocity, m/s^2	100	200	100	200	100	200	100	200
Height, mm	300	100	300	100	300	100	300	100
RMS in X, m/s^2	0,0016	0,0038	0,0012	0,0028	0,0016	0,0044	0,0010	0,0042
RMS in Y, m/s^2	0,0012	0,0041	0,0008	0,0032	0,0013	0,0046	0,0014	0,0024
Freq., in X, Hz	8,652	15,602	7,837	10,233	8,674	16,842	9,109	10,194
Freq., in Y, Hz	5,655	17,289	5,000	10,261	6,989	17,039	5,050	11,000

The most precise configuration is characterized by a velocity of 100 mm/s and a height from the robot base of 300 mm which indicates the stiffer robot posture. Here frequencies are between 5 Hz and 10 Hz. On the other hand, the least accurate scenario was observed at a velocity of 200 mm/s and a height from the robot base of 100 mm. Here frequencies are between 10 Hz and 20 Hz. From all the analysis, there was a common higher frequency peaks at around 50 Hz which can be interpreted as an electrical system noise and lower than 5 Hz peaks which may come from the mechanical part of the industrial robot.

4 Conclusions

During the analysis of industrial robotic tool path following, system research was designed and described. During the methodology chapter, description of the paths, the equipment used in experiments are provided, both software and hardware was characterized. During the experimental phase of the research a total of 16 different measurement cases were performed changing the moving velocity and path heigh for each shape path. Standardized acceleration in X axis and in Y axis, discrete Fourier transformation of standardized acceleration data in X axis and in Y axis were analysed. From the analysed time domain data, the most accurate path parameters can be recognized.

The most precise configuration is characterized by a velocity of 100 mm/s and a height from the robot base of 300 mm which indicates the stiffer robot posture. On the other hand, the least accurate scenario was observed at a velocity of 200 mm/s and a height from the robot base of 100 mm. After analysing the discrete Fourier transformations of all the measurements, it can be stated that there are mechanical main frequencies in the system between 5 Hz and 20 Hz depending on the robot tool height position and movement velocity. Higher frequency peaks at 50 Hz can be interpreted as an electrical system noise.

As industries continue to develop automation, the importance of vibration analysis of industrial robots is becoming more and more important as it is shaping the future of intelligent and resilient manufacturing environments. Further analysis will be carried out in the future using additional tools with camera for the visual detection of the followed path. The methodology presented in this paper and future work will be combined to train a machine learning algorithm to optimize robot tool path following.

Acknowledgement. This project has received funding from the European Social Fund (project No P-ST-23-83) under a grant agreement with the Research Council of Lithuania (LMTLT).

References

1. Han, H., Lin, Y., Gu, L., et al.: Vibration analysis based condition monitoring for industrial robots. Mech. Mach. Sci. **105**, 186–195 (2020). https://doi.org/10.1007/978-3-030-75793-9_19
2. Ariano, A., Perna, V., Senatore, A., et al.: Simulation and experimental validation of novel trajectory planning strategy to reduce vibrations and improve productivity of robotic manipulator. Electronics (Switzerland) **9**, 581 (2020). https://doi.org/10.3390/ELECTRONICS9040581
3. Jin, J., Gans, N.: Parameter identification for industrial robots with a fast and robust trajectory design approach. Robot. Comput. Integr. Manuf. **31**, 21–29 (2015). https://doi.org/10.1016/J.RCIM.2014.06.004
4. Xiong, G., Ding, Y., Zhu, L.M.: Stiffness-based pose optimization of an industrial robot for five-axis milling. Robot. Comput. Integr. Manuf. **55**, 19–28 (2019). https://doi.org/10.1016/J.RCIM.2018.07.001

5. Xie, H., Li, W.L., Zhu, D.H., et al.: A systematic model of machining error reduction in robotic grinding. IEEE/ASME Trans. Mechatron. **25**, 2961–2972 (2020). https://doi.org/10.1109/TMECH.2020.2999928
6. Nguyen, V., Johnson, J., Melkote, S.: Active vibration suppression in robotic milling using optimal control. Int. J. Mach. Tools Manuf **152**, 103541 (2020). https://doi.org/10.1016/J.IJMACHTOOLS.2020.103541
7. Zhu, D., Feng, X., Xu, X., et al.: Robotic grinding of complex components: a step towards efficient and intelligent machining – challenges, solutions, and applications. Robot. Comput. Integr. Manuf. **65**, 101908 (2020). https://doi.org/10.1016/J.RCIM.2019.101908
8. Wahballa, H., Duan, J., Dai, Z.: Constant force tracking using online stiffness and reverse damping force of variable impedance controller for robotic polishing. Int. J. Adv. Manuf. Technol. **121**, 5855–5872 (2022). https://doi.org/10.1007/S00170-022-09599-X/FIGURES/26
9. Hu, J., Pagilla, P.R.: A novel force and motion control strategy for robotic chamfering of gears. IFAC-PapersOnLine **53**, 8710–8715 (2020). https://doi.org/10.1016/J.IFACOL.2020.12.282
10. Wu, K., Li, J.: Prediction of the eigenfrequency of industrial robots based on the ANN model. In: ACM Cloud and Autonomic Computing Conference, pp. 1595–1598 (2021). https://doi.org/10.1109/CAC53003.2021.9728068
11. Shaker, W.K., Klimchick, A.: Simulation study for robot-based single point incremental forming. In: International Conference on Industrial Engineering, Applications and Manufacturing, pp. 730–736 (2023). https://doi.org/10.1109/ICIEAM57311.2023.10138999
12. Bi, G., Borys, S., Kaczmarek, W., et al.: Experimental Study of the Vibration of the Spot Welding Gun at a Robotic Station (2022). https://doi.org/10.3390/app122312209
13. Tiryaki, M.E., Zhang, X., Pham, Q.C.: Printing-while-moving: a new paradigm for large-scale robotic 3D printing. In: 2019 IEEE/RSJ International Conference on Intelligent Robots and Systems (IROS), pp. 2286–2291 (2018). https://doi.org/10.1109/IROS40897.2019.8967524
14. Garbev, A., Atanassov, A.: Comparative analysis of RoboDK and robot operating system for solving diagnostics tasks in off-line programming. In: 2020 International Conference Automatics and Informatics, ICAI 2020 – Proceedings (2020). https://doi.org/10.1109/ICAI50593.2020.9311332
15. Moeller, C., Schmidt, H.C., Koch, P., et al.: Real time pose control of an industrial robotic system for machining of large scale components in aerospace industry using laser tracker system "real time pose control of an industrial robotic system for machining of large scale components in aerospace industry using laser tracker system. SAE Int. J. Aerosp. **10** (2017). https://doi.org/10.4271/2017-01-2165
16. Yamaguchi, N., Aiuchi, K., Morita, N., et al.: Calibration of robotic woodworking machinery using a motion capture system. In: Calibration of Robotic Woodworking Machinery Using a Motion Capture System 2022 IEEE/SICE International Symposium on System Integration (SII) (2022). https://doi.org/10.1109/SII52469.2022.9708733
17. Denkena, B., Bergmann, B., Lepper, T.: Design and optimization of a machining robot. Procedia Manuf **14**, 89–96 (2017). https://doi.org/10.1016/J.PROMFG.2017.11.010

Adaptive Neural Network Control for Mobile Robot with Mecanum Wheels: Experimental Validation

Zenon Hendzel[1] and Maciej Kołodziej[2(✉)]

[1] Faculty of Mechanical Engineering and Aeronautics, Department of Applied, Mechanics and Robotics, Rzeszow University of Technology, Rzeszow, Poland
zenhen@prz.edu.pl

[2] Doctoral School of the Rzeszow University of Technology, Rzeszów, Poland
d515@stud.prz.edu.pl

Abstract. In this paper, neural network (NN) adaptive tracking control for uncertain nonlinear systems is proposed to attenuate the effects caused by unmodeled dynamics, disturbances and variable operating conditions of a wheeled mobile robot with mecanum wheels. A NN with basic sigmoid functions is used to compensate for the nonlinearity and variable operating conditions of the robot. There are two features in our approach. The first one is that the drive structure of a Mecanum wheeled mobile robot with four wheels has 3 degrees of freedom and four drive modules, it is an object of an over-actuated type, the missing moment is determined on the basis of the power balance of the drive systems. The second one is that experiments are performed to clarify the effectiveness and advantage of the proposed method.

Keywords: mobile robot · mecanum wheels · disturbance attenuation · experimental validation

1 Introduction

The advent of Mecanum wheel-based applications represents a groundbreaking leap in the field of mobile robotics engineering [11]. These wheels offer unparalleled movement capabilities, allowing omnidirectional travel and opening new avenues in robotics and automation [1]. Their ability to navigate through tight spaces and complex environments places them at the forefront of advanced transportation system development [6]. An effective control approach is crucial for accurately following a set trajectory, particularly over long distances or in complex environments. Mecanum Wheeled Mobile Robots (MWMRs) face challenges such as nonholonomic constraints, internal dynamics, sensor feedback, and disturbances from external loads, leading to a range of unpredictable uncertainties. Thus, implementing a more accurate and robust method for tracking trajectories can significantly enhance the operational efficiency of MWMRs. Traditional control system design approaches, such as optimal control and various algebraic methods, rely on knowing the mathematical model of the system and its parameters. However,

R. Szewczyk et al. (Eds.): AUTOMATION 2024, LNNS 1219, pp. 213–222, 2024.
https://doi.org/10.1007/978-3-031-78266-4_19

accurately defining a system's model and evaluating its parameters is often challenging, particularly when the system's characteristics change during operation. The challenges associated with motion control and regulation have been thoroughly examined and have garnered significant attention from numerous researchers in the area of mobile robotics. In paper [3], the synthesis of the NN control algorithm is presented, based on the integration of Lyapunov stability theory and the Hamilton-Jacobi inequality. In study [10], the authors introduce a synthesis of a neural control algorithm using Lyapunov's stability theory. Paper [12] presents an online actor-critic synchronous learning algorithm. Study [9] describes a robust nonsingular terminal sliding mode based on Lyapunov stability theory.

This article presents the verification of theoretical considerations regarding NN control of MWMR. To tackle the control task of a nonlinear object, Lyapunov's stability theory has been applied. For compensating nonlinearities, Random Vector Functional Link (RVFL) [7] networks with sigmoidal activation functions were used. The algorithm also incorporates a PD controller. The research focuses on a four-wheeled mobile robot, Husarion Panther [8], equipped with Mecanum wheels. Husarion Panther is an industrial-grade robot used often as a mobile platform [5]. Section 2 formulates the kinematics and dynamics equations for MWMR. Section 3 describes the control algorithm. Section 4 contains the results of the experimental verification of the control algorithm. Section 5 provides a summary.

2 Kinematics and Dynamics of MWMR

The kinematic structure and actual construction of the analyzed MWMR are depicted in Fig. 1.

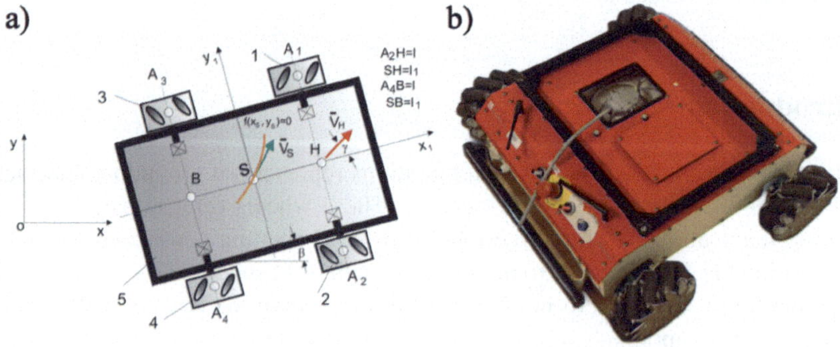

Fig. 1. a) robot diagram, b) real robot

This model's essential components include frame 5 and the driving units. The driving unit consists of wheels 1 to 4. These wheels spin around their own axes at angular velocities $\dot{\varphi}_{(.)}$, maintaining their position relative to the frame. Rollers are positioned around the circumference of the wheels, angled at $\alpha = \pi/4$ [rad] relative to the axis of the driving wheel. The center of mass of the frame is indicated by point S. Utilizing the system's geometry and classical mechanics methods, a kinematic description of a mobile wheeled robot can be derived as outlined in references [3].

$$\dot{x}_s\cos(\beta-\alpha)+\dot{y}_s\sin(\beta-\alpha)-\dot{\beta}(l\cos\alpha+l_1\sin\alpha)=\dot{\varphi}_1(R+r)\cos\alpha \tag{1}$$

$$\dot{x}_s\cos(\beta+\alpha)+\dot{y}_s\sin(\beta+\alpha)+\dot{\beta}(l\cos\alpha+l_1\sin\alpha)=\dot{\varphi}_2(R+r)\cos\alpha \tag{2}$$

$$\dot{x}_s\cos(\beta+\alpha)+\dot{y}_s\sin(\beta+\alpha)-\dot{\beta}(l\cos\alpha+l_1\sin\alpha)=\dot{\varphi}_3(R+r)\cos\alpha \tag{3}$$

$$\dot{x}_s\cos(\beta+\alpha)+\dot{y}_s\sin(\beta+\alpha)-\dot{\beta}(l\cos\alpha+l_1\sin\alpha)=\dot{\varphi}_4(R+r)\cos\alpha \tag{4}$$

where: β is the angle of temporary rotation of the frame, $\dot{\varphi}_{(.)}$ are angular velocity vectors of wheels.

The derived equations enable us to solve the inverse kinematics problem, which involves defining the motion trajectory for point S of the MWMR within the control system. To establish the dynamic motion equations for the mobile wheeled robot (Fig. 1), we utilized Maggi's equations, useful in the control system's synthesis. The general formulation of Maggi's equations is as follows [3, 4]:

$$M(q)\ddot{q}+C(q,\dot{q})\dot{q}+F(\dot{\varphi})+\tau_d(t)=u \tag{5}$$

where

$$M(q)=\begin{bmatrix} a_1 sin\beta+a_2 cos(\beta-\alpha) & a_2 sin(\beta-\alpha)-a_1 cos\beta & -a_3 \\ a_2 cos(\beta+\alpha)-a_8 sin(\beta-\alpha) & a_2 sin(\beta+\alpha)-a_8 cos(\beta-\alpha) & -a_4 \\ a_1 cos\beta+a_2 cos(\beta-\alpha) & a_2 sin(\beta-\alpha)+a_1 sin(\beta-\alpha) & a_3 \end{bmatrix}$$

$$C(q,\dot{q})=\begin{bmatrix} -a_2 sin(\beta-\alpha)\dot{\beta} & a_2 cos(\beta-\alpha)\dot{\beta} & 0 \\ -a_2 sin(\beta+\alpha)\dot{\beta} & a_2 cos(\beta+\alpha)\dot{\beta} & 0 \\ -a_2 sin(\beta-\alpha)\dot{\beta} & a_2 cos(\beta-\alpha)\dot{\beta} & 0 \end{bmatrix}$$

$$F(\dot{\varphi})=\begin{bmatrix} a_5\ sgn\dot{\varphi}_1 \\ a_6\ sgn\dot{\varphi}_3 \\ a_7\ sgn\dot{\varphi}_4 \end{bmatrix} \quad q=\begin{bmatrix} x_s \\ y_s \\ \beta \end{bmatrix} \quad u=\begin{bmatrix} M_1 \\ M_3 \\ M_4 \end{bmatrix} \quad \tau_d(t)=\begin{bmatrix} \tau_{d1} \\ \tau_{d3} \\ \tau_{d4} \end{bmatrix}$$

the vector $a=[a_1,\ldots,a_8]^T$ contains parameters resulting from the geometry, mass distribution, and the MWMR resistance to motion. $\tau_d(t)$ is disturbance signal satisfying the constraint $\|\tau_d(t)\|\leq b$, $b=const>0$.

From the dynamic model of the MWMR, only three driving torques can be determined. The missing torque M_2 is obtained through a power balance of the drive systems

$$M_2=a_d(|M_3|+|M_4|-|M_1|)sgn(\dot{\varphi}_{2d}) \tag{6}$$

where a_d is a positive design parameter that mitigates differences in the drive systems, $\dot{\varphi}_{2d}$ – is desired angular velocity of wheel 2.

3 Control Algorithm

In adaptive control methods for tracking motion in wheeled mobile robots, the control task is solved by introducing an approximation of the nonlinear functions present in the compensating element for the object's nonlinearities. It is assumed that the nonlinear function $f(x_r)$ is a combination of known base functions $Y(x_r)$ and the object's parameters $a_{(.)}$. To approximate the nonlinearities in the mathematical model, RVFL neural networks were used, featuring the following structure [7]

$$y = W^T S\left(V_x^T x\right) = W^T \phi_x(\overline{x}) \tag{7}$$

where the activation function $S(.)$ in RVFL networks is commonly implemented as a sigmoid function. V_x represents the fixed weights of the input layer, chosen randomly, and W are the weights of the output layer. The tracking error $e \in \mathbb{R}^3$, the sliding surface s, and the auxiliary vector v were defined as follows

$$\begin{aligned} e &= q_d - q \\ s &= \dot{e} + \Lambda e \\ v &= \dot{q} - \Lambda e \end{aligned} \tag{8}$$

The where Λ is a positively defined diagonal matrix. By inserting Eq. 8 into Eq. 5, the nonlinear function can be defined as:

$$f(x_r) = M(q)\dot{v} + C(q, \dot{q})v + F(\omega) \tag{9}$$

where $x_r = \left[v^T, \dot{v}^T, \dot{q}^T, q^T, \dot{\varphi}^T\right]^T$,

$$M\dot{s} = -\mathrm{u} - C(\dot{q})s + f(x_r) \tag{10}$$

for the approximation of the nonlinear function, a neural network described by Eq. 7 was utilized, which is expressed as follows

$$f(x_r) = W^T \phi_x(x_r) + \varepsilon \tag{11}$$

ε is approximation error satisfying the constraint $\|\varepsilon\| \leq \varepsilon_N$, $\varepsilon_N = const > 0$, the approximation of the function $f(x_r)$ is given by

$$\hat{f}(x_r) = \widehat{W}^T \phi_x(x_r) \tag{12}$$

where $\widehat{W}$ is the estimate of the weights of the ideal neural network. Utilizing Eq. (13), the control law is adopted in the form

$$u = \widehat{W}^T \phi_x(x_r) + K_D s \tag{13}$$

The second component of Eq. (13) is a PD-type controller, $K_D = K_D^T > 0$, substituting Eqs. (11) and (13) into Eq. (10), the following was obtained

$$M\dot{s} + C(\dot{q})s + K_D s = \tilde{f}(x_r) + \tau_d(t) \tag{14}$$

where $\tilde{f}(x_r)$ is the approximation error of the function $f(x_r)$, amounting to

$$\tilde{f}(x_r) = f(x_r) - \hat{f}(x_r) = W^T\phi_x(x_r) - \widehat{W}^T\phi_x(x_r) + \varepsilon = \tilde{W}^T\phi_x(x_r) + \varepsilon \tag{15}$$

where $\tilde{W} = W - \widehat{W}$ is NN weights estimation error, inserting Eq. (14) into Eq. (15), Eq. (16) was derived

$$M\dot{s} + C(\dot{q})s + K_D s = \tilde{W}^T\phi_x(x_r) + \tau_d(t) + \varepsilon \tag{16}$$

To develop the weight update algorithm for $\widehat{W}$, Lyapunov's stability theory was employed. Defined a function as a proposed Lyapunov function:

$$V = \frac{1}{2}s^T M s + \frac{1}{2}tr\left(\tilde{W}^T\eta^{-1}\tilde{W}\right) \tag{17}$$

where $\eta = \eta^T > 0$ is the design matrix, and the derivative of the function V along the system trajectory (16) with respect to time is given by

$$\dot{V} = s^T M\dot{s} + \frac{1}{2}s^T\dot{M}s + tr\left(\tilde{W}^T\eta^{-1}\dot{\tilde{W}}\right) \tag{18}$$

substituting from Eq. (16), and considering that the matrix $\dot{M} - 2C(\dot{q})$ is a zero matrix for $\beta(t) = 0$, the result obtained is

$$\dot{V} = -s^T K_D s + tr\left\{\tilde{W}\left[\eta^{-1}\dot{\tilde{W}} + \phi_x(x_r)s^T\right]\right\} + s^T[\tau_d(t) + \varepsilon] \tag{19}$$

adopting the network weight learning algorithm in the form of

$$\dot{\tilde{W}} = \eta\phi_x(x_r)s^T \tag{20}$$

Yielded

$$\dot{V} = -s^T K_D s + s^T[\tau_d(t) + \varepsilon] \leq -K_{D_{min}}\|s\|^2 + (\varepsilon_N + b)\|s\| \tag{21}$$

where $K_{D_{min}}$ is the smallest eigenvalue of the matrix K_D, and $\varepsilon_N + b$ is a constant value. Thus, $\dot{V} < 0$ if the following condition is met:

$$\psi = \left\{s\colon \|s\| > \frac{\varepsilon_N + b}{K_{D_{min}}} \equiv b_s\right\} \tag{22}$$

From relation (22), it follows that the sliding surface s is uniformly ultimately bounded within the set ψ, with a practical boundary denoted as b_s. This approach to the synthesis of neural control allows the control system with the PD regulator to function correctly until the NN begins to adapt the weight values.

4 Verification Tests

The verification experiment was conducted using a rapid prototyping station, which includes a dSpace 1103 control and measurement card [2], a PC with MATLAB/Simulink R2012a software, and a Husarion Panther robot with Mecanum wheels. The card generates a PWM signal that is executed by the robot's drive system. Incremental encoders are used to measure the wheels' angular velocity, with the rotation angle being determined through numerical integration. Husarion Panther is equipped with four BLDC 80 PMB800K.80RBL motors, each having a power of 473 [W] (900 [W] peak power), coupled with a 1:30 planetary gearbox and an incremental encoder. The robot is powered by a 36 [V] battery with a capacity of 740 [Wh]. The manufacturer has modified the robot to allow direct access to its actuation systems, bypassing the internal control system. The masses and basic dimensions of the robot are shown in Table 1.

Table 1. Husarion Panther parameters

l[m]	l_1[m]	m_r[kg]	m_k[kg]	R[m]	r[m]
0.34	0.22	45	2.4	0.09	0.01

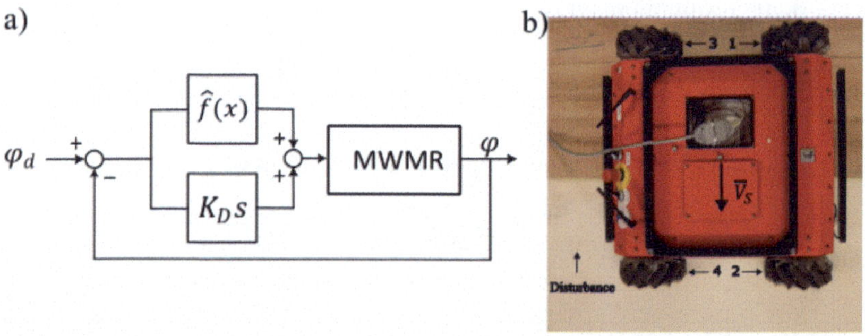

Fig. 2. a) A schematic of the control algorithm, b) MWRM encountering a disturbance

Based on the results of numerous experimental studies, the motion of a robot with a given configuration and the direction of the velocity vector of point S causes the greatest resistance to motion. Therefore, the results of experimental studies of this case have been included in the work. The experiment hypothesizes that the robot will move from the starting point $x_{S_0} = 1$ [m], $y_{S_0} = 1$ [m] at a speed of $v_S = 0.3$ [m/s] along a linear trajectory, under the following conditions: the initial angle of the robot's frame $\beta = 0$ [rad], the angular velocity of the robot's frame $\dot{\beta}(t) = 0$ [rad/s], and the velocity of point S vector V_S angle is -π/2. During the experiment, at approximately $t = 11$ [s] for wheels 2 and 4, and $t = 13.5$ [s] for wheels 1 and 3, there is a change in the frictional resistance due to the robot moving onto a disturbance in the form of a polyurethane foam, designated T25 and measuring 5 [mm] in thickness (see Fig. 2b). Value of point

S velocity is approximated to achieve smooth velocity profiles, formulated as follows:

$$v_S = v_S^*\left(\frac{1}{1+e^{-c(t-b)}} - \frac{1}{1+e^{-c(t-b_1)}}\right) \tag{23}$$

where: $b = 5$ [s] – starting phase time, $b_1 = 17$ [s] – braking phase time, $c = 5$ [1/s] – design parameter.

The control signal of the PD regulator is determined based on the measurement of tracking errors. PD controller parameters are $\Lambda = diag[2, 2, 2]$. To compensate for the nonlinearity of the robot and its variable operating conditions, a neural network comprising 6 bipolar sigmoidal neurons was used, learning rate of NN $\eta = diag[20, 20, 20]$. The weights of the input layer V_x were randomly selected from the range -0.2 to 0.2, while the output layer weights W were initialized with zero conditions. For approximating the nonlinear function $f(x_r) \in \mathbb{R}^3$, a separate neural networks are utilized, however, based on the robot's dynamic model, it is possible to establish only three NN's. The missing compensation signal for wheel 2 is determined from relation (6), where $a_d = 0.99$. The obtained patterns of output layer weights of wheel 1 NN are shown in Fig. 3d.

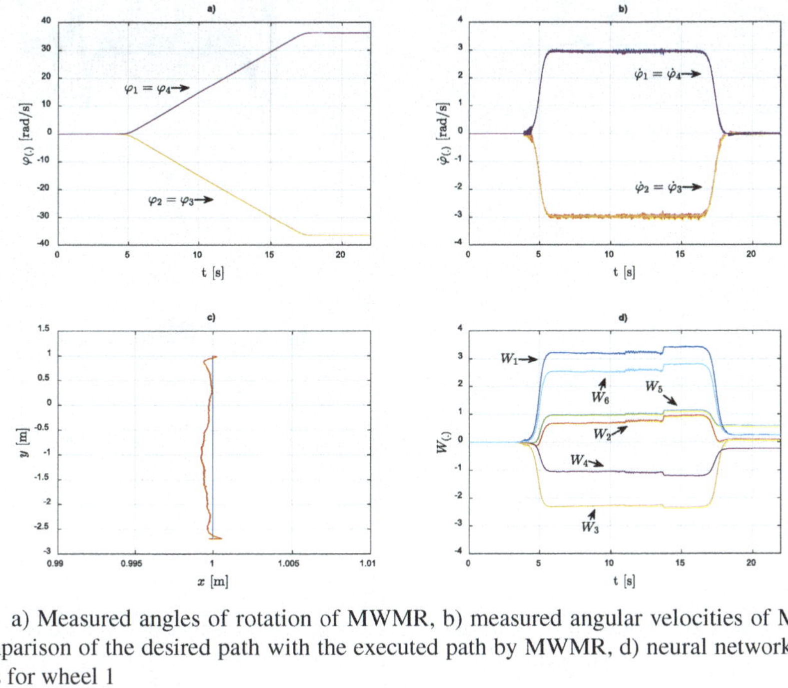

Fig. 3. a) Measured angles of rotation of MWMR, b) measured angular velocities of MWMR, c) comparison of the desired path with the executed path by MWMR, d) neural network weight signals for wheel 1

The angular parameters obtained are depicted in Fig. 3. Subfigure 3a illustrates the achieved rotation angles of the wheels of the MWMR, Fig. 2b displays the angular velocities of the MWMR wheels. The initial configuration of the robot frame and the assumption that the angular velocity of the robot's frame $\dot{\beta}(t)$ remains unchanged result in the robot moving in a negative direction along the Y-axis of the global coordinate system while executing the given path with a velocity vector angle of $-\pi/2$. This specified motion path leads to the angular parameters of wheels 1 and 4 having positive values, while those of wheels 2 and 3 are negative. The path of point S executed by the robot, marked in orange, is presented relative to the given trajectory in Fig. 3c.

For evaluating the quality of the tracking control system of the MWMR, the following set of indicators was introduced:

- maximum wheel rotation angle error $e_{(.)} = \left(\varphi_{(.)d} - \varphi_{(.)}\right)$[rad], $e_{max(.)} = \left|e_{(.)}\right|$ [rad],
- RMSE (Root Mean Square Error) for the wheel rotation angle errors $\varepsilon_{(.)} = \sqrt{\left(\frac{1}{n}\right)\sum_{k=1}^{n} e_{(.)k}^{2}}$[rad], where k is the number of subsequent discrete measurements,
- RMSE for the angular velocity errors of the wheel $\dot{\varepsilon}_{(.)} = \sqrt{\left(\frac{1}{n}\right)\sum_{k=1}^{n} \dot{e}_{(.)k}^{2}}$ [rad/s],
- maximum distance $d_k = \sqrt{(x_S - x_{dS})^2 + (y_S - y_{dS})^2}$ [m], $d_{max} = max(d_k)$ [m].

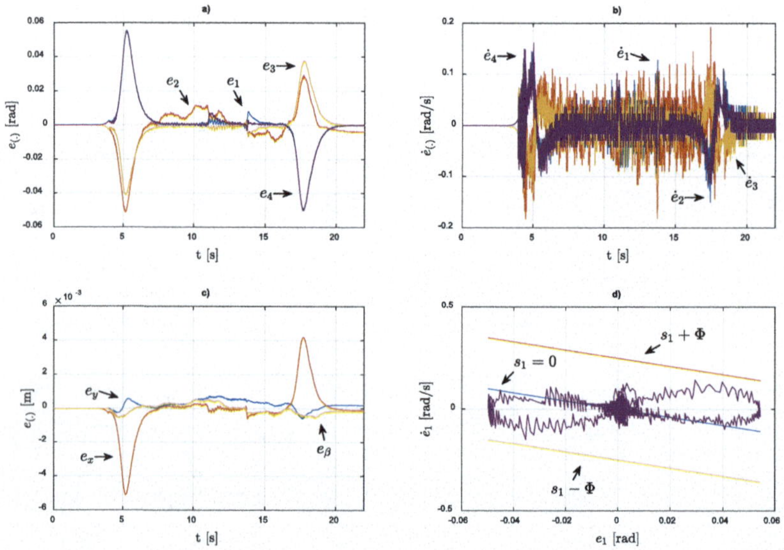

Fig. 4. Tracking errors, a) MWMR angle of rotation errors, b) MWMR angular velocity errors, c) MWMR linear errors, d) phase trajectory of the error of wheel 1 in vicinity of the sliding manifold $s_1 = 0$

Figure 4 display the errors in rotation angle(Fig. 4a) and angular velocity(Fig. 4b) for the wheels during verification. The most significant error in rotation angle tracking of value ± 0.06 [rad] is observed at the start of the movement, particularly in the acceleration phase and just after it. This is attributed to the abrupt rise of the control signal and the inertia of the MWMR, and with the neural network weights initialized to zero. At approximately $t = 11$ [s] for wheels 2 and 4, and at $t = 13.5$ [s] for wheels 1 and 3, a minimal increase in the rotational angle error is observed, along with a noticeable peak in the angular velocity error trajectory. Linear errors, as shown in Fig. 4c, exhibit the highest values in the e_y signal, attributed to the robot's movement along Y axis.

Errors in the X-axis(e_x), as well as the rotational angle error of the robot's frame, e_β, are noticeably lower. The linear error trajectories display peaks that coincide with those in the angular trajectories (Fig. 4a), particularly at moments when disturbances occur on individual wheels. Vicinity $\pm\Phi = 0.25$ of the sliding manifold $s_1 = 0$ was introduced to effectively illustrate the dynamics of the error trajectory on the phase plane for wheel 1, as shown in Fig. 4d. The values of the adopted quality indicators are presented in Table 2.

Table 2. Values of quality indicators

	e_{imax}[rad]	ε_i[rad]	$\dot{\varepsilon}_i$[rad/s]	d_{max}[m]
1	0.054	0.013	0.030	0.005
2	0.051	0.011	0.039	
3	0.051	0.010	0.028	
4	0.054	0.013	0.031	

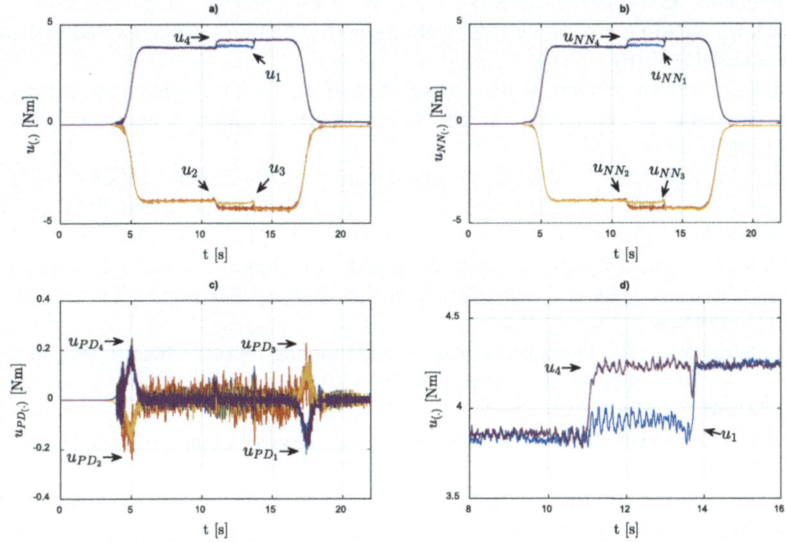

Fig. 5. Control signals, a) total control, b) NN control, c) PD control, d) enlarged total control

Control signals are shown in Fig. 5. During the startup phase of MWMR, the PD control signal is predominant (Fig. 5c). However, around $t = 7$ [s], it gets overshadowed by the signal generated by neural networks (Fig. 5b), due to the NN weights initially being set to zero. The adaptation process of the neural network weights leads to an increase in the value of control signals produced by the network, which then start to dominate the total control signal value. Similarly, during the occurrence of a parametric disturbance, there is a brief rise in the PD signal value, which diminishes as the network weights adjust to the new conditions. The outcomes of the solution for the robust control of MWMR motion using neural networks, which evolves over time. The total control signal for wheels 1 and 4 is enlarged in Fig. 5d at the moment of occurrence of a parametric disturbance.

5 Summary

This paper presents the verification of a novel approach to tracking control of MWMR with unspecified descriptions, considering variable operating conditions. Neural networks with sigmoidal activation functions were used for approximating nonlinearities. Validation results have shown that the tracking performance of the proposed solution is quite satisfactory, even with disturbances and unknown non-linear dynamics in an overactuated system.

References

1. Adascalitei, F., Doroftei, I.: Practical applications for mobile robots based on Mecanum wheels - a systematic survey. Romanian Rev. Precis. Mech. Opt. Mechatron. 21–29 (2011)
2. dSpace, DS1103 PPC Controller Board. Hardware Installation and Configuration. Padeborn: dSpace GmbH (2010)
3. Hendzel, Z.: Robust neural networks control of omni-mecanum wheeled robot with hamilton-jacobi inequality. J. Theor. Appl. Mech. **56**(4), 1193–1204 (2018). https://doi.org/10.15632/jtam-pl.56.4.1193
4. Hendzel, Z., Kołodziej, M.: Parametric identification of the mathematical model of a mobile robot with mecanum wheels. In: Szewczyk, R., et al. (eds.) Automation 2023. LNNS, vol. 630, pp. 107–117 (2023). https://doi.org/10.1007/978-3-031-25844-2_10
5. Junge, K., Pires, C., Hughes, J.: Lab2Field transfer of a robotic raspberry harvester enabled by a soft sensorized physical twin. Commun. Eng. **2**, 40 (2023). https://doi.org/10.1038/s44172-023-00089-w
6. Kanjanawanishkul, K.: Omnidirectional wheeled mobile robots: wheel types and practical applications. Int. J. Adv. Mechatron. Syst. **6**, 289 (2015). https://doi.org/10.1504/IJAMECHS.2015.074788
7. Lewis, F.L., Jagannathan, S., Yesilderek, A.: Neural Network Control of Robot Manipulators and Nonlinear System. Taylor & Francis, London (1999)
8. Panther - outdoor AMR. (n.d.). Husarion | Autonomous Mobile Robots Made Simple. https://husarion.com/manuals/panther
9. Sun, Z., Xie, H., Zheng, J., Man, Z., He, D.: Path-following control of mecanum-wheels omnidirectional mobile robots using nonsingular terminal sliding mode. Mech. Syst. Signal Process. **147**, 107128 (2020). https://doi.org/10.1016/j.ymssp.2020.107128
10. Szeremeta, M., Szuster, M.: Neural tracking control of a four-wheeled mobile robot with mecanum wheels. Appl. Sci. **12**, 5322 (2022). https://doi.org/10.3390/app12115322
11. Taheri, H., Zhao, C.: Omnidirectional mobile robots, mechanisms and navigation approaches. Mech. Mach. Theory **153**, 103958 (2020). https://doi.org/10.1016/j.mechmachtheory.2020.103958
12. Zhang, D., Wang, G., Wu, Z.: Reinforcement learning-based tracking control for a three mecanum wheeled mobile robot. IEEE Trans. Neural Netw. Learn. Syst. 1–8 (2022). https://doi.org/10.1109/TNNLS.2022.3185055

Adaptive Controller Using Genetic Algorithm for Autonomous Wheeled Mobile Robot

Paweł Penar(✉) and Zenon Hendzel

Faculty of Mechanical Engineering and Aeronautics, Department of Applied, Mechanics and Robotics, Rzeszow University of Technology, Rzeszow, Poland
{ppenar,zenhen}@prz.edu.pl

Abstract. This paper presents the application of a genetic algorithm (GA) to the problem of tracking and stabilizing the motion of a wheeled mobile robot (WMR) operating under variable conditions. The proposed approach is to apply GA to adaptive control. Its novelty is the transformation of the parameter adaptation law into an iterative form, which allows the application of GA in real time. In addition, by introducing a supervisory control into the control signal, the asymptotic stability of the closed-loop control system can be obtained. This is confirmed by the simulation carried out, which indicates that the control system achieves a high degree of motion accuracy under varying operating conditions.

Keywords: Adaptive control · genetic algorithms · supervisory control

1 Introduction

Control methods for nonlinear objects are not trivial, as evidenced by the constant interest of researchers [1]. One of the techniques (based on the mathematical model of the controlled object) used in this area is the compensation of nonlinearity implemented by the feedback linearization [2, 3]. Its disadvantage is sensitivity to parametric disturbances, which can be understood as variable operating conditions (in the case of mobile robots, variable operating conditions can be interpreted as a change in the mass carried or a change in the conditions of motion). Their occurrence affects the quality of control. This has to do with the deterioration of the representation of the controlled object by the model, which consequently worsens the compensation of nonlinearity [3]. Hence the need to introduce a mechanism for estimating model parameters into the control system. Therefore, control theory develops methods of adaptive control in which the control algorithm is designed so that it is possible to adapt the parameters under changing operating conditions [4, 5]. Such a system includes a mechanism that, based on the response of the object, updates the values of the vector of parameter evaluations on an ongoing basis so as to ensure the desired behavior of the closed system.

The next section of this work sets the problem to be solved, which is the application of a genetic algorithm in the parameter adaptation procedure. Its solution is described in the next three sections including: description of the controlled object (Sect. 3), presentation of genetic algorithms (Sect. 4) and synthesis of the optimal adaptive control algorithm

R. Szewczyk et al. (Eds.): AUTOMATION 2024, LNNS 1219, pp. 223–232, 2024.
https://doi.org/10.1007/978-3-031-78266-4_20

(Sect. 5). The paper concludes with a presentation of simulation results (Sect. 6) and a summary (Sect. 7).

2 Problem Statement

The paper applies adaptive control to one of the tasks of mobile robotics, the task of tracking. It is implemented for varying operating conditions, which, with respect to the analyzed example, are understood as a change in the resistance to motion. Due to the task at hand, the goal of the control algorithm will be to reduce the error of realization of the desired trajectory, which is widely discussed in the paper [3].

The parameter adaptation law was chosen so that the closed-loop control system remains stable. Therefore, as shown in works [3, 4], it is in the form of a dynamic system, which was derived by analyzing stability in the Lapunow sense. In this work, the parameter adaptation law implemented in continuous time was written in iterative form, which allowed the use of genetic algorithms. The use of the iterative law of parameter adaptation means that their adaptation is understood as the problem of finding the maximum of the objective function that is a function of the estimated parameters. When implementing this type of adaptation, it is necessary to ensure the stability of the control system. Hence the need to show that the derivative of the Lapunov function for a closed system remains negative.

3 Wheeled Mobile Robot

The kinematics and dynamics of robots, especially mobile robots, have been widely described in the literature [6–8]. Therefore, the description presented below will be limited only to those issues that are relevant in the context of the described control algorithms.

Figure 1 shows a schematic of a 2-wheeled mobile robot Pioneer 2-DX, consisting of two driving wheels designated by 1 and 2, a castor wheel 3 and a frame 4. Wheels 1 and 2 rotate around their own axes, which do not change position relative to the frame 4. The driving wheels are driven by independent drives consisting of electric motors and gears. The radius of wheels 1 and 2 is r, and their angles of self-rotation are α_1 and α_2, respectively.

The description of the kinematics of any point of a 2-wheeled mobile robot requires the equations of kinematics, which are widely described in works [6, 8]. The motion of the selected characteristic point of the WMR along a given motion path can be understood as a combination of two basic types of motion: motion of the selected point of the robot along a straight line track with a given velocity v_A, and motion along a loop-shaped motion path of radius R with a given velocity v_A. A path of the WMR consists of a sequence of motions: along a straight line, along a loop-shaped track and then along a straight line. The angular velocities of its driving wheels can be defined by the equation:

$$\dot{\alpha}_i = \frac{v_A}{r} + (-1)^i \dot{\beta} \frac{l_1}{r} \left(\frac{1}{1 + e^{c_\beta (t - b_{1\beta})}} - \frac{1}{1 + e^{c_\beta (t - b_{2\beta})}} \right), i = 1,2 \tag{1}$$

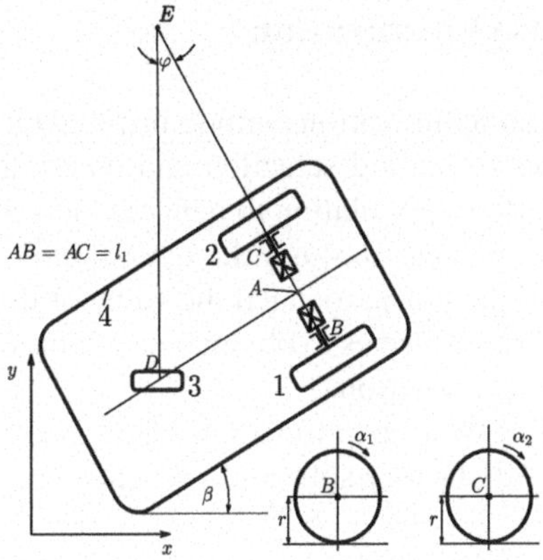

Fig. 1. Scheme of the Pioneer 2-DX 2-wheeled mobile robot

wherein $\dot{\alpha}_1, \dot{\alpha}_2$ are the angular velocities of the driving wheels of the WMR, $b_{1\beta}, b_{2\beta}$ define the start and end times of movement along a loop-shaped track with a transition term whose length depends on the value of c_β. Using a velocity profile in the form [7]

$$v_A = v_A^* \left(\frac{1}{1 + e^{c(t-b_1)}} - \frac{1}{1 + e^{c(t-b_2)}} \right) \tag{2}$$

it is possible to approximate the velocity v_A in the three stages of motion, i.e. starting, performing established motion with velocity v_A^* and braking. The coefficient c affects the rate of change of velocity in the starting and braking phases, v_A^* is the velocity in the established motion phase, while the values b_1, b_2 indicate the time of the starting and braking phases.

The WMR dynamic equation can be written as in [7, 8]

$$\boldsymbol{M}(\boldsymbol{a})\ddot{\alpha} + \boldsymbol{C}(\dot{\alpha}, \boldsymbol{a})\dot{\alpha} + \boldsymbol{F}(\dot{\alpha}, \boldsymbol{a}) = \boldsymbol{u} \tag{3}$$

where $\dot{\boldsymbol{\alpha}} = \left[\dot{\alpha}_1 \; \dot{\alpha}_2 \right]$ the vector of angles of rotation of the wheels driving the WMR and $u = [M_1, M_2]^T$ is the vector of torques driving these wheels. The matrices **M,C** can be written as

$$\boldsymbol{M}(\boldsymbol{a}) = \begin{bmatrix} a_1 + a_2 + a_3 & a_1 - a_2 \\ a_1 - a_2 & a_1 + a_2 + a_3 \end{bmatrix}, \boldsymbol{C}(\dot{\alpha}, \boldsymbol{a}) = \begin{bmatrix} 0 & 2a_4(\dot{\alpha}_2 - \dot{\alpha}_1) \\ -2a_4(\dot{\alpha}_2 - \dot{\alpha}_1) & 0 \end{bmatrix}$$

whereby the grouped parameters $a_1 \ldots a_4$ are elements of the parameter vector a having physical interpretations given in [7], for example. The next parameters i.e. a_5, a_6 are related to the resistance to motion, where the resistance vector is defined $F(\dot{\alpha}, a) = \left[a_5 sign(\dot{\alpha}_1), a_5 sign(\dot{\alpha}_2) \right]^T$. Due to the discontinuous nature of the *sign* function, it was decided to approximate it by hyperbolic tangent functions, i.e. $sign(\alpha_i) \approx tanh(c_F \dot{\alpha})$ where c_F is the project coefficient.

4 Genetic Algorithm Optimization

Genetic algorithm (GA) is an optimization method inspired by the process of biological evolution, i.e. the principles of natural selection and inheritance [9]. It starts with the generation of an initial (random) population to generate new generations of individuals in subsequent stages of the process through the operation of selection, crossover and mutation [10]. With each of them is associated the value of the fitness function [11]. In next iterations, the GA algorithm strives to obtain better solutions understood as striving for the maximum of the fitness function.

Evaluation of the value of the fitness function, selection, crossover and mutation are repeated iteratively until a satisfactory solution is reached or a defined stop condition is met.

5 Synthesis Optimal Adaptive Control

In the adaptive control method, the control algorithm is designed so that it can independently modify its properties with changing operating conditions. In the case of adaptive control of the tracking motion of a mobile 2-wheeled robot, it is necessary to determine the control law $\mathbf{u}$ and the adaptation law of the parameter vector $\hat{\boldsymbol{a}}$, such that for the initial conditions $[\boldsymbol{\alpha}(0), \dot{\boldsymbol{\alpha}}(0)] \in \boldsymbol{\Omega} = [(\boldsymbol{\alpha}, \dot{\boldsymbol{\alpha}}) : \boldsymbol{\alpha}_d(0) = \boldsymbol{\alpha}(0), \dot{\boldsymbol{\alpha}}_d(0) = \dot{\boldsymbol{\alpha}}(0)]$ converges to the manifold $\boldsymbol{\Omega}_d$ definded as: $\boldsymbol{\Omega}_d = [(\boldsymbol{\alpha}, \dot{\boldsymbol{\alpha}}) : \boldsymbol{\alpha} = \boldsymbol{\alpha}_d, \dot{\boldsymbol{\alpha}} = \dot{\boldsymbol{\alpha}}_d]$.

This paper presents a new approach in adaptive control, in which a genetic algorithm is used to estimate the parameters of a mathematical model and thus obtain an estimate of a nonlinear robot function.

Defining the tracking error $\boldsymbol{e} = \boldsymbol{\alpha}_d - \boldsymbol{\alpha}$ we obtain:

$$\dot{\mathbf{e}} = \dot{\boldsymbol{\alpha}}_d - \dot{\boldsymbol{\alpha}}, \ddot{\boldsymbol{e}} = \ddot{\boldsymbol{\alpha}}_d - \ddot{\boldsymbol{\alpha}} \tag{4}$$

Let's introduce additional signals $\boldsymbol{v}$, $\dot{\boldsymbol{v}}$ definded as

$$\boldsymbol{v} = \dot{\boldsymbol{\alpha}} - \boldsymbol{\Lambda}\boldsymbol{e}, \dot{\boldsymbol{v}} = \ddot{\boldsymbol{\alpha}} - \boldsymbol{\Lambda}\dot{\boldsymbol{e}} \tag{5}$$

and generalized error

$$\boldsymbol{s} = \dot{\boldsymbol{e}} + \boldsymbol{\Lambda}\boldsymbol{e} \tag{6}$$

where $\boldsymbol{\Lambda}$ is a diagonal design matrix with positive elements of λ_i. By differentiating (6) and considering (3), we obtain a description of the closed-loop control system as a function of the generalized error $\mathbf{s}$

$$\boldsymbol{M}(\boldsymbol{a})\dot{\boldsymbol{s}} = -\boldsymbol{C}(\dot{\boldsymbol{\alpha}}, \boldsymbol{a})\boldsymbol{s} + \boldsymbol{f}(\boldsymbol{x}) - \boldsymbol{u} \tag{7}$$

where the nonlinear function $\mathbf{f}(\mathbf{x})$ is given as

$$\boldsymbol{f}(\boldsymbol{x}) = \boldsymbol{M}(\boldsymbol{a})(\ddot{\boldsymbol{\alpha}} - \boldsymbol{\Lambda}\dot{\boldsymbol{e}}) + \boldsymbol{C}(\dot{\boldsymbol{\alpha}}, \boldsymbol{a})(\dot{\boldsymbol{\alpha}} - \boldsymbol{\Lambda}\boldsymbol{e}) + \boldsymbol{F}(\dot{\boldsymbol{\alpha}}, \boldsymbol{a}) \tag{8}$$

The vector $\boldsymbol{x} = \left[\boldsymbol{e}^T\ \dot{\boldsymbol{e}}^T\ \dot{\boldsymbol{\alpha}}^T\ \ddot{\boldsymbol{\alpha}}^T\right]^T$ includes signals to determine $\mathbf{f}(\mathbf{x})$. In the synthesis of adaptive control, it is assumed that the function $\mathbf{f}(\mathbf{x})$ is linear due to the parameters, which can be written as

$$f(x) = \boldsymbol{M}(\boldsymbol{a})\dot{\boldsymbol{v}} + \boldsymbol{C}(\dot{\boldsymbol{\alpha}}, \boldsymbol{a})\boldsymbol{v} + \boldsymbol{F}(\dot{\boldsymbol{\alpha}}, \boldsymbol{a}) = \mathbf{Y}_{\mathbf{V}}(\dot{\boldsymbol{\alpha}}, \mathbf{v}, \dot{\mathbf{v}})\mathbf{a} \tag{9}$$

By analogy, the estimate of the function $\mathbf{f}(\mathbf{x})$ is written as

$$\widehat{f}(x) = \widehat{\boldsymbol{M}}(\boldsymbol{a})\dot{\boldsymbol{v}} + \widehat{\boldsymbol{C}}(\dot{\boldsymbol{\alpha}}, \boldsymbol{a})\boldsymbol{v} + \widehat{\boldsymbol{F}}(\dot{\boldsymbol{\alpha}}, \boldsymbol{a}) = \boldsymbol{Y}_V(\dot{\boldsymbol{\alpha}}, \boldsymbol{v}, \dot{\boldsymbol{v}})\widehat{\boldsymbol{a}} \tag{10}$$

where $\hat{\boldsymbol{a}}$ is the estimates of the vector of unknown parameters. We will use a standard genetic algorithm to determine the control that compensates for the nonlinearities and changing operating conditions of the mobile robot. Typically, the parameter adaptation law is applied in continuous time. As is known [12], the genetic algorithm uses a discrete approximation of the problem under analysis. Therefore, let's write the evaluation of the unknown parameters of the nonlinear function (9) in iterative form

$$\hat{\boldsymbol{a}}_{GA}^{k} = \hat{\boldsymbol{a}}_{GA}^{k-1} + \left[\boldsymbol{P}\boldsymbol{Y}_{\boldsymbol{v}}(\dot{\boldsymbol{q}}, \boldsymbol{v}, \dot{\boldsymbol{v}})\boldsymbol{s}^T\right]^{k-1} \tag{11}$$

where $t = kh$, h is the discret step, $\mathbf{P}$ is the gain matrix. Let's define the parameter estimation error as

$$\boldsymbol{E} = \hat{\boldsymbol{a}}_{GA}^{k-1} + \boldsymbol{P}[\boldsymbol{Y}_{\boldsymbol{v}}(\dot{\boldsymbol{\alpha}}, \boldsymbol{v}, \dot{\boldsymbol{v}})]^{k-1} - \hat{\boldsymbol{a}}_{GA}^{k} \tag{12}$$

and a function in quadratic form

$$J = argmin_{\hat{a}_{GA}^{k}}\left(\boldsymbol{E}^T\boldsymbol{Q}\boldsymbol{E}\right) \tag{13}$$

where $\boldsymbol{Q} = \boldsymbol{Q}^T > 0$ is the project matrix. The fitness function is assumed to be of the form

$$J^* = \underset{\hat{a}_{GA}^{k}}{argmax}(c/(J + c)) \tag{14}$$

where $c > 0$ is the project constant. Take the estimate of the nonlinear function (10) as the compensation control

$$\boldsymbol{u}_{GA} = \boldsymbol{Y}_V(\dot{\boldsymbol{\alpha}}, \boldsymbol{v}, \dot{\boldsymbol{v}})\hat{\boldsymbol{a}}_{GA} \tag{15}$$

Our task is to ensure the stability of the closed-loop system, without changing the projected control signal to compensate for nonlinearities and varying operating conditions, $\boldsymbol{u}_{\mathrm{GA}}$. This means that the control signal $\mathbf{u}$ must be projected to ensure the stability of the closed-loop control system, i.e. the generalized error s must be limited, i.e. $|s_i| \le \Phi_i \forall t > 0$ where Φ_i, i = 1,2, assumes constant values. For this purpose, supervisory control $\boldsymbol{u}_{\mathrm{GA}}$ is added to the control signal $\boldsymbol{u}_s$, generated by an additional supervisory system. It is different from zero when the state vector achieves the boundaries of the set of $\{s_i: \ |s_i| \leq \Phi_i, i = 1{,}2\}$. For the problem posed in this way, let us introduce the law of control

$$u = u_{\mathrm{GA}} + Ks + Iu_s \tag{16}$$

where $I = \mathrm{diag}[I_1, I_2]$ is a matrix with elements I_i, which take the value equal to 1 for $|s_i| \geq \Phi_i$ and $I_i = 0$ for $|s_i| < \Phi_i$. Substituting (15) into (7), we get a description of a closed system of the form

$$M\dot{s} = -Cs - Ks + f(x) - u_{\mathrm{GA}} - Iu_s \tag{17}$$

Let's define a positive function $L = \frac{1}{2}s^T Ms$. For the case when $|s_i| \geq \Phi_i$ we get the following

$$\dot{L} = s^T M\dot{s} + \frac{1}{2}s^T \dot{M}s = -s^T Ks + \frac{1}{2}s^T(\dot{M} - 2C)s + s^T[f(x) - u_{\mathrm{GA}}] - s^T u_s \tag{18}$$

Assuming that $\dot{M} - 2C$ is a skew-symmetric matrix, the derivative of the scalar function (18) takes the form

$$\dot{L} \leq -s^T Ks + \sum_{i=1}^{2} |s_i|[|f(x))| - |u_{\mathrm{GA}i}|] - \sum_{i=1}^{2} s_i u_{\mathrm{s}i} \tag{19}$$

If we choose a supervisory control signal in the form of

$$u_{si} = [F_i + |u_{\mathrm{GA}i}| + \eta_i]\mathrm{sgn}s_i \tag{20}$$

assuming

$$|f_i(x)| \leq F_i \quad i = 1, 2 \quad F_i > 0, \ \eta_i > 0 \tag{21}$$

we obtain:

$$\dot{L} < -s^T Ks - \sum_{i=1}^{2} \eta_i |s_i| \tag{22}$$

The derivative of the Lapunov function is negatively determined, hence the closed system (17) is asymptotically stable. In the control signal (16) there is a step function I, which activates the supervisory control signal u_s at the boundary $|s_i| = \Phi_i$. In order to eliminate the high switching frequency of the executive system, the approximation of the **s** function in the neighbourhood of Φ_i according to the formula [7]

$$\begin{matrix} \frac{s_i}{\Phi_i} & |s_i| < \Phi_i \\ \mathrm{sgn}(s_i) & |s_i| \geq \Phi_i \end{matrix} \quad \mathrm{i} = 1{,}2 \tag{23}$$

The introduced approximation leads to weakening of the asymptotic stability condition of the system. An example diagram of a control system with genetic parameter adaptation is shown in Fig. 2.

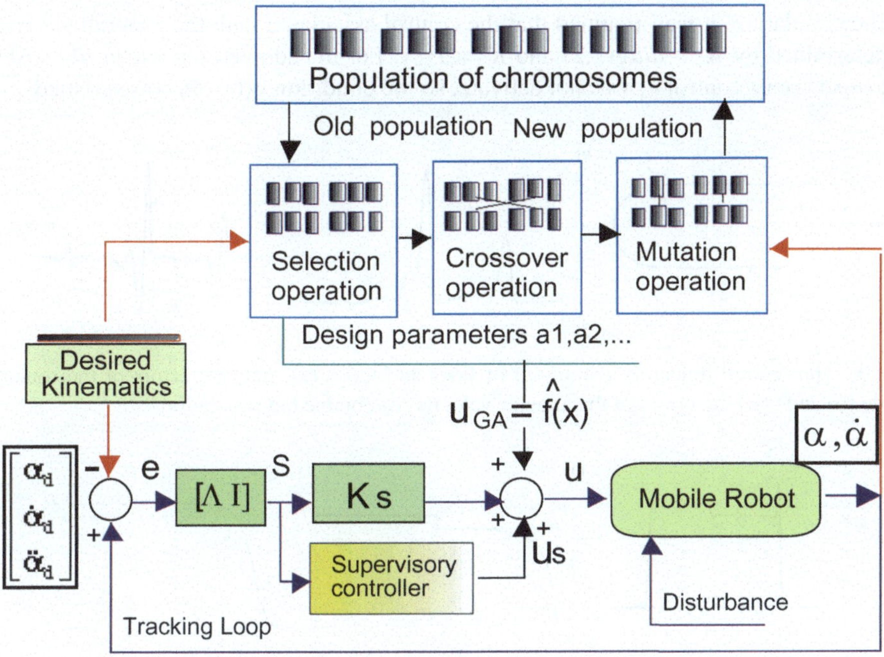

Fig. 2. Architecture of adaptive control using GA.

6 Simulation Example

Assuming a WMR model of the form (3) with the parameters $a_1 = 0.0691$, $a_2 = 0.0511$, $a_3 = 0.0291$, $a_4 = 0.0014$, $a_5 = a_6 = 4.3001$ and assuming variable operating conditions, the tasks of tracking the desired trajectory of the straight-loop-straight type (Fig. 3a) were realized using MATLAB/Simulink software and GAOT libraries [3]. Based on the desired trajectory, the angular velocities of the driving wheels and other parameters determined from relations (1) and (2) are assumed: $v_A = 0.4[\text{m/s}]$, $c = 10[-]$, $b_1 = 5[\text{s}]$, $b_2 = 20.5[\text{s}]$, $c_\beta = 5[-]$, $b_{1\beta} = 10[\text{s}]$, $b_{2\beta} = 15.5[\text{s}]$, $l_1 = 0.163[\text{m}]$, $r = 0.0825[\text{m}]$. The change in operating conditions is understood as a change in the resistance to motion, which corresponds to a change in the value of the parameters a_5, a_6. In addition, it was assumed that the resistance to motion on each wheel is equal, so $a_5 = a_6$. A control of the form (21) was used to implement the tracking task, with a genetic algorithm used to determine u_{GA}.

The initial population was assumed to consist of eight individuals representing the values of the parameters a_5, a_6 randomize from the range [4, 6]. At each discrete step ($h = 0.01[\text{s}]$), three steps of the GA algorithm consisting of selection (ranking selection function based on the normalized geometric distribution was used), crossover (arithmetic crossover was used) and mutation (Non uniform mutation was used), leading to maximization of the fitness function of the form (14), are performed. Using the GA algorithm it was assumed: $P = I_{2\times2}$, $Q = 100I_{2\times2}$, c = 5. In addition, it should be noted that the values of parameter estimates generated by the GA algorithm are filtered.

In the simulation, it was assumed that the control associated with the generalized error is determined by $\mathbf{\Lambda} = diag(2{,}2)$ and $K = I_{2\times2}$. For the adopted parameter $\Phi_i = 0.9$, the supervisory control u_s was not activate, so the condition $|s_i| < \Phi_i$ was satisfied.

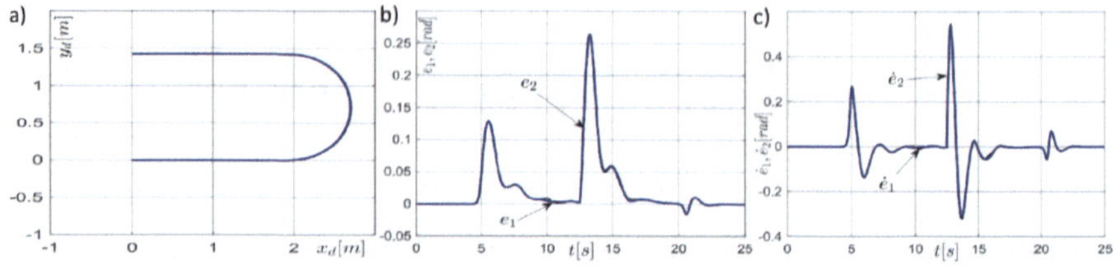

Fig. 3. The desired trajectory composed of lines and loops (a); tracking errors of the rotation angle (b) and tracking errors of the angular velocity (c) obtained in simulation.

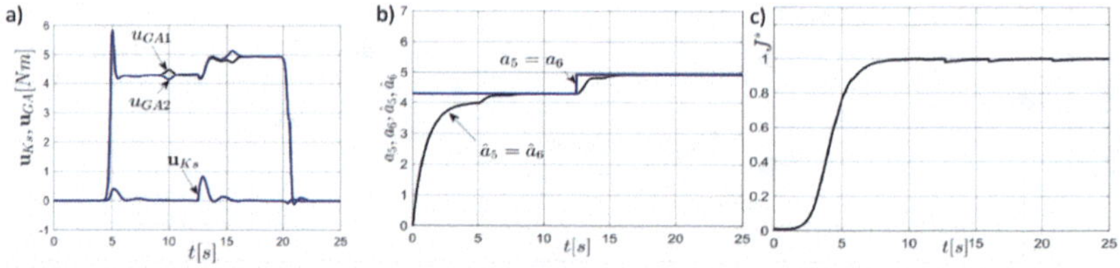

Fig. 4. The control waveform of the $\boldsymbol{u_{GA}}$ and $\boldsymbol{u_{Ks}}$(a); the waveform of adaptation of the parameters estimates to the parameters of the WMR (b); the waveform of the fitness function of the GA algorithm (c).

Figures 3b and 3c show the tracking error of rotation angles and angular velocities for MRK drive wheels, respectively. The error values are small, indicating good tracking quality. This is due to the correct adaptation of the parameters that are the basis for determining adaptive control. This adaptation is carried out by the GA algorithm. The increase in the modulus of error values occurs during the starting phase of motion and is also related to the timing of the change in operating conditions, which was realized by changing the values of the parameters a_5, a_6 by 15 percent.

The quality of the realization of the tracking task is determined by the control system. The waveform of the adaptive control u_{GA} and the generalized error proportional control, i.e. u_{Ks}, is shown in Fig. 4a. It shows that the control u_{GA} dominates over u_{Ks}. The control signal u_{Ks} increases its amplitude only in the case of the start phase of motion and during the change in the resistance to motion of the WMR. This occurs at time $t = 13[s]$. The decrease in the value of the control signal generated by u_{Ks} is related to the takeover of control by u_{GA}, which is related by the process of adaptation of the parameters $\widehat{a_5} = \widehat{a_6}$ by the GA algorithm running online. This process is shown in Fig. 4b. On the other hand, Fig. 4c shows the value of the fitness function J^* that has been filtered. The decrease in the value of the fitness function seen in the case of variable conditions is due to worsening of the quality of the estimation of parameters related to motion resistance. Next steps of the GA algorithm are realized through selection, mutation and crossover search to maximize fitness function, which allows adaptation of parameter estimates to new values, as shown in Fig. 4b.

7 Conclusions

In this paper, a real-time MRK tracking control was synthesized. The adopted solution uses adaptive control in which parameter adaptation was implemented by the GA algorithm. By introducing an additional supervisory signal, the asymptotic stability of the closed-loop control system was proven. This was confirmed in simulation.

In addition, the simulation shows that the adapted parameters tend to the reference value and also adapt to the change. On the basis of the small tracking error values, the quality of the tracking control implementation should be considered good, and the obtained results prompt further research into the adopted solution. Their direction can be at least twofold. Verification on a real object and the study of the relationship between the quality of parameter adaptation and the selection, crossover and mutation algorithms used in the GA algorithm.

References

1. Iqbal, J., Ullah, M., Khan, S.G., Khelifa, B., Ćuković, S.: Nonlinear control systems - A brief overview of historical and recent advances. Nonlinear Eng. **6**(4), 301–312 (2017). https://doi.org/10.1515/nleng-2016-0077
2. Gasparyan, O.N.: On application of feedback linearization in control systems of multicopters. In: Sergey, Yu., Misyurin, V.A., Avetisyan, A.I. (eds.) Advanced Technologies in Robotics and Intelligent Systems, pp. 343–351. Springer, Cham (2020). https://doi.org/10.1007/978-3-030-33491-8_41
3. Giergiel, M., Hendzel, Z., Żylski, W.: Modeling and control of mobile wheeled robots. PWN (2013). (in Polish)
4. Astrom, K.J., Wittenmark, B.: A survey of adaptive control applications. In: Proceedings of 1995 34th IEEE Conference on Decision and Control (1995). https://doi.org/10.1109/CDC.1995.478986
5. Astolfi, A., Karagiannis, D., Ortega, R.: Nonlinear and Adaptive Control with Applications, vol. 187 (2008). https://doi.org/10.1007/978-1-84800-066-7
6. Żylski, W.: Kinematics and dynamics of mobile wheeled robots. Publishing House Rzeszów Univ. of Technology (1996). (in Polish)
7. Szuster, M., Hendzel, Z.: Intelligent Optimal Adaptive Control for Mechatronic Systems, vol. 120. Studies in Systems, Decision and Control (2018). https://doi.org/10.1007/978-3-319-68826-8
8. Giergiel, J., Żylski, W.: Description of motion of a mobile robot by Maggie's equations. J. Theor. Appl. Mech. **43**(3) (2005)
9. Carr, J.: An introduction to genetic algorithms. Sr. Proj. **1**(40) (2014)
10. Katoch, S., Chauhan, S.S., Kumar, V.: A review on genetic algorithm: past, present, and future. Multimed. Tools Appl. **80**(5) (2021). https://doi.org/10.1007/s11042-020-10139-6
11. Lambora, A., Gupta, K., Chopra, K.: Genetic algorithm- a literature review. In: 2019 International Conference on Machine Learning, Big Data, Cloud and Parallel Computing (COMITCon) (2019). https://doi.org/10.1109/COMITCon.2019.8862255

12. Deb, K.: An introduction to genetic algorithms. Sadhana (1999)
13. Houck, C.R.: A genetic algorithm for function optimization: a matlab implementation. ACM Trans. Mathmatical Softw. (1996)

Search and Detection of People in the Water Using YOLO Architectures: A Comparative Analysis from YOLOv3 to YOLOv8

Nataliya Bilous(✉), Vladyslav Malko, and Nazarii Moshenskyi

Kharkiv National University of Radio Electronics, Kharkiv, Ukraine
nataliya.bilous@nure.ua

Abstract. The rapid development of computer vision and deep learning technologies has significantly improved the accuracy and speed of object recognition in a variety of applications, including security, surveillance, and search and rescue. One of the key challenges in this area is the detection of people in water areas, which is crucial for improving water safety and emergency response. In this research, a detailed comparative analysis of YOLOv3 to YOLOv8, is performed to evaluate their ability for effective detection people in the water. The analysis focuses on assessing the accuracy of each version's identification of people in the water, the speed of real-time image processing, the ability to adapt to different water conditions, and the required computing resources for effective operation. The purpose of the research is to perform a detailed comparative analysis of YOLO architectures, from YOLOv3 to YOLOv8, for evaluating their ability to effectively detect people in the water area. The research not only assesses current capabilities, but also suggests directions for future innovation to improve the efficiency and reliability of detecting and tracking people on water.

Keywords: YOLO · object detection · deep learning

1 Introduction

In the modern world, the speed and accuracy of object recognition using computer vision are key in a variety of applications, from automated video surveillance to security systems and search and rescue operations. One of the critical tasks in this context is recognizing people in the water, which is crucial for preventing accidents, improving water safety, and responding effectively to emergencies. The development of deep learning algorithms, in particular You Only Look Once (YOLO) architectures, has significantly advanced the ability to recognize objects quickly and accurately in real time. Since its first introduction, YOLO has gone through several iterations of enhancements, each bringing significant improvements in accuracy, speed, and generalization ability. From YOLOv3 to the most recent version, YOLOv8, each iteration offers unique innovations that have the potential to significantly improve the effectiveness of human recognition in the water. However, to achieve the best results, it is important to have a thorough understanding of the features, benefits, and limitations of each version.

R. Szewczyk et al. (Eds.): AUTOMATION 2024, LNNS 1219, pp. 233–255, 2024.
https://doi.org/10.1007/978-3-031-78266-4_21

2 Purpose of the Research

The purpose of the research is to perform a detailed comparative analysis of the YOLO architectures, from YOLOv3 to YOLOv8, to assess their ability to effectively detect people in the water area. The analysis focuses on several key aspects: how accurately each version can identify people in the water, how fast it processes images for real-time use, its ability to adapt to different water areas, and the computing resources required to operate efficiently.

It is planned to identify what improvements have been made to each new version of the YOLO architecture and how these changes affect their ability to detect people in the water. This includes analyzing the impact of external conditions, such as illumination and obstacles on the water, on the performance of the models. The main idea is to understand which version of YOLO is best suited for this specific task, providing the optimal combination of accuracy, speed, and efficiency in different conditions. In addition, the research will help identify potential areas for further improvements in deep learning and computer vision technologies aimed at improving human safety on the water. We aim not only to assess existing capabilities, but also to consider how future innovations can improve the process of detecting and tracking people on the water, making it more efficient and reliable.

3 General Statement of the Problem

To address the challenge of efficiently searching for and detecting people on water, this research focuses on analyzing and comparing different versions of YOLO architectures, which are known for their ability to recognize objects quickly and accurately in images. The importance of this analysis is to determine the optimal model that can adapt to the unique conditions of the water area, where factors such as water surface glare, water turbidity, and other variables can significantly affect the identification process. The research will evaluate each version of YOLO by the criteria of recognition accuracy, processing speed, ability to generalize to different conditions, and efficiency of computing resources. This will not only help identify the most appropriate model for the task of detecting people on water, but also identify possible areas for further research and improvement in this area.

In addition to the technical analysis, the research will also include the development of recommendations for customizing and adapting the selected YOLO models to specific water environment conditions. This will include investigating the impact of external conditions on model performance and developing techniques to optimize their performance, taking into account the challenges inherent in the water area. Thus, this research aims not only to solve the actual problem of detecting people on water using YOLO architectures, but also to contribute to the development of computer vision technologies, ensuring increased safety on water.

4 Review the Literature

Alwani [1] described a methodology for recognizing actions performed by people using joint angles and skeleton in-formation. Unlike classical methods that focus on the body contour, the presented approach was based on joint angle measurements obtained from time series of skeleton data captured by depth sensors. In turn, Chen [2] proposed a system for real-time intelligent object detection designed specifically for drones, based on Field-Programmable Gate Array (FPGA) technology and integrated into the drone. The proposed system can detect objects at a speed of 8 frames per second. In [3] Gallego used unmanned aerial vehicles equipped with multispectral cameras to search for people as part of maritime rescue operations. The aligned multispectral images were used to train a convolutional neural network (CNN) aimed at body detection. The evaluation results showed that the best classification performance obtained when combining the Green, Red Edge, and NIR channels. In particular, it was found that the precise localization method is the most suitable, showing accuracy similar to the "sliding window" method, but with a spatial localization of approximately 1 m. Debapriya Maji [4] introduced YOLO-pose, a new method for detecting joint points and estimating the 2D position of multiple people in an image without using heat maps, based on the well-known YOLO object detection system. He obtained a new quality standard on the validation and test datasets above 90%, outperforming all existing methods without the need for additional tests and methods to improve results during testing. In [5] Bilous and Kalugin developed algorithms based on computer vision technologies to determine the position of the human body. The combination of BlazePose and the weighted distance method was found to be the best suited for position recognition, providing high ac-curacy and stability in a range of challenging scenarios. As a result, the authors conclude that the weight distance meth-od showed the best results among the position comparison methods. Daniel Hernández [6] focused on the active deployment of Unmanned Aerial Vehicles (UAVs) in various contexts and areas of application. Special attention was given to equipping these devices with a high level of intelligence and autonomy to perform complex tasks. He proposed applying deep learning techniques for the semantic segmentation of images. Experimental results demonstrated the feasibility of performing high-quality image processing of UAVs in real time using solutions developed based on Deep Neural Networks (DNNs). Xin Li [7] focused on traditional approaches to human position estimation based on heat maps. He presented an end-to-end lightweight network for human position estimation that utilizes a multiscale coordinate attention mechanism based on the Yolo-Pose network. The average accuracy of the proposed network on the COCO 2017 validation dataset increased by 4.8% while reducing the number of network parameters and computations. Nowadays, the challenge is to accurately and quickly locate people stranded on the surface of the water in flood conditions to provide them with immediate assistance. In his paper, José Gomes da Silva Neto [8] presents the initial stage of a broader research focused on the use of RGB images to analyze human behavior. At this stage, the researcher created a prototype system that combines hardware and software and can detect human postures based solely on RGB data. The choice of hardware fell on the NVIDIA Jetson Nano because of its relatively high computational efficiency compared to alternatives such as Raspberry Pi and Arduino. While searching for optimal algorithms for determining human position suitable for running on the computationally limited Jetson Nano platform, such significant developments as HyperPose, TensorRT Pose Estimation, and tf-pose estimation were identified, the latter of which was chosen for use in the project. However, the results of performance tests in terms of frames per second (FPS) showed that Jetson Nano has rather poor performance when using the selected algorithm, especially in comparison with similar systems such as NVIDIA Jetson TX2 and NVIDIA Jetson Xavier. In their article, Rakova A. O., and Bilous N. V. [9] consider the importance of tracking the direction of a person's face, which is an indicator of attention. Such data can be useful in various aspects of everyday life, including human-computer interaction, teleconferencing, virtual reality, and 3D audio rendering. Furthermore, head position detection can serve as a means of comparing the exercises performed by a person with standardized exercises. Systems based on depth cameras, which are often used for this purpose, have serious drawbacks, such as reduced accuracy in direct sunlight and the need for additional equipment. Therefore, more and more attention is being paid to recognition from 2D images, which eliminates the problems associated with depth cameras and allows them to be used both indoors and outdoors. The authors propose a landmark method that reduces the set of recorded vectors to the minimum number needed to describe head movements. They also analyze and compare existing face vector detection methods in terms of their use in the proposed approach. The results showed that the landmark method can significantly reduce the set of head direction vectors describing the movement. According to the results of the research, regression-based methods provided significantly higher accuracy and independence from lighting and partial face occlusion, so they were chosen to obtain the head direction vector in the landmark method. The findings of the research confirm the suitability of the landmark method for tracking human movements and demonstrate that methods for determining the human head vector using a 2D image can compete in accuracy with RGBD-based methods, while having fewer limitations in use. Javier Smith [10] in his research focuses on the problem of human position estimation in thermal images using convolutional neural networks and Vision Transformer architectures. The author adapts eight position estimation methods developed for visible images to the thermal domain. Due to the lack of large, labeled thermal image datasets, it is necessary to use training transfer between the visible and thermal domains, as well as a database to fine-tune the networks in the thermal domain. Dhanushree M. [11] in his article considers the problem of detecting people in the flood zone - a typical natural disaster for monsoon countries, including India. It is noted that effective detection of people in flooded areas is critical for rescue, crisis and other operational services. Since weather conditions such as rain, clouds, and fog seriously complicate the detection process, the author introduces data augmentation techniques that simulate these conditions. The author also presents the HOG-based Robust Human Object Detection (HOG_based_RHOD) algorithm, which efficiently and reliably detects people in photographs of flooded areas under various weather conditions. In their article [12], Oleh Hramm and Nataliya Bilous developed a flexible and customizable solution for cell segmentation using the Hough transform for circles and the watershed algorithm. This approach allows to optimize image processing for different datasets, providing high performance even in the presence of noise and artifacts. Considerable attention is paid to parameter tuning, which allows the algorithm to be precisely adapted to specific segmentation tasks. The results show that the developed solution has an average error of 2.69% on a sample of test images, which is an important contribution to the field of biological object image processing. Xin Wu [13] proposes a novel approach to utilize lidar data in deep learning. They focus on processing low-resolution 360° images acquired from lidar sensors. The research shows that with adequate preprocessing, lidar data can be efficiently processed by existing deep learning models for object detection and semantic segmentation.

Valarmathi [14] proposes the use of the YOLOv3 (You Only Look Once) algorithm to detect people and recognize their actions in search and rescue operations during natural disasters. This method, used with drones, allows for more efficient and faster analysis of disaster zones than traditional methods. The YOLOv3 algorithm demonstrates high accuracy (94.9%) and is able to process im-ages in 0.40 ms, which is far superior to existing methods. Li Tan's research [15] improves YOLOv4 for target detection in UAV imagery by using new techniques to address the challenges of target attenuation and small object detection. It introduces a receptive field block for better feature extraction, a lightweight attention mechanism for representing features at different scales, and soft non-maximum suppression to reduce ocularity misses. The paper is organized into five chapters covering its contributions, related work, methodology, experimental validation, and conclusions, significantly advancing UAV-based monitoring and detection applications. The paper by Farzaneh Dadrass Javan [16] presents a modification of the YOLOv4 algorithm for drone recognition using visual data. The research focuses on the challenges associated with the small size of drones, their confusion with birds, the presence of hidden areas, and crowded backgrounds. The authors modified the YOLOv4 architecture, in particular the number of convolutional layers, to extract semantic features more accurately. Experimental results showed an improvement in recognition accuracy and speed compared to the baseline YOLOv4 model. Hung-Cuong Nguyen [17] explores the application of 3D human pose estimation in sports, robotics, and healthcare, focusing on fast and accurate methods. The author introduces the YOLOv5-HR-TCM (YOLOv5-HRet-Temporal Convolution Model), which is based on a 2D to 3D approach for estimating human postures in three dimensions. The model is designed to optimally perform each stage of the estimation process: face detection, 2D pose estimation, and 3D pose estimation. YOLOv6 [18], although not an official member of the YOLO series, was inspired by the original YOLO detectors. Its primary aim is to cater to industrial applications. The main distinction from YOLOv5 lies in YOLOv6's adoption of an anchor-free method, boosting detection speed by 51%. Additionally, it incorporates the EfficientRep backbone and Rep-PAN neck and separates the regression and classification branches. For smaller models, YOLOv6 employs SIoU loss, whereas larger models utilize IoU loss. Several articles used both YOLOv6 and YOLOv5 detection models trained on various datasets. Chenhao [19] proposes an experiment that compares both models for obstacle detection for visually impaired people. In the paper, they utilized seven different YOLO detection models. They made use of a custom dataset which consists of 15 classes, 7938 labels and 2205 of them are of the class "Person". YOLOv5 reached an overall precision of 78.1%, recall of 68.2%, and mAP@0.5 of 74.2%. YOLOv6 - an overall precision of 78.5%, recall of 71.4%, and mAP@0.5 of 78.4%. When selecting the right model scientists should consider different conditions like training time, inference speed, precision, and model size. According to Horvat et al. [20] the duration of training is smaller for smaller

YOLOv5 models (YOLOv5m6 and YOLOv5s) but larger model YOLOv5x trains longer than the largest YOLOv6 model (YOLOv6-M6) – 17h 20m for 300 epochs and 11h 18m for 300 epochs respectively. Although inference time results for each v6 model are much better than corresponding v5 models of the same size, the NMS stage neglects all the improvements and results in at least twice longer total processing time. YOLOv5x processes video at twice as many frames per second as the YOLOv6-M6 model with the same complexity. According to detection results, YOLOv5 is better at handling unbalanced labels but YOLOv6 is better in object localization. YOLOv6-M reached better results in identifying the class "Person" on the COCO2017 dataset (60.3% precision, 46.0% recall, and mAP@.5 – 49.1%). These results outperform all YOLOv5 models at the detection stage. Wei Yang et al. detected vulnerable road users [21] such as motorcyclists, pedestrians, and bicyclists. They picked suitable images from different public datasets (from car cameras to regular images) and labeled them. Comparing different models they proposed their solution based on YOLOv5 as the original version performs badly on small targets. The result shows that, although the original YOLOv5 reached better FPS (98 FPS) than YOLOv6 (70.6 FPS) but still has lower accuracy – mAP@.5 85.7% when YOLOv6 mAP@.5 – 96.6%. In this experiment, only YOLOv5 exceeded 90 FPS value while all other models are far lower. Also, YOLOv5 had fewer parameters and GFLOPs compared to other models. In their research, Jinghui Yan et al. [22] developed an improved underwater object detection model based on the YOLOv7 algorithm integrated with a CBAM attention mechanism and a fast spatial pyramidal pooling through stages (SPPFCSPC) module. This research solves the problems of recognizing blurry and small objects in complex underwater environments by improving the accuracy and speed of detection. The ACFP-YOLO model has shown high accuracy compared to other state-of-the-art algorithms on URPC datasets and underwater debris detection, which is confirmed by numerous experiments and analysis. This research makes an important contribution to the development of underwater object recognition technologies. The research by Yalin Zeng et al. [23] presented the YOLOv7-UAV algorithm aimed at improving object detection in drone images. The authors make improvements to YOLOv7, including optimizing the architecture and using new methods to improve small object detection. Experiments have shown a 27% increase in detection speed compared to YOLOv7, as well as improved accuracy on the VisDrone2019 and TinyPerson datasets. These results confirm the effectiveness of YOLOv7-UAV for analyzing aerial images. Chen Haiwei and Zhou Guohui in their paper [24] describe in detail an approach to improving the YOLOv8 model for identifying student behavior. They propose an innovative module, C2f_Res2block, which combines elements from Res2Net and YOLOv8, and introduce EMA and MHSA mechanisms to improve the accuracy of identifying student behavior in the classroom. These innovations contribute to the overall accuracy of the model, as demonstrated by the 4.2% improvement in mAP scores. This research is important for the development of effective systems for monitoring and analyzing human activity. In the paper Gang Wang et al. [25] the authors developed a UAV-YOLOv8 model based on the YOLOv8 algorithm. They implemented an advanced loss function, an efficient attention mechanism, and a multi-level feature fusion network. These optimizations help to improve the detection of small objects and reduce detection misses, which is especially important in UAV aerial photography scenarios. The research showed that the new model has better detection accuracy compared to the baseline YOLOv8 model and other popular models. The authors developed the UAV-YOLOv8 model based on the YOLOv8 algorithm. They implemented an advanced loss function, an efficient attention mechanism, and a multi-level feature fusion network. These optimizations help to improve the detection of small objects and reduce detection misses, which is especially important in UAV aerial photography scenarios. The research showed that the new model has better detection accuracy compared to the baseline YOLOv8 model and other popular models. J. Berndt in his paper [26] researches the accuracy of using the YOLOv8 convolutional neural network for detecting people in nadir aerial images, which is important for search and rescue operations. In this research, a unique dataset of nadir images with different ground spacing distances (GSD) was created to train YOLOv8. The impact of GSD on detection accuracy is analyzed, and it is found that networks trained on images with lower GSD (higher resolution) perform better. The results emphasize the dependence of neural network performance on the GSD of the training data, which is key to achieving high detection accuracy in aerial search and rescue scenarios.

5 Review of the YOLO Architectures

YOLO (You Only Look Once) architectures represent a revolutionary approach to computational vision, offering fast and efficient real-time object recognition. Since its first introduction, YOLO has gone through several significant enhancement iterations, each bringing new improvements and optimizations, making it increasingly accurate and fast. Below is an overview of the major versions of YOLO, from YOLOv3 to YOLOv8, highlighting their key features and improvements.

5.1 YOLOv3

YOLOv3 [27], introduced in 2018, is one of the key versions of the You Only Look Once (YOLO) architecture for real-time object recognition. This version made a significant step forward from its predecessors, offering enhancements that improved recognition accuracy and speed, as well as the model's ability to identify objects of different sizes. Main features of YOLOv3:

- YOLOv3 uses three different scales of output frames, which allows the model to better detect small, medium, and large objects. This feature is critical for applications where objects may appear at different distances from the camera.
- As the basis for feature extraction, YOLOv3 uses the Darknet-53 network, which is deeper and has better performance than the Darknet-19 used in YOLOv2. Darknet-53 includes 53 convolutional layers, which enables more efficient detection of complex features in images.
- YOLOv3 introduces a new approach to predicting object classes and their sizes, using logistic regression for each object class, which allows for high classification accuracy.
- Despite the increased accuracy and ability to detect objects of different sizes, YOLOv3 remains a high-performance model capable of running in real time on modern hardware.

YOLOv3 has significantly improved the ability of computer vision systems to recognize objects accurately and quickly in images or videos. However, like any deep learning model, it has its limitations, particularly in cases where objects are highly overlapping or where there are many small objects in the scene. The introduction of YOLOv3 was an important milestone in the development of object recognition algorithms, laying the groundwork for further innovations in this area. This model demonstrated that it is possible to achieve high accuracy and speed at the same time, which paved the way for the development of subsequent versions of YOLO, each of which continues to improve these key aspects.

5.2 YOLOv4

YOLOv4 [28], introduced in 2020, continues the innovative path set by previous versions of the YOLO architecture. This version focuses on achieving the highest possible accuracy and speed of object recognition on a variety of hardware, including power-limited devices. YOLOv4 introduces several innovations and optimizations that make it one of the most powerful models for computer vision applications.

Main features of YOLOv4:

- One of the key goals of YOLOv4 was to ensure high performance not only on powerful GPUs, but also on less powerful hardware. This makes YOLOv4 available for a wider range of applications, including embedded systems and mobile devices.
- YOLOv4 integrates the latest advances in deep learning and computer vision, including techniques such as mish-activation, Cross-Stage Partial networks (CSPNet), Self-Adversarial Training (SAT), and others. These innovations help improve recognition accuracy and speed.
- YOLOv4 implements an automatic process for finding optimal hyperparameters and network structures, which maximizes the model's efficiency without the need for manual tuning.
- Through the use of advanced training and optimization techniques, YOLOv4 demonstrates an improved ability to detect objects in challenging environments such as low light, overlapping objects, and high object density in the image.

YOLOv4, with its high accuracy and speed, is a significant contribution to the field of computer vision. However, like any complex system, it has challenges, particularly in cases of extremely small or difficult objects to detect. Nevertheless, YOLOv4 remains one of the most popular and widely used models for real-world object recognition tasks, setting the standard for future developments in this field.

5.3 YOLOv5

YOLOv5 [29], launched in 2020, is not an official continuation of the YOLO line from its creator Joseph Redmon, but it quickly gained popularity in the community for its affordability, high performance, and ease of use. It was developed and published as an open-source project that continues the evolution of YOLO for object recognition.

Main features of YOLOv5:

- The YOLOv5 has been optimized to run quickly on a variety of hardware, including power-limited devices, making it highly flexible for a variety of applications.
- The ease of setup and use of YOLOv5 makes it accessible even to those new to computer vision, offering ample opportunity for customization and experimentation.
- YOLOv5 offers several model options with different numbers of parameters, from lightweight to heavier versions, allowing you to choose the optimal balance of speed and accuracy depending on your project needs.
- Improved accuracy: Despite its speed, the YOLOv5 delivers high recognition accuracy, rivaling the heavier models in the other YOLO series.

YOLOv5 continues to innovate in the area of fast and accurate object recognition, while delivering significant improvements in ease of use and accessibility. However, as with any technology, its performance can vary depending on specific application conditions, such as the variety of objects, lighting conditions, and the level of object overlap. YOLOv5 has become a popular choice for developers and researchers looking for an effective solution for a wide range of object recognition tasks, from smart video surveillance to autonomous driving systems. It has set new standards for the ratio of speed, accuracy, and ease of use in computer vision.

Although YOLOv5 is known for its high performance and efficiency in real-time object recognition, it is important to note that its training process requires significant computing resources. This is due to several factors. YOLOv5, like most deep neural networks, contains millions of parameters that need to be optimized during the training process. This optimization requires a significant amount of computation, especially with large datasets that are typically used to train computer vision models. To achieve high accuracy and overall model generalization capability, YOLOv5 is trained on large and diverse datasets. Processing and analyzing such a large number of images requires significant computing power, especially to ensure real-time training efficiency. To optimize the training process of deep learning models, in particular YOLOv5, it is recommended to use specialized hardware such as graphics processing units (GPUs). GPUs significantly speed up training due to their ability to process large amounts of data in parallel. However, the cost and availability of powerful GPUs can be a barrier for some researchers and developers. YOLOv5 training requires a large amount of RAM and video memory to store intermediate data such as model weights, gradients, and image batters. This increases the overall computational resource requirements needed for effective training. Despite these challenges, YOLOv5 remains one of the most effective and widely used models for real-time object recognition, offering high accuracy and speed. Optimization of resource utilization and hardware development continue to help reduce these limitations, making YOLOv5 training and deployment increasingly affordable.

5.4 YOLOv6

YOLOv6 [18, 30] is a modern version of the famous YOLO series of architectures for real-time object detection. Developed by the Meituan team, YOLOv6 has made a number of innovative improvements to the architecture and training methodology aimed at improving accuracy and processing speed without significantly increasing computational costs.

Main features of YOLOv6:

- Bidirectional Concatenation Module (BiC). This module improves object localization by enabling the model to integrate information from different layers of the network more efficiently. It provides a performance boost with minimal speed degradation, increasing the overall accuracy of the system.
- Anchored training strategy (AAT). AAT is used to combine the benefits of anchored and unsupervised approaches, optimizing the learning process and increasing the efficiency of inference. This allows YOLOv6 to achieve high accuracy while maintaining processing speed.
- Improvements in the model's spine and neck architecture allow YOLOv6 to achieve impressive results on large datasets such as COCO, especially at high input resolutions.
- Self-distillation strategy. This strategy aims to improve the performance of smaller models by reinforcing an additional regression branch during training and removing it during output. This helps to avoid reducing the processing speed without losing accuracy.
- YOLOv6 provides flexibility in backbone selection, offering support for CSPDarknet, EfficientNet, and ResNet, among others. Users can tailor the backbone to their needs by replacing CSPDarknet53 with any other CNN architecture that is compatible with YOLOv6 input and output data sizes.

With its innovative features, YOLOv6 is being used in a wide range of applications, from video surveillance and security systems to smart image analysis and autonomous transportation systems, where high accuracy of real-time object detection is required. YOLOv6 is a significant step forward in object detection algorithms, offering high performance, accuracy and flexibility to meet the needs of modern computer vision applications.

5.5 YOLOv7

YOLOv7 [31] is one of the newest versions in the YOLO series of real-time object recognition architectures. YOLOv7 continues the tradition of innovation established by previous versions, offering significant improvements in accuracy and speed while maintaining high efficiency.

Main features of YOLOv7:

- YOLOv7 introduces a number of technological innovations that significantly improve the accuracy of object recognition, including improved image processing algorithms and optimized network architecture.
- Despite the increased accuracy, YOLOv7 maintains a high processing speed, enabling it to be used in applications that require instant response, such as video surveillance, autonomous driving, and interactive systems.
- YOLOv7 runs efficiently on a wide variety of hardware, including both high-performance GPUs and less powerful computer systems, making it available for a wide range of applications.
- The model's architecture is designed to be easily adapted to specific object recognition needs, allowing users to customize the model for optimal performance depending on the task at hand.

YOLOv7 is setting new standards in object recognition accuracy and speed, but like any advanced technology, it has its challenges. The main ones include the need for significant computing resources to train the model, especially when working with large datasets, as well as the need for deep machine learning knowledge to effectively adapt and customize the model for specific tasks. YOLOv7 continues to advance the field of computer vision by offering powerful tools for developers and researchers interested in real-time object recognition. Its high performance and adaptability make it an ideal choice for a wide range of applications, from industrial monitoring to interactive technology development.

5.6 YOLOv8

Ultralytics YOLOv8 [32] represents a significant advancement in object detection, instance segmentation, and image classification technologies. This version provides unsurpassed accuracy and high speed, allowing for real-time object detection without delays, distinguishing it from previous versions.

Main features of YOLOv8:

- YOLOv8 provides users with the ability to customize the model architecture to meet specific needs and requirements. This flexibility is key to adapting the model to a variety of use cases, from simple applications to complex image analysis systems.
- With the introduction of new adaptive training capabilities such as loss function balancing and the Adam optimizer, YOLOv8 improves training speed and delivers better accuracy and faster model convergence. These innovations contribute to the overall improved model performance.
- The model has advanced image analysis capabilities, allowing not only to detect objects, but also to recognize actions, color, texture, and establish connections between objects. This opens up new horizons for applications in the security, medical, retail, and other industries.
- New data augmentation techniques help the model better cope with image variations, such as low resolution or occlusion, improving the accuracy of real-world object detection.

YOLOv8 offers support for multiple backbones, allowing users to choose between CSPDarknet, EfficientNet, ResNet, and others, or even customize their own CNN architecture. This provides great flexibility and the ability to adapt the model to specific tasks. YOLOv8, like any other deep learning model, has its limitations that can affect its effectiveness in certain scenarios. Potential limitations of YOLOv8:

- Detecting very small objects: Although YOLOv8 has improved its ability to detect small objects compared to previous versions, it may still have difficulty with very small or distant objects, especially in environments with a lot of noise or clutter.
- Dependence on data quality: The performance of YOLOv8 is highly dependent on the quality and variety of data used in training. Insufficiently representative or limited training data can lead to poor model generalisation and poor performance on new or unfamiliar images.

- Handling collisions and overlapping objects: In scenes with a high level of overlap or collision between objects, YOLOv8 may have difficulty accurately detecting individual objects, especially if the objects have similar characteristics or are in close clusters.
- Difficult weather conditions: The performance of YOLOv8 may be adversely affected by difficult weather conditions, such as rain, fog, or snow, which may reduce the visibility of objects in images or videos.
- Computational requirements: Despite improvements in computational efficiency, YOLOv8 still requires powerful computing resources for real-time training and inference, especially when dealing with high resolution or large amounts of input data.
- Overall adaptability: While YOLOv8 is highly adaptive to a variety of object detection tasks, there are specific scenarios or domains where it may require significant modifications or tweaks to achieve optimal performance.

Understanding these limitations is important for choosing the most appropriate architecture for a particular task and for improving the model through further research and development. YOLOv8 can be adapted to specific tasks through fine-tuning using a specialized dataset. This process allows the model to "remember" the knowledge gained during previous training and adapt it to new conditions, improving the accuracy of detecting specific objects. Fine-tuning and full training of YOLOv8 can be performed through Python code or via the command line interface, providing flexibility and accessibility to a wide range of users. These innovations and enhancements make YOLOv8 an ideal tool for developers and researchers looking for effective computer vision solutions for a wide variety of applications.

6 Comparative Analysis of YOLOv8 with Other Neural Network Models

The YOLO architecture is known for its ability to quickly detect objects in images with high accuracy. The latest version, YOLOv8, has made significant improvements in speed and accuracy, making it a potentially ideal choice for applications such as searching for and detecting people on water. To evaluate the performance of YOLOv8, it is important to compare it to other advanced object detection models.

Faster R-CNN is a model that combines high accuracy with the ability to handle complex scenes efficiently. Despite its high accuracy, Faster R-CNN has a significantly slower detection rate than YOLOv8, which can be a limiting factor for real-time applications such as water body monitoring. YOLOv8 outperforms Faster R-CNN in terms of processing speed while providing comparable accuracy, making it more suitable for rapid emergency response. SSD (Single Shot Multibox Detector) is distinguished by its ability to detect objects quickly with reasonable accuracy. It uses fixed anchor frames to detect objects of different sizes at different scales, making it effective for a wide range of applications. However, YOLOv8 performs better when detecting small objects and in challenging environments, such as water environments with variable visibility and light refraction. EfficientDet is one of the newest object detection models that is optimised for maximum efficiency without significant loss in accuracy. This model uses an efficient scalable architecture and an optimised loss function to detect objects of different scales with high accuracy. Despite its optimisation for efficiency, YOLOv8 outperforms EfficientDet in terms of processing speed, which is a key factor for real-world applications where time is of the essence. In addition, YOLOv8 performs better in detecting people in complex aquatic environments, where the model's performance is determined not only by accuracy but also by its ability to quickly adapt to dynamic conditions. For a structured comparison of YOLOv8 with other advanced object detection models, you can create a table that displays the key characteristics of each architecture (Table 1).

Table 1. Structured comparison of YOLOv8 with other neural network models

Feature	YOLOv8	Faster R-CNN	SSD	EfficientDet
Speed	Very High	Low	High	High
Accuracy	High	Very High	Moderate	Very High
GPU Resources	Moderate	High	Moderate	Low
Complexity	Moderate	High	Low	Moderate
Use Case	Real-Time	Detailed Analysis	Broad Range	Optimization
Features	Fast Detection	Precise Localization	Scalability	Efficiency

YOLOv8 sets new standards for speed and accuracy in object detection in challenging environments, such as people detection on water. Its comparison with other advanced models such as Faster R-CNN, SSD, and EfficientDet highlights its advantages in processing speed and adaptability. While each of the models reviewed has its own strengths depending on the specific application, YOLOv8 stands out as a particularly powerful tool for fast and accurate real-time object detection. This makes it ideal for applications where a quick response is crucial, especially in search and rescue missions on the water.

7 Experiments

7.1 Data Preparation

In the experiment to compare versions of YOLO architectures for detecting people in the water, three key datasets were used: COCO2017, SARD (Search and Rescue Dataset), and SeaDronesSee. These datasets were chosen to provide a broad coverage of potential scenarios for recognizing people in different environments, including water. The COCO2017 [33] dataset, with its extended set of 118,000 training and 5,000 validation images, was used as the basis for the overall training of the models for object recognition. Given the goal of the experiment, special attention was paid to images containing scenes with water and people in these contexts. The SARD [34] dataset, containing 1,981 manually labeled images extracted from video footage of simulated search and rescue scenarios, was used to train the models on specific conditions. The images included a variety of terrain types and human activity, allowing the models to adapt to a wide range of detection scenarios. Specializes in search and rescue scenarios, including the detection of people in difficult conditions. This dataset helps to determine how well YOLO models can identify people in critical situations on the water where speed and accuracy are key. SeaDronesSee [35] includes images of marine areas captured by unmanned aerial vehicles. With 8,930 training, 1,547 validation, and 3,750 test images, SeaDronesSee facilitated model training and evaluation on a large amount of data from maritime scenarios. The drone imagery provided a unique perspective for detecting people in the water.

In general, data preparation prior to model training included thorough normalization, augmentation, and specific processing to optimize the images for the specific conditions of the experiment. This provided a comprehensive framework for training, evaluating, and comparing the performance of different versions of YOLO in the task of detecting people on water, providing the models with the ability to adapt to a variety of scenarios and conditions, from search and rescue operations to water monitoring.

7.2 Training YOLO Architectures

The process of training the YOLO architectures for the task of detecting people on water was carried out considering the unique challenges posed by this specific task. Thanks to the detailed COCO2017, SARD, and SeaDronesSee datasets, a solid foundation was created for effective model training. This process involved several key steps aimed at maximizing the performance and accuracy of each version of YOLO. Before the training, the models were adapted to the specifics of the task of detecting people on water. This meant selecting and optimizing images from datasets that best represent potential scenarios that the model could encounter in the real world. Data augmentation was also applied to create a variety of conditions, such as changes in lighting, weather conditions, and other factors that affect recognition. The training of each version of YOLO was customized to ensure optimal performance. Given the wide variety of scenarios present in the datasets used, training parameters such as learning rate, batch size, and number of epochs were carefully chosen for each model. This allowed us to strike a balance between the training speed and the ability of the models to generalize learning on a variety of data. During the training process, the performance of the models was evaluated using validation datasets. This made it possible to identify potential problems with overtraining or undertraining and to adjust training parameters in a timely manner. The key metric for evaluation was average precision (AP), which provided information about the models' ability to accurately localize and identify people in the water. At this stage, detailed optimization of the architectures was performed. For training the YOLO architecture, we used the NVIDIA TESLA A-100-80GB GPU. The choice of this powerful GPU was driven by the need to provide high computational power, which is critical for efficient training of deep neural networks, especially when working with large datasets and complex architectures such as those characterizing YOLO models. NVIDIA TESLA A-100-80GB delivers exceptional performance with its Ampere architecture, which is specifically designed to accelerate deep learning operations. With 80GB of GPU memory, this processor is capable of processing huge amounts of data, allowing you to train complex models with high speed and efficiency. Using the TESLA A-100 not only significantly reduced model training time, but also allowed us to run experiments with high resolution images and complex architectures without compromising on accuracy and performance. This GPU also helped to improve the efficiency of the data augmentation process and the implementation of additional training techniques such as fin-tuning and additional hyperparameter tuning. In addition, the use of such a powerful computing platform made it possible to analyze in detail the behavior of each version of YOLO in different conditions and with different configurations, determining the optimal parameters for maximizing the performance of the models. This, in turn, contributed to a deeper understanding of the mechanisms of object detection and opened up new opportunities for further improvement of machine vision technologies. Thus, the use of NVIDIA TESLA A-100-80 GB in the experiment with human detection on water using YOLO architectures has significantly improved the quality and speed of the research, demonstrating the importance of choosing a powerful computing platform for training modern deep learning models. A special optimization formula was used to adjust and fine-tune the neural network parameters, ensuring optimal performance:

$$E(\theta) = \sum_{i=1}^{N} L(yi, f(xi; \theta)) \quad (1)$$

where $E(\theta)$ – general error on the dataset, L - loss function that determines the difference between the actual and predicted values, yi - is the actual value of the label for the i-th sample, $f(xi; \theta)$ - is the predicted value of the model for the i-th sample with parameters θ, N - number of samples in the dataset.

The training process is based on the YOLO architecture using Adam optimizer. To determine the accuracy of the model, F1-Score is used. Table 2 presents the comparative performance of different versions of YOLO architectures for object recognition after 30 epochs of training. It includes such metrics as Precision, Recall, average accuracy at a threshold of 0.5 (mAP@0.5), and F1-Score for each version of the model from YOLOv3 to YOLOv8.

Table 2. YOLO architectures performance.

Architecture	Precision	Recall	mAP@0.5	F1-Score
YOLOv3	0.67	0.68	0.675	0.675
YOLOv4	0.81	0.83	0.82	0.82
YOLOv5	0.73	0.73	0.73	0.73
YOLOv6	0.62	0.62	0.62	0.62
YOLOv7	0.68	0.68	0.68	0.68
YOLOv8	0.87	0.89	0.88	0.88

YOLOv6, performing lower on key metrics such as precision, recall, and mAP@0.5, stands out as a model that needs further tuning to improve its ability to identify true positives. The recommendation of Li et al. [18] to increase the number of training epochs from 300 to 400 as a way to improve performance and mAP is valuable because it is based on empirical data and can help in detecting more objects in images with higher accuracy. At the same time, the high memory consumption of the YOLOv5GPU during training, especially with the 64 batch size of 73.2 GB, indicates significant resource requirements. This can be a challenge for researchers and developers, especially in resource-constrained environments, and requires consideration of alternative optimization strategies or the use of more efficient neural network architectures to reduce computing resource requirements without significant performance loss.

Based on this data, after 30 epochs of training, YOLOv8 showed the highest overall performance, recording the best mean accuracy (mAP) of all YOLO versions. The results of the training showed that each version of YOLO has its own strengths and weaknesses in the context of the task of detecting people in the water. Based on this data, additional model optimizations were made to further improve their performance. This included fine-tuning the models using specific datasets and adapting the architectures to the specific challenges identified during the experiment. Thus, the training process of the YOLO architectures was deeply integrated with the unique requirements of the waterborne detection task, demonstrating the importance of careful data preparation, proper training setup, and an adaptive approach to model evaluation and optimization.

8 Results

After performing thorough experiments using different versions of the YOLO architectures for the task of recognizing people on water, significant data was obtained that reflects the performance of each model. The models were trained on a balanced dataset including images from COCO2017, SARD, and SeaDronesSee to evaluate their performance in a variety of waterborne detection environments. The analysis of the training results after 30 epochs showed that the YOLO models demonstrate different levels of performance, in particular in such key metrics as accuracy, recall, average accuracy at a threshold of 0.5 (mAP@0.5) and F1-score. The YOLO architecture classifies objects using a single neural network pass to simultaneously predict multiple classes and locations of objects in an image. The model divides the image into a grid and for each cell predicts the bounding box, the probability of an object being in the box, and the object's

Fig. 1. Human detection by YOLOv7

classification. This allows YOLO to detect and classify objects quickly and with high accuracy, making it ideal for real-time applications. Figure 1 shows the classification of people on the water using surfboards with YOLOv7.

Figure 2 shows the classification of objects using YOLOv7. As we can see, the YOLOv7 architecture has detected all the people in the frame and highlighted them with a dark blue rectangle with the class "person".

Fig. 2. Object detection by YOLOv7 when human partially submerged on water.

An analysis of "misclassified objects" in the context of people-on-water detection using YOLO architectures revealed that this type of error is an important aspect for optimizing model performance. Misclassified objects can include cases where the model falsely identifies inanimate objects as people or misses people by classifying them as part of the environment. Reducing the number of such errors requires improving training algorithms, increasing the diversity of training data, and optimizing data preprocessing. Figure 3 shows the incorrectly classified "person" object in the frame. Instead of the correct class, YOLO categorized the person as "bird".

In the context of using YOLO for object detection, misclassification can occur due to several factors. Among them is the limited ability of the model to effectively distinguish between objects on complex backgrounds or in conditions of object overlap. Insufficient data diversity can lead to a high specificity of the model, which works well only on data like the training set. Also, the quality of data annotation plays an important role: incorrectly labeled objects in the training set can distort training. In addition, limitations of the YOLO architecture, such as loss of detail due to reduced input image size or suboptimal anchor frame size, can also cause classification errors. To reduce the number of misclassified objects in YOLO architectures, it is recommended to increase the variety and volume of training data, including images from different perspectives and under different lighting conditions. It is important to improve data preprocessing, in particular, the use of augmentation techniques to simulate potential operating conditions. It is also necessary to optimize the architecture, tune hyperparameters, and use techniques to reduce the loss of details on small objects, for example, by introducing more accurate detection mechanisms.

Fig. 3. Misclassified partially submerged human

The YOLOv8 architecture proved to be the most effective among all versions of YOLO models in the task of detecting people on water. This version not only showed high accuracy in detecting persons, but was also able to build skeletons on the detected people. This feature greatly expands the analysis capabilities, allowing for further motion analysis and detection of critical states, such as determining whether a person is alive. This approach opens up new horizons for the development of rescue systems and water safety monitoring. Figure 4 shows the result of recognizing a person in the water using YOLOv8. The person was highlighted with a red rectangle and key points were overlaid for further processing and motion analysis. If the detected rectangle with a person moved and changed its coordinates, it assumed that the person was moving and potentially alive.

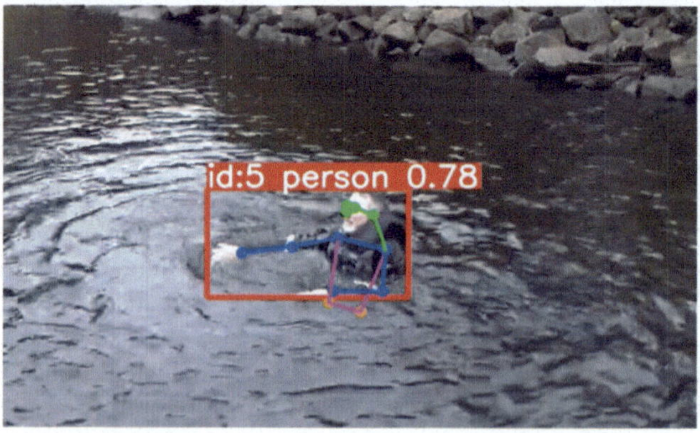

Fig. 4. Object detection when human partially submerged on water

YOLOv8 excels at detecting multiple people in a video frame, demonstrating the ability to build skeletons quickly and accurately for each detected person. Figure 5 shows the result of capturing several people on the surface of the water with YOLO. After detecting a person in the water, a skeleton was built for each of them to track joint movements.

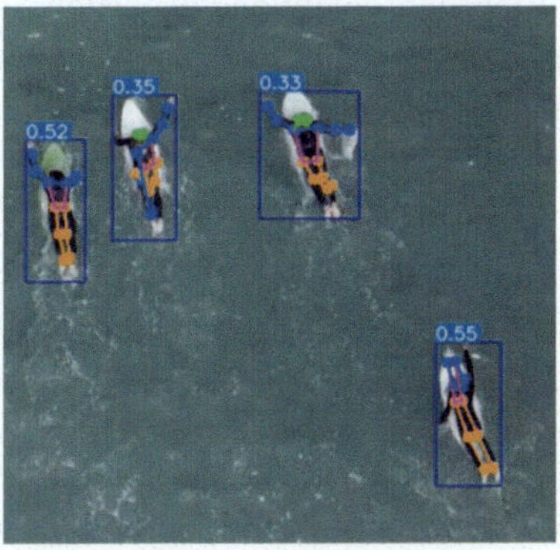

Fig. 5. Multiple human detection by YOLOv8

YOLOv8 is highly efficient in real-time, delivering high frame rates even on devices with limited processing power. This expands its applicability in a wide range of environments and on a variety of devices. A comparative analysis of object detection errors using different versions of the YOLO architecture is performed. This analysis provides a deeper understanding of how each version copes with the challenges of object identification in a variety of environments. By looking at typical errors such as partial detection, misclassification, or complete object misses, we can determine which aspects of the architecture need to be optimized to improve detection accuracy. Table 3 provides a detailed analysis of the errors made by different versions of the YOLO architectures during the human detection experiments. It shows the percentage of cases where objects were partially detected, misclassified, or completely missed by each version of the models from YOLOv3 to YOLOv8.

Table 3. Analysis of detection errors in different versions of YOLO architectures

Architecture	Partially Detected	Misclassified	Missed
YOLOv3	16%	9%	8%
YOLOv4	10%	7%	6%
YOLOv5	11%	8%	6%
YOLOv6	21%	12%	11%
YOLOv7	10%	6%	6%
YOLOv8	8%	4%	4%

It can be concluded that YOLOv8 significantly outperforms the other versions in all respects, indicating its high efficiency and accuracy in detecting people in the water. On the other hand, YOLOv6 showed the highest percentage of errors, which emphasizes the need for further optimization. When the YOLOv8 model fails in challenging scenarios, especially when dealing with low light conditions, occlusion, and different water textures, it can have a significant impact on its performance. Low light conditions pose challenges for any computer vision system due to reduced visibility and contrast between objects and their backgrounds. This can cause YOLOv8 to lose its ability to correctly identify people in the water, as the features the model looks for to classify and localise objects become less distinct. Such conditions can also distort the characteristics of an object, making its silhouette blurry or incomplete. Occlusion presents another serious problem when objects are partially or completely blocked by other objects in the image. This can be particularly common in dynamic water environments where people may be partially obscured by waves or other objects. In such cases, YOLOv8 may misinterpret a person's shape or position, resulting in false detections or misses. Water textures add another layer of complexity, as they can vary significantly from calm water to stormy water with large waves and splashes. Sun glare on the surface of the water or changes in the colour of the water from different angles can create a large number of false positives, where the model may detect objects that are not actually there, or vice versa, fail to detect a person in need of help. The diversity and dynamics of the aquatic environment requires the model to be able to adapt to a wide range of visual conditions, which is a significant challenge.

All of these factors highlight the importance of developing more robust deep learning methods that can better adapt to complex detection environments. Optimising models like YOLOv8 to perform in such conditions is possible by expanding training data, incorporating image enhancement techniques to increase visibility in low light conditions, developing algorithms that better recognise and distinguish objects during occlusion, and using more sophisticated texture analysis techniques to identify people in different water conditions.

9 Discussion

The analysis of the results of the experiment with detecting people on water using different versions of YOLO architectures opens up interesting prospects and challenges for further research. It was found that the improvement of the architecture and training methods from YOLOv3 to YOLOv8 significantly increased the effectiveness of the models in detecting people in the water area. The results of YOLOv8 are especially impressive, which emphasizes the progress in the development of deep neural networks and their application in specific detection tasks. YOLOv8 demonstrated the best results, which indicates significant progress in object detection technology. This version effectively solves the problem of detecting people in the water, demonstrating high accuracy and response. This success can be attributed to improvements in the image processing algorithm, including better recognition of the scene context and optimization to handle a variety of lighting conditions and backgrounds. On the other hand, YOLOv6 showed lower performance, which emphasizes the importance of further optimization and adaptation of models for specific tasks. This opens up a discussion about the need to balance the versatility of models with their specialization for specific use cases. One of the key challenges identified during the experiments is the high computational resource requirements, especially for versions such as YOLOv5. This requires the development of optimization strategies and trade-offs between model performance and the availability of resources for training and implementation. At the same time, the results indicate a significant potential for further improvement of YOLO models through fine-tuning and contextualization. Future research could focus on developing specialized versions of YOLO that are optimized for detecting people in complex environments, such as water areas with a variety of influencing factors. The results of the experiment are important for the development of security and surveillance systems, especially in the context of improving the efficiency of search and rescue operations on the water. Improvements in human detection algorithms can contribute to faster identification of victims and more effective emergency response. In general, the discussion of the results emphasizes the importance of further research in the field of computer vision and the development of intelligent systems capable of adapting to complex and dynamic real-world conditions. The development and optimization of YOLO architectures for specific object detection tasks opens up new opportunities for improving safety and efficiency in various applications.

10 Conclusions

The promising results obtained during the experiments indicate that the YOLOv8 architecture is not only theoretically sound, but also practically applicable. Its high accuracy and low error rate make it an invaluable tool for search and rescue missions, where fast and correct decision-making is of paramount importance. Given its performance, it can make a significant contribution to saving lives in emergencies by effectively locating and identifying people in need of assistance. Thus, the results confirm the effectiveness, reliability, and practicality of the proposed architecture. Due to its high precision, accuracy, recall, and F1-Score, as well as minimal errors, the architecture proves to be better than previous models such as YOLOv3, YOLOv4, YOLOv5, YOLOv6, and YOLOv7. Its practical application in real-world search and rescue missions makes it a significant contribution to the field, opening up prospects for future implementations and developments in emergency response systems. The findings lay a solid foundation for further research and development in this area, paving the way for more advanced and reliable rescue technologies in the future. The modular nature of the technology also means that it can be adapted to different types of disasters and emergencies, from forest fires to earthquakes. In essence, the inherent scalability of this project ensures that as technology evolves, the architecture can be adapted and expanded, making it a long-term solution for disaster response on a global scale.

Acknowledgements. The work was carried out under the auspices of the Software Engineering Department and the research laboratory "Information Technologies in Learning and Computer Vision Systems" at the Kharkiv National University of Radio Electronics. The authors extend their profound appreciation to the scientists from the Technical University of Applied Sciences Wildau,

Brandenburg, Germany, and the Volkswagen Foundation for their invaluable support and contributions. Server used for research is med-IoT-KI, TH Wildau, project 85053663 "AI-supported IoT environment for medical examinations" MWFK program to promote the infrastructure for research, development and innovation (InfraFEI) with funds from the European Regional Development Fund (ERDF). This international collaboration has not only paved the way for groundbreaking research but also fostered an enriching academic and investigative exchange among the participating institutions.

References

1. Alwani, A.A., Chahir, Y., Goumidi, D.E., Molina, M., Jouen, F.: 3D-posture recognition using joint angle representation. In: Laurent, A., Strauss, O., Bouchon-Meunier, B., Yager, R.R. (eds.) Information Processing and Management of Uncertainty in Knowledge-Based Systems. Springer International Publishing, Cham, pp. 106–115 (2014)
2. Chen, C., Min, H., Peng, Y., Yang, Y., Wang, Z.: An intelligent real-time object detection system on drones. Appl. Sci. **12**, 10227 (2022)
3. Gallego, A., Pertusa, A., Gil, P., Fisher, R.B.: Detection of bodies in maritime rescue operations using unmanned aerial vehicles with multispectral cameras. J. Field Robot. **36**, 782–796 (2019)
4. Maji, D., Nagori, S., Mathew, M., Poddar, D.: YOLO-pose: enhancing YOLO for multi person pose estimation using object keypoint similarity loss (2022)
5. Bilous, N.V., Ahekian, I.A., Kaluhin, V.V.: Determination and comparison methods of body positions on stream video. RIC **52** (2023)
6. Hernández, D., Cecilia, J.M., Cano, J.-C., Calafate, C.T.: Flood detection using real-time image segmentation from unmanned aerial vehicles on edge-computing platform. Remote Sens. **14**, 223 (2022)
7. Li, X., Guo, Y., Pan, W., Liu, H., Xu, B.: Human pose estimation based on lightweight multi-scale coordinate attention. Appl. Sci. **13**, 3614 (2023)
8. Neto, J.G.D.S., Teixeira, J.M.X.N., Teichrieb, V.: Analyzing embedded pose estimation solutions for human behaviour understanding. In: Anais Estendidos do Simpósio de Realidade Virtual e Aumentada (SVR Estendido 2020). Sociedade Brasileira de Computação, Brasil, pp. 30–34 (2020)
9. Rakova, A.O., Bilous, N.V.: Reference points method for human head movements tracking. RIC **0**, 121–128 (2020)
10. Smith, J., Loncomilla, P., Ruiz-Del-Solar, J.: Human pose estimation using thermal images. IEEE Access **11**, 35352–35370 (2023)
11. M D, S C, C.M. B,: Robust human detection system in flood related images with data augmentation. Multimed Tools Appl **82**, 10661–10679 (2023)
12. Hramm, O., Bilous, N., Ahekian, I.: Configurable cell segmentation solution using Hough circles transform and watershed algorithm. In: 2019 IEEE 8th International Conference on Advanced Optoelectronics and Lasers (CAOL). IEEE, Sozopol, Bulgaria, pp 602–605 (2019)
13. Wu, X., Li, W., Hong, D., Tao, R., Du, Q.: Deep learning for UAV-based object detection and tracking: a survey. IEEE Geosci. Remote Sens. Mag. **10**, 91–124 (2022)
14. Valarmathi, B., et al.: Human detection and action recognition for search and rescue in disasters using YOLOv3 algorithm. J. Electr. Comput. Eng. **2023**, 1–19 (2023)
15. Tan, L., Lv, X., Lian, X., Wang, G.: YOLOv4_Drone: UAV image target detection based on an improved YOLOv4 algorithm. Comput. Electr. Eng. **93**, 107261 (2021)
16. Dadrass Javan, F., Samadzadegan, F., Gholamshahi, M., Ashatari Mahini, F.: A modified YOLOv4 deep learning network for vision-based UAV recognition. Drones **6**, 160 (2022)

17. Nguyen, H.-C., Nguyen, T.-H., Scherer, R., Le, V.-H.: Unified end-to-End YOLOv5-HR-TCM framework for automatic 2D/3D human pose estimation for real-time applications. Sensors **22**, 5419 (2022)
18. Li, C., et al.: YOLOv6: a single-stage object detection framework for industrial applications (2022). https://doi.org/10.48550/ARXIV.2209.02976
19. He, C., Saha, P.: Investigating YOLO models towards outdoor obstacle detection for visually impaired people (2023). https://doi.org/10.48550/ARXIV.2312.07571
20. Horvat, M., Jelečević, L., Gledec, G.: Comparative analysis of YOLOv5 and YOLOv6 models performance for object classification on open infrastructure: insights and recommendations (2023)
21. Yang, W., Tang, X., Jiang, K., Fu, Y., Zhang, X.: An Improved YOLOv5 algorithm for vulnerable road user detection. Sensors **23**, 7761 (2023)
22. Yan, J., et al.: Underwater object detection algorithm based on attention mechanism and cross-stage partial fast spatial pyramidal pooling. Front. Mar. Sci. **9**, 1056300 (2022)
23. Zeng, Y., Zhang, T., He, W., Zhang, Z.: YOLOv7-UAV: an unmanned aerial vehicle image object detection algorithm based on improved YOLOv7. Electronics **12**, 3141 (2023)
24. Chen, H., Zhou, G., Jiang, H.: Student behavior detection in the classroom based on improved YOLOv8. Sensors **23**, 8385 (2023)
25. Wang, G., Chen, Y., An, P., Hong, H., Hu, J., Huang, T.: UAV-YOLOv8: a small-object-detection model based on improved YOLOv8 for UAV aerial photography scenarios. Sensors **23**, 7190 (2023)
26. Berndt, J., Meißner, H., Kraft, T.: On The Accuracy of YOLOV8-CNN regarding detection of humans in nadir aerial images for search and rescue applications. Int. Arch. Photogramm. Remote Sens. Spatial Inf. Sci. XLVIII-1/W2–2023, 139–146 (2023)
27. Redmon, J., Farhadi, A.: YOLOv3: an Incremental Improvement (2018). arXiv preprint: arXiv:1804.02767
28. Bochkovskiy, A., Wang, C.-Y., Liao, H.-Y.M.: YOLOv4: optimal speed and accuracy of object detection (2020)
29. Jocher, G.: Ultralytics YOLOv5 (2020). https://doi.org/10.5281/zenodo.3908559
30. Li, C., et al.: YOLOv6 v3.0: a full-scale reloading (2023)
31. Wang, C.-Y., Bochkovskiy, A., Liao, H.-Y.M.: YOLOv7: trainable bag-of-freebies sets new state-of-the-art for real-time object detectors (2022). arXiv preprint: arXiv:2207.02696
32. Jocher, G., Chaurasia, A., Qiu, J.: Ultralytics YOLOv8 (2023)
33. Lin, T.-Y., et al.: Microsoft COCO: common objects in context (2014). https://doi.org/10.48550/ARXIV.1405.0312
34. Sambolek, S., Ivasic-Kos, M.: Search and rescue image dataset for person detection - SARD (2021). https://doi.org/10.21227/AHXM-K331
35. Varga, L.A., Kiefer, B., Messmer, M., Zell, A.: SeaDronesSee: a maritime benchmark for detecting humans in open water (2021). https://doi.org/10.48550/ARXIV.2105.01922

Measuring Techniques and Systems

Applications of Machine Learning Methods for Positioning in Scanning Electrochemical Microscope

Vadimas Ivinskij[1(✉)], Vijitashwa Pandey[2], and Inga Morkvėnaitė-Vilkončienė[1,3]

[1] Vilnius Gediminas Technical University, 10223 Vilnius, Lithuania
{vadimas.ivinskij,inga.morkvenaite-vilkonciene}@vilniustech.lt
[2] Oakland University, Rochester, MI 48309, USA
[3] State Research Institute Center for Physical Sciences and Technology, Vilnius, Lithuania

Abstract. A typical scanning electrochemical microscope (SECM) consists of a three-axis positioning system, usually with stepper motors and potentiostat. Working at a micro and nanoscale, the most consistent problem is vibrations, which an antivibration table can solve. However, this is not a solution for some devices placed in small environments. Therefore, we suggest reducing vibrations by machine learning methods, making SECM more modular. In this paper, we compare and evaluate the use of neural networks within similar systems that have specialized architecture requirements, the use of samples that require a specialized approach to attribute extraction concerning feature fidelity for learning algorithms, control-machine learning combination methods for static and dynamic applications in electrochemical systems as well as application of neural network functions as stand-alone packages. The results were taken for the application of prototyping a tangible and portable electrochemical scanning device system. Potential improvements, such as micro stepping instead of time delay between steps, are considered. The proposed methods in this paper are aimed at countering the disturbances caused by transient processes during actuation and control tuning using non-evasive methods for cheaper solution alternatives.

Keywords: Scanning electrochemical systems · Machine learning · Control systems

1 Introduction

Instrumentation of artificial intelligence (AI) methods is used alongside automation technologies to provide an otherwise unavailable approach to control of specialized systems [1]. Several types of software typically accompany the instrumentation of a machine learning algorithm with the trained model and neural network architecture prepared in advance, with the firmware for use on the expected data types. Applying such methods to the electrochemical systems usually presupposes the necessity of precision in the range of 0.1–1.0 μm and a control method that can deal with typically small and gradually indistinguishable disturbances from a specific working environment [2,

R. Szewczyk et al. (Eds.): AUTOMATION 2024, LNNS 1219, pp. 259–271, 2024.
https://doi.org/10.1007/978-3-031-78266-4_22

3]. The standard procedure to minimize outside disturbances for systems based on a combination of electrochemical interactions and associated kinematics is a generalized damping ratio, ζ factor reduction method either in a system or component scale.

An example of more direct approaches to reducing vibrations in such systems is granite slab fixtures, foundations, and anti-vibrational tables or armatures, which are highly cumbersome as an additional anti-damping system. System parameter modification using non-tangible modification methods and machine learning can be used to implement the abovementioned methodologies in an approach that offers more modular electrochemical positioning system solutions. Neural network combinations can be performed to generate feature-extracted information with which we can examine and modify the data morphologically to detect, localize, and replace missing data, similar to traditional control systems. Partially, this would lead to a control system that is portable and less expensive than the damping methods mentioned above. However, these specialized methods are less accurate than more expensive system modifications or popular pre-trained machine learning or computer vision-based neural network models due to limited data and training time. The problems mentioned above can be circumvented by developing proprietary software using combination neural networks (NNs, CNNs, DNNs), perceptron functions, segmentation techniques (Masks, Template Matching), and optimization methods (Simplex based augmentation, Back-propagation, Random Forest) to develop separate algorithms for various instances of expected data related to acquisition parameters for machine learning models [4–6].

These methodologies and their applications can be seen in use in microscopy and optical devices that measure electrochemical activity, such as scanning electrochemical microscopes (SECMs), where sample scanning is processed and fed back into the control of the linear axis actuators. These tasks can, under normal circumstances, be performed inadequately. De-localized mixed samples or unknown samples, as well as outside disturbances, require several system implementations and constant steady scanning, which consumes more time while the probe finds the reactive regions of the given sample. We can achieve similar behavior in an electrochemical-based system without heavy equipment or slow scanning methods by combining conditional statements as machine learning-trained models with system parameter control methods [7].

Solutions for limitations can be categorized into a) positioning of the device using perceptron deep neural networks (DNNs) for classification and detection in the x and y axis, as well as optimization and error reduction methods for finding local maxima as the height of the z-axis, finally concatenating these three values to get a 3D map of the sample which would be similar to the samples produced by typical ultra-micro-electrode (UME) scanning methods; and b) using different categories of photoelectric and capacitive MEMS sensors like the ADXL345-EP 3*axis, high resolution, 3 mm $\times$ 5 mm $\times$ 1 mm micro accelerometer or a device embedded BOSCH BMA150 2–8G I2C/SPI 12LGA accelerometer which output candela, lumen, lux or oscillation magnitude values respectively as conditional operators in the neural network architecture algorithm to detect varying environmental, vibrational and brightness conditions which will in turn switch between control settings of the positional systems operational modes and the neural network models automatically [8].

The SECM can easily be equipped with a top-mounted microscope camera for computer vision model data acquisition and a potentiostat for machine learning model data acquisition followed by data pre-processing and feature extraction using neural network model-based training with an experimentally determined number of layers and their activation functions, loss functions, and back-propagation algorithm. This would allow the system to automatically classify, track, localize, compensate, and replace missing topographical data and rectify the positional inaccuracies from environmental and vibrational disturbances [9, 10].

This article proposes applying machine learning and automatic control with a tangible electrochemical system in one modular package that automates the sample scanning, detection, and positioning of the camera/electrode and counteracting outside disturbances. Machine learning code acts as an automated tool for measurements, consequently resulting in data augmentation of the prepared samples of yeast cell varieties [10].

2 Machine Learning

Machine learning can be divided into categories that solve different morphological feature discrimination methods. These methods separate the data received into sections responsible for their exclusive morphological operations. These operations can also be divided into several machine learning instrumentation categories responsible for the solutions of detection and analytical problems associated with the system and those responsible for the automatic control of the device.

Classification problems for a single class with discretely distinguishable features simplify the process of identifying the label of the object in the detected sample from which we can extract the positional coordinates as the endpoints of the area of the remaining thresholded feature mask. The clear separation and feature range can be seen in Fig. 1, shown below.

Additionally, we can see from graph b) in Fig. 1 that the explicit value range shown is a prime feature candidate for the most distinct feature in the image. We would need more confirmation before selecting this feature –We can divide the thresholding methods into five categories and analyze the thresholded values to see if they match the graph in Fig. 1b). The proposed methods can be categorized into five separate average pixel brightness value calculation cases:

1. Convert the image to greyscale and return average pixel brightness values.
2. Convert the image to greyscale and return RMS pixel brightness values.
3. Average pixels, and then transform to localized brightness values.
4. RMS of pixels then transforms to localized brightness values.
5. Calculate localized brightness values of pixels and then return the average.

Average pixel brightness was determined using several steps, each giving a different result. The results of each method are shown in Table 1 below.

As can be seen from Table 1, the values float around the 155 to 160 intensity mark. Considering the similar environment of the data acquisition methods, these values match with the values from the histogram (b) in Fig. 1 from previous experiments.

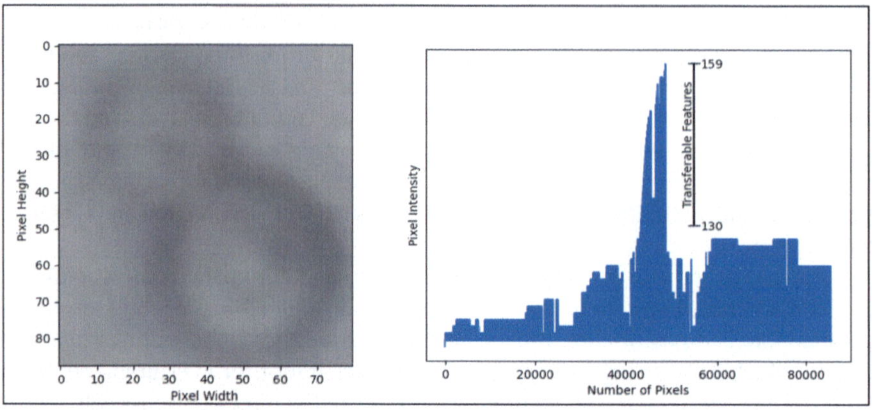

(a) (b)

Fig. 1. The above figure shows the single yeast cell sample under the magnification of the SECM probe camera that shows a) unprocessed double yeast cell 3–5 μm wide, b) the processed image histogram equalization fast Fourier transform that shows the frequency of the most prevalent features of each pixel intensity. From graph b), it can be seen that most of the juxtaposed pixels of the thresholded frequencies lie in the 140–155 value range. Notice the axis is flipped due to the integrated development environments (IDEs) inherent plotting starting point at the top left corner.

Table 1. Averaged brightness or intensity values of the same image output from five different methods. (ITU-R 601–2 luma transform).

Method Nr	Average Intensity Value
Method 1	158.048
Method 2	158.648
Method 3	158.243
Method 4	158.856
Method 5	158.248

The objective for determining our machine learning model parameters using first principles is to determine the perceptron and activation function architecture. Neural networks use layers of neurons, also known as perceptrons, to train the network to optimize feed-forward and back-propagation parameters. The perceptron has a mathematical model similar to a line function ($y = mx + c$). However, in our case value, the magnitude, x, and the bias, shown here as value, c, are unknown and initially taken from our threshold calculation and analysis and then optimized. The machine learning model parameters can be seen from the features of the appearance of the pixels, separated by the perceptrons activation threshold linear function appearance on the graphs in Fig. 2 shown below.

The perceptron separation consideration for non-linearity is not the only prevalent property while extracting information from a sample. Generally, the neural network

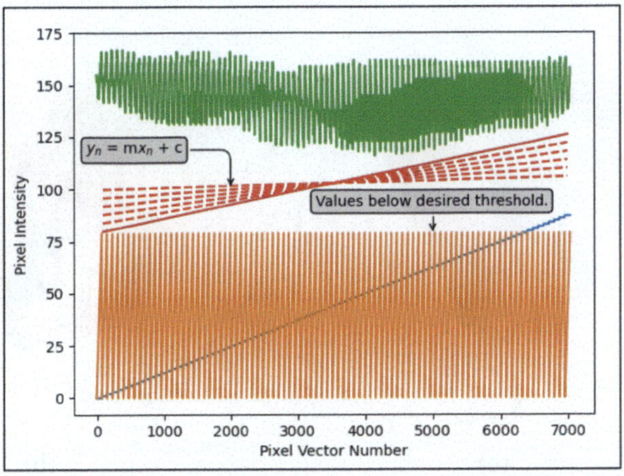

Fig. 2. Graph showing the clear separation of threshold pixel intensity (green) feature space and background values (orange, blue) with a straight line (red) representing the perceptron threshold separation algorithm as a function of a slope and intercept similar to linear regression. The side view of the yeast cell region of interest from the cropped image with the red line linearly separating feature values gives the assumption of the necessary number of perceptrons and activations based on variability. (Color figure online)

and deep layer model must consider the juxtaposition before the final algorithm for the neuron architecture is decided. The complex feature plane contrast between different pixel parameters morphology and topography can be manipulated to justify expanding the deep layer and the activation threshold number. Looking at the 3D-mapped image of the pixel brightness intensities, we can determine if the sample geometry contrast can be adjusted and used for localization and detection. An example is shown in Fig. 3 below.

Graph (a) from Fig. 3 above shows that the values must be taken from the deep "valleys" of the pixel intensity map to have a distinct enough juxtaposition for correct localization and appropriate labeling classification. At such conditions and surface morphology, the data is insufficient for accurate feature extractions and needs to be modified to produce the desired output in the hidden layers of the neural network. Several techniques can help us make the desired results using normalization methods and additional medial equalization pixel morphology, as shown in Fig. 6 below.

3 Theoretical Control Method

The control models used are based not on the individual component and housing equivalent circuit construction but rather on the step response of the shaft load during rotation by an incoming voltage control signal. Reactive damping methods and theoretical system time domain control scheme property manipulation methods increase the stepper motors' time delay between executing discrete rotational commands between pulses. The chosen motor and its parameters were selected from the "Standa" company data.

Micro-stepping is considered a more consistent control method for future implementation due to advanced digital motor shaft angle control (Table 3).

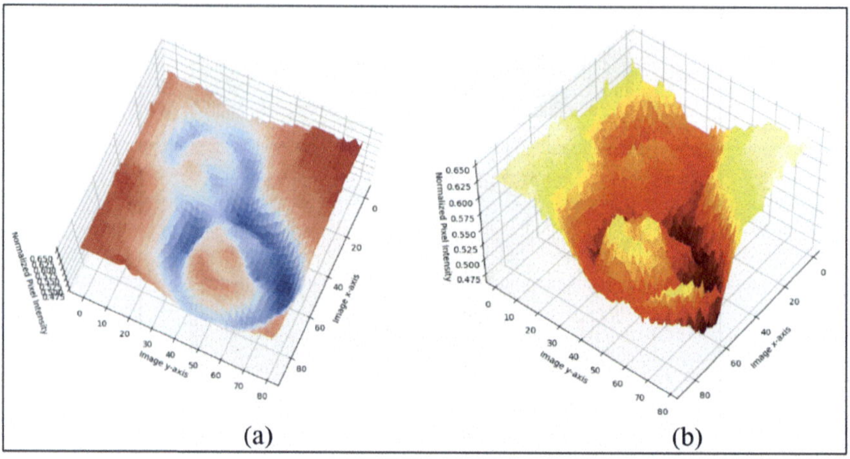

Fig. 3. Pixel input space cool-warm and heat maps of the pixel values mapped in 3D space. Graph (a) shows the possible threshold area view of the yeast cell input space for object detection data points, and (b) shows the heat map with the most prominent features shown as minimums and maximums in valleys and peaks of the 3D input space

Table 2. Chosen Stepper Motor parameter overview.

Model name	Wiring scheme	Design Voltage, V	Phase Current, A	Holding Torque, Nm	Detent Torque, mNm	Steps/Revolution
Model 28	Bipolar	4.56	0.67	0.106	3.75	200 (1.8°)

Table 3. Control stepper motor system theoretical response comparison based on time delay, t_d variation for increased damping equating to amplitude reduction.

Step response, #	Natural frequency, ω_n rad/s	Damped frequency, ω_d rad/s	Damping ratio, ζ	% Overshoot, M_p	Time delay, t_d ms
Graph 1	1.999	1.977	0.15	62.091	0.581
Graph 2	0.71	0.6	0.3	37.232	0.636
Graph 3	1.917	1.534	0.6	9.5	0.741
Graph 4	5.279	2.301	0.9	0.152	0.309

We can tell from Table 2 that the time delay and the damping ratio impact overshoot, representing the vibrational disturbances of the surrounding environment and the vibrations due to inertia in the joints of the connection between the motor and linear screw stages. As seen from theoretical data, the problems mentioned above can be compensated either with delay or some physical frictional component motion reduction. However, secondary harmonics in resonance and the physical disturbance to the device structure can harm the system due to increased discrepancy in mixing physical components. Therefore, an increase in the damping coefficient through the increase in time delay is more favorable to the precision working of the system.

Time delay, as seen from the programmable code, can be easily implemented in controlling the stepper motors between each control step. However, it remains challenging to insert the delay into the operational movement of the drive itself, as seen from step response plots in Fig. 4 below, without physically altering the device with something like automatic brakes or additional frictional material. Here, we can implement a theoretical concept in stepper motors called micro-stepping, in which we can divide the angle of each step into potentially 255 smaller components. However, instead of dividing them into pieces, we divide them into operational timing algorithms, similar to a timing graph.

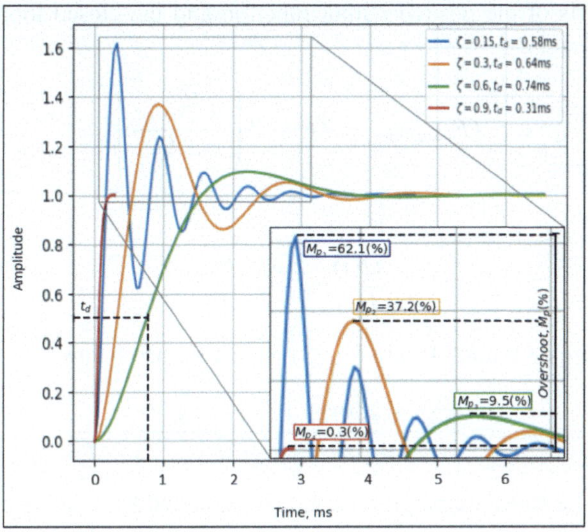

Fig. 4. Graphs showing the influence of various time delays on the second-degree system response

The last graph, d) in Fig. 4, shows the most decrease in amplitude, which would be the apparent choice of parameters for our system considering the lowest values for overshoot (actually under-shoot) and steady state. However, that is not the case since the system response shows an under-damped trend which would result in loss of inertia and torque, thus preventing the stepper motor from turning the linear screw drive and thus preventing the movement of stages. The micro-stepping method, mentioned before, will make the stepper motor run slower and more accurately; however, to increase a linear drive's speed, as an example for rapid motion, we would have to increase pulse frequency, which requires a different control algorithm complicating the overall system control. In terms of time delay introduced into the system response shown above, also known as second order plus time delay (SOPTD) system response, an appropriate method to deliver such delay other than physical damping method would require consideration. An increase in gain or signal delay by adding a dead-time term to the end of our system transfer function that would serve as a time delay is shown in Eq. 1 below.

$$TF = \frac{\omega_n^2}{s^2 + 2\zeta\omega_n s + \omega_n^2} e^{-T_d s} \tag{1}$$

Rewriting the above transfer function into pole form we get Eq. 2, shown below with time constants, τ and gain, K.

$$TF_{CL}(s) = \frac{Ke^{-T_d s}}{(\tau_1 s + 1)(\tau_2 s + 1)} = \frac{Ke^{-T_d s}}{\tau^2 s^2 + 2\zeta\tau s + 1} \tag{2}$$

Applying direct synthesis method to Eq. 2 and assuming the response of the system is satisfactory with first order kinematic approximation, as we do not wish for the linear screw shaft to oscillate, we can write the combined approximation for a standard PID with separate parts of the desired output function and the closed loop plant function shown below in Eq. 3.

$$TF_{controller} = TF_{CL}{}^{-1} \left[\frac{\frac{O(s)}{I(s)}}{1 - \frac{O(s)}{I(s)}} \right] \tag{3}$$

where,

$$\frac{O(s)}{I(s)} = \frac{e^{-T_d s}}{\tau s + 1} \tag{4}$$

Substituting Eqs. (2) and (4) into Eq. (3) we get Eq. (5) below,

$$TF_{controller} = \frac{\tau^2 s^2 + 2\zeta\tau s + 1}{Ke^{-T_d s}} * \frac{e^{-T_d s}}{(2\zeta\omega_n)^{-1} s + T_d s}, \zeta \neq 0 \tag{5}$$

Simplifying the above expression we get an expanded form in Eq. 6:

$$TF_{controller} = \frac{\tau^2 s^2 + 2\zeta\tau s + 1}{K} * \frac{1}{(2\zeta\omega_n)^{-1} s + T_d s} = \frac{1}{K} * \frac{1}{((2\zeta\omega_n)^{-1} + T_d)s} + \frac{2\zeta}{K} * \frac{1}{(2\zeta\omega_n)^{-1} + T_d} + \frac{1}{K} * \frac{\tau^2 s}{(2\zeta\omega_n)^{-1} + T_d} \tag{6}$$

Comparing Eq. 6 with the standard PID controller form from [11], we get each component of the proportional, integral and derivative gain as a separate equation as shown below.

$$TF^{ideal}_{controller} = K_c \left(1 + \frac{1}{\tau_I s} + \tau_D s \right) \tag{7}$$

$$TF_{controller} = \frac{1}{K} * \frac{2\zeta}{(2\zeta\omega_n)^{-1} + T_d} * (1 + \frac{1}{(2\zeta\tau)s} + \frac{\tau^2}{2\zeta\tau} s) \tag{8}$$

$$K_c = \frac{1}{K} * \frac{2\zeta}{2(\zeta\omega_n)^{-1} + T_d}, \tau_I = 2\zeta\tau, \tau_D = \frac{\tau}{2\zeta} \tag{9}$$

As we can see from the above Eqs. (7–9), the exponential factor adds a discrete time delay to the overall system response, however the delay only shifts the response to the right, more specifically it is an offset in the time vector rather than an introduction of a longer time delay into the response curve. Simulating a more damped response is necessary, therefore the micro-stepping method was chosen to achieve system joint level oscillation reduction by dividing the pulse width modulated (PWM) control signal or angles of rotation intro smaller components.

A combined software application used as an all-in-one model has been developed to process several stages during the devices' object detection and positional movement process stages in scanning mode. The results of the most significant theoretical results in pre-processing object topography of the feature and input space, and the concept of a control scheme for the execution of the overall system have been shown and described in the following section.

4 Results

The results for theoretical calculations and theoretical research with equivalent models were performed using Python programming language in an integrated development environment (IDE) using the PyCharm software package. The combination control and neural network machine learning algorithm architecture was developed with dynamic data acquisition and system adaptability to detect objects under various environmental and artificial conditions. This control method integrates machine learning with control methods using modified system parameters for cost reduction and feature implementation. A theoretical control approach is shown in Fig. 5, the diagram below.

The setup consists of a 200X–2000X Zoom Stereo Industry Microscope Camera, Coaxial Light Lamp Monocular 40 mm/50 mm Ring C-Mount Glass Lens, and 4K Ultra HD 30fps HDMI and USB Output Digital Microscope Camera with SONY imx334 1/1.8″ CMOS Sensor mounted on top of a three-axis stage structure connected to proprietary drivers and firmware. The drive has a Raspberry Pi board as the central controller for the triple motor linear screw stage positioning setup. The initial data is fed manually to set up the desired model control method, which gives the training data for the neural network model. Once the neural network is developed and its topography is corrected, we send the data through several modules that determine the coordinates and area of the detected object. These are then fed into a localizer model to superimpose the detected areas as a mask onto the desired image and draw bounding boxes. The received bounding box coordinates and their width and height are sent as a modified g-code string to the driver to move select stages. While the data moves, several sensors can trigger conditional variables with pre-determined coefficients, which are used to switch specific detection models for increased efficiency and detection chance against outside disturbances and unnecessary bioactivity.

The edges of the threshold function, as seen in Fig. 6 (b), show a value of 0.6; however, this is due to the non-normalized axis of the graph. In reality, the edges of the function would have the value of 0 and 1 otherwise in the decision algorithm for classification. However, additional operations have to be applied to measure the correct axis height from the intensities; otherwise, the device will be selected for an incorrect

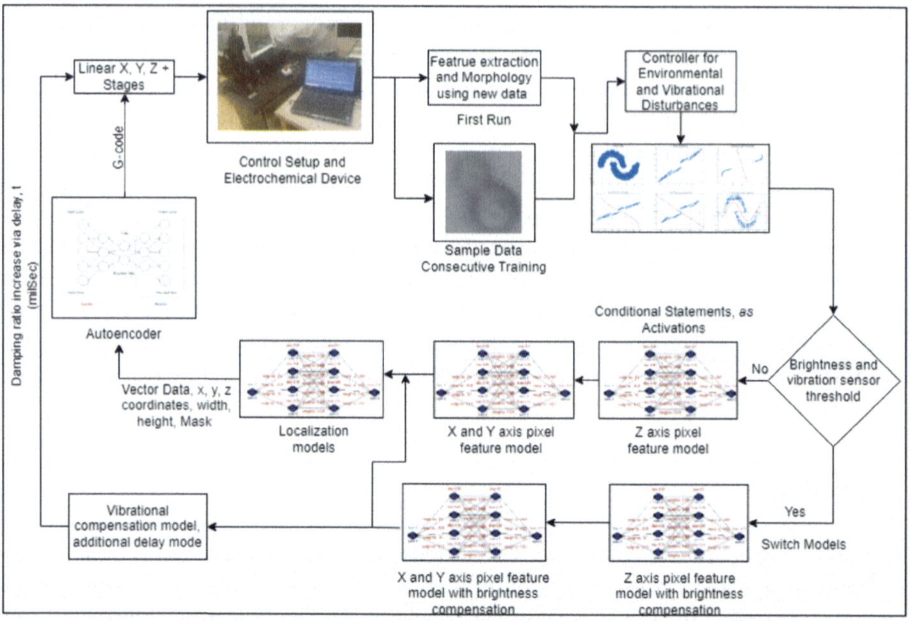

Fig. 5. Control algorithm consisting of a principled experimental setup with an electrochemical positional system and top-mounted microscope from which the initial data was gathered.

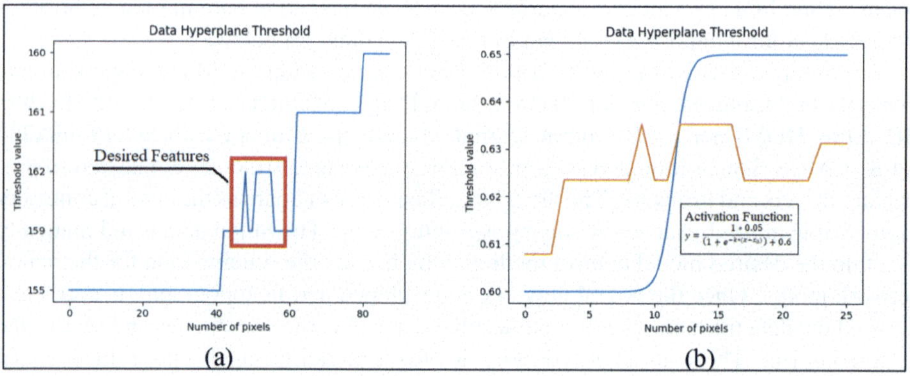

Fig. 6. Desired learned feature space decision boundary – (a) and desired are with sigmoid activation threshold function – (b) showing the possible fit for a sigmoidal curve for the selected feature data. It can be seen that the discrete data points of the pixel intensity will be picked up, theoretically, by the smooth sigmoidal curve from pixels 7 to 18, and the rest of the fill falls on the ends of the function where it will turn to zero, which will enable us to filter out the desired feature data.

constant height. A trial and error approach to find a constant coefficient for the actual height and pixel intensity in feature space is possible but impractical; therefore, it was decided to use a kernel feature filter to average and smooth the input space of the values, as shown below.

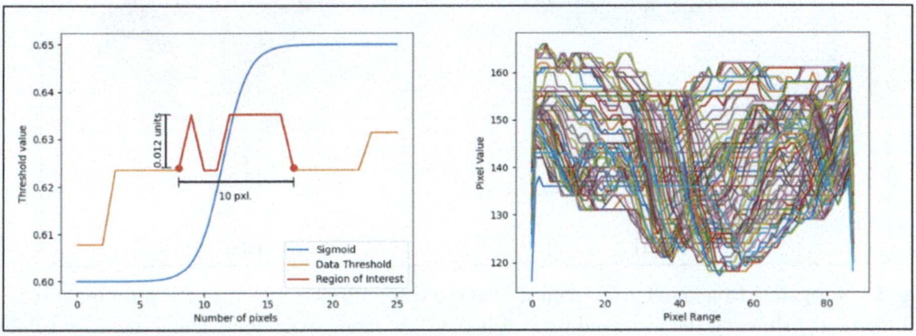

Fig. 7. Feature space histogram, (a) highlighting desired pixel feature spectrum (red) and intensity value as height, $h_{f_{intensity}}$. . Input space two dimensional view of each pixel channel shape – (b). (Color figure online)

It can be seen that the extreme valleys and canyons in the input 3D space of graph (b) in Fig. 7 are entirely predominant in maxima and minima values, and the surrounding area smoothed using a filtering method. It can also be noted that the pixel intensity value per feature, $h_{f_{intensity}}$ of 0.012 is not to scale with the number of pixels and their overall dimensions, giving the notion of disproportionate height, which is critical for constant height scanning methods. The algorithm developed using the Python language uses the array functions exclusively to manipulate the input space satisfactorily using coefficients to vary the level of feature suppression and rounding error reduction. This results in a smoother and more realistic surface for the data transfer of the actual height of the constant scanning mode in the system's positional control, which will trigger during data acquisition.

It can be seen from Fig. 8 above that every pixel intensity channel in the array except for two outlying values has been smoothed over with the received discrete measurements. This allows us to derive the sample's height and, consequently, the axis's position when using the specific stages for constant height measurement in the electrochemical device. The coefficient of the number of iterations for each pixel averaged magnitude, S_n chosen value was five and threshold 20. The threshold range seems to be between approximately 140 and 160 pixel intensity non-normalized. A kernel filter was used to go over each pixel in the given image using a function for rounding from the "Numpy" Python library. The feature space in Fig. 8 above corresponds to the input space in Fig. 7 (b).

5 Conclusions and Summary

The methodologies and control algorithm concepts of the instrumentation of machine learning and control system combinations directly support the need for cheaper and more modular alternatives for electrochemical scanning technology research. If turned into practice, these concepts would enable faster and more precise scanning and better

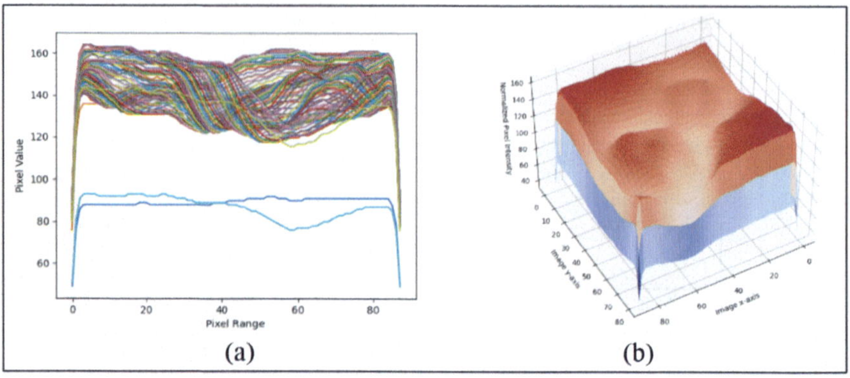

Fig. 8. Normalized averaged pixel intensity feature space surface map – (a) and pixel input space 3D map – (b) showing the averaged pixel values of the input space with actual intensity values showing a smoothed out area for increased positional accuracy.

control system robustness. In general, as mentioned before, the traditional methods for control and optimization would need to be replaced by machine learning and dynamic control methods to be beneficial.

Follow-up research will focus on determining methods for concrete feature extraction for more concise feature manipulation and try implementing more unique techniques in learned feature decisions, such as the ones used for the so-called Fourier Neural Network models. The use of experimental research will lead to determining the continued trend in the development of future neural network models and their architecture. The algorithms in machine learning and control methods mentioned in this article will be used to not only provide solutions to existing practical problems but to create open-source software from publicly generated databases from which continued data can be applied for more broad neural network feature generated structures and models. As mentioned, this will contribute to a more portable device design and reduce the costs associated with practical applications of tangible electrochemical system development.

The most challenging feature in machine learning algorithms today is applying the control and machine learning or computer vision and implementing them while running multiple algorithms for different models, which require separate technological utilities such as gate arrays to function. These artificial intelligence models can be further improved with the addition of force sensors and other kinetically sensitive input transducers to determine and conditionally correct the displacement caused by the absolute coordinate error for increased precision of the positional stages as well as provide better feedback for the g-code coordinate generation for example in the scanning probe microscopes.

Acknowledgements. This project received funding from Taipei Tech – Vilnius Tech Joint Research Collaborative Seed Grant Program 2023.

References

1. Li, Z., et al.: Machine-learning-based positioning: a survey and future directions. IEEE Netw. **33**(3), 96–101 (2019)
2. Makulavičius, M., Dzedzickis, A., Bučinskas, V., Subaciute-Zemaitiene, J., Morkvenaite-Vilkonciene, I.: Theoretical simulations of scanning electrochemical microscope positioning system. automation 2022: new solutions and technologies for automation, robotics and measurement techniques. AUTOMATION 2022. Adv. Intell. Syst. Comput. **1427** (2022)
3. Lazenby, R.A., McKelvey, K., Unwin, P.R.: Hopping intermittent contact-scanning electrochemical microscopy (HIC-SECM): visualizing interfacial reactions and fluxes from surfaces to bulk solution. Anal. Chem. **85**(5), 2937–2944 (2013)
4. Esmaeel, A., Elmelegy, T., Abdelgawad, M.: Multi-purpose machine vision platform for different microfluidics applications. Biomed. Microdev. **33**(3), 21 (2019)
5. Leal-Junior, A.G., Campos, V., DÃaz, C., Andrade, R.M., Frizera, A., Marques, C.: A machine learning approach for simultaneous measuring magnetic field position and intensity with fiber Braag grating and magnetorheological fluid. Opt. Fiber Technol. **56**(), (2020)
6. Zhu, R., et al.: ScratchDet: training single-shot object detectors from scratch. In: Proceedings of the IEEE/CVF Conference on Computer Vision and Pattern Recognition, pp. 2268–2277 (2019)
7. Leemann, S.C., et al.: Demonstration of machine learning-based model-independent stabilization of source properties in synchrotron light sources. Phys. Rev. Lett. **123**(19), 6 (2019)
8. Pang, J., Chen, K., Shi, J., Feng, H., Ouyang, W., & Lin, D.: Libra R-CNN: towards balanced learning for object detection. In: Proceedings of the IEEE/CVF Conference on Computer Vision and Pattern Recognition, pp. 821–830 (2019)
9. Wang, J., Liang, K., Zhang, N., Yao, H., Ho, T.-Y., Sun, L.: Automated calibration of 3D-printed microfluidic devices based on computer vision. Biofluidics **15** (2021)
10. Soviany, P., Ionescu, R.T.: Optimizing the trade-off between single-stage and two-stage deep object detectors using image difficulty prediction. In: 20th International Symposium on Symbolic and Numeric Algorithms for Scientific Computing (SYNASC), Timisoara, Romania, pp. 209–214 (2018)
11. İçmez, Y., Can, M.S.: Smith predictor controller design using the direct synthesis method for unstable second-order and time-delay systems. Processes **11**(3), 941 (2023)

Navigation Support System for Rotating Missiles

Dawid Florczak(✉), Dariusz Miedziński, and Robert Głębocki

Faculty of Power and Aeronautical Engineering, Warsaw University of Technology, Nowowiejska 24 Street, 00-665 Warsaw, Poland
{dawid.florczak.dokt,dariusz.miedzinski2.dokt,robert.glebocki}@pw.edu.pl

Abstract. The key problem in the navigation of rotating missiles based on inertial navigation using MEMS sensors is the problem of precisely calculating the rotational speed. Limitations in the accuracy of rotational speed measurement are the main cause of errors in inertial navigation. This problem constitutes a significant barrier to the improvement and control of traditional, unguided rotating missile designs. As a potential solution, an innovative system supporting rotational speed calculation is proposed, which would measure changes in the missiles surroundings. The work presents the issue of developing a system supporting the precise calculation of the rotational speed of a rotating missile. The issues related to the use of inertial navigation systems based on MEMS sensors and the problems resulting from the use of this type of sensors in rocket technology are presented in detail. A dedicated research stand is presented along with the results of preliminary laboratory tests, which indicate the possibility of obtaining data on the missiles inclination angle by analyzing the light intensity in its surroundings.

Keywords: navigation · rotating missiles · MEMS sensors · electromagnetic sensors

1 Introduction

To understand the current use of missile artillery systems in armed conflicts, it is important to examine the characteristics of these conflicts. A key factor here is the transformation of modern battlefields, especially in the context of intelligence gathering. The use of unmanned aerial vehicles, both small ones similar to consumer drones and larger ones resembling civilian light aircraft, has played an important role. In addition, the availability of satellite information, including images across various spectrum ranges, has increased significantly due to lower satellite launch costs and the popularity of satellite services. This has enabled rapid acquisition of detailed information about hostile units, which, in turn, has enhanced the effectiveness of precision missile artillery strikes.

D. Florczak, D. Miedziński and R. Głębocki—These authors contributed equally to this work.

R. Szewczyk et al. (Eds.): AUTOMATION 2024, LNNS 1219, pp. 272–281, 2024.
https://doi.org/10.1007/978-3-031-78266-4_23

The cost of producing guided missiles is much higher than their unguided counterparts, but their economic value is justified by their effectiveness. Generally, missile artillery is used for mass attacks on enemy units to break the front and allow attacks by specialized units. However, due to the proximity of enemy troops, units equipped with missile artillery must be prepared for a response from the enemy's artillery. The development of technology and the ability to precisely determine the position of the enemy, combined with improved propellant materials that have extended the range of missile artillery, have led to a change in their strategic use. This artillery is now equipped with advanced technological effectors capable of directly striking distant important targets, such as ammunition depots, temporary camps, bridges, command centers, and other strategic objects. However, the navigation of the projectiles remains a challenge, especially those flying with high rotational speeds, which affect the spatial orientation of the missiles.

The key problem in the navigation of rotating objects using navigation based on MEMS sensors is the problem of increasing error over time, resulting from the nature of the operation of gyroscopes, the indications of which are burdened with an error increasing per unit of time. It is also worth emphasizing that problems arising at high rotational speeds of rotating bodies also concern not only gyroscopes but also accelerometers. In particular, the key aspects influencing the decrease in the accuracy of determining spatial orientation are the following elements:

- *Gyroscopic Drift*
  Measurements of high angular velocities are subject to drift (slow increase of errors per unit of time) resulting from, among others, imperfections in the production process, temperature fluctuations and mechanical stresses resulting from the forces.
- *Centrifugal Effects*
  High angular velocities can expose MEMS sensors to centrifugal forces. As a result, the forces that appear in the measuring system may deform or cause stress directly on the sensor elements. Due to the above, it is possible for changes in sensor characteristics to occur, causing measurement inaccuracies and contributing to the overall error in the inertial navigation system.
- *Vibration and Mechanical Resonance*
  The operation of the inertial navigation system in an object exposed to vibrations and mechanical resonances resulting from both the translational and rotational motion of the missile is also subject to errors resulting from the very nature of the motion. High-speed spinning can cause vibrations and mechanical resonances in MEMS sensors. These vibrations can disrupt the functioning of accelerometers and gyroscopes, leading to inaccurate readings. Mechanical resonances can cause temporary instability and errors (additional noise) on the sensor communication lines.

- *Limited Dynamic Range*
  MEMS sensors often have limitations in terms of dynamic range (the range of accelerations or angular velocities they can accurately measure). Particularly in the case of rocket technologies, the obvious fact is the issue of purchasing sensors whose measurement range corresponds to the real values of rotational speeds possible during the movement of the missile. Sensors operating within the limits of measurement values are exposed to saturation and loss of accuracy.
- *Cross-Axis Sensitivity*
  MEMS sensors may have lateral sensitivity, which means that accelerations or rotations in one axis may significantly affect measurements in other axes. Due to the physics of missile motion, it is understandable that rotational speeds occur in more than one body rotation axis, resulting from, for example, precession motion. The occurrence of high rotational speeds in such a system may exacerbate lateral sensitivity, leading to additional errors in the calculated navigation parameters.
- *Environmental Conditions*
  Extreme conditions occurring on rotating missiles, also caused by high rotational speeds, imply the emergence of factors that have a particular impact on the accuracy of inertial navigation systems. In particular, elevated temperature and mechanical stress as described above can further increase sensor drift and cause other errors to occur. At this point, it is worth emphasizing that the issue of factors affecting the accuracy of inertial navigation based on MEMS sensors is a multi-layered aspect. Eliminating one factor causing a given type of error may not significantly improve the final accuracy of the navigation system. For this reason, a multi-aspect approach to improving the accuracy of the measurement system is necessary.

In order to limit the impact of errors that increase over time in inertial navigation based on MEMS sensors, calibration techniques are often used. In particular, a popular solution is to connect sensors with other navigation systems such as satellite navigation systems (e.g. GPS, Beidou, Glonass, Galieo). However, due to the ease of jamming satellite navigation systems, currently these systems are not considered for military applications. Moreover, a relatively common solution is also the use of advanced filtering algorithms to increase accuracy and reliability, the popularity of which is growing due to the strong development of microprocessors and the increase in available computing power. Additionally, the designed systems are subject to advanced strength optimization in order to reduce errors occurring due to the mechanical properties of the structure.

The solution to reducing individual components of errors that accumulate over time and errors resulting from other operating factors of the inertial navigation system used in rotating missiles may be an innovative support system - calibrating navigation based on sensors recording changes in factors around the rotating missile. The issue of navigation using electromagnetic waves in visible light is widely used in astro navigation satellites, which calculate spatial orientation by searching the sky for the brightness of the star, which has been presented in publications: [1,2]. However, it should be emphasized that work in the field of inertial navigation based on MEMS sensors is unique in both Poland and Europe. In case of our country, few actions are being taken to develop a national inertial navigation technology, which has been presented in publications: [3–5]. On a European and global scale, the issue of navigation in missiles and sounding rockets is widely addressed by global R&D centers for the development of rocket technologies, as presented in the works: [6–8]. However, it should be emphasized that these works did not strictly address the issue of developing an inertial navigation support system based on sensors of environmental changes around the missile.

2 Navigation Solution Impact on Missile's Accuracy

Surface-to-surface (STS) missiles can be equipped with two types of control effectors: aerodynamic and gasdynamic, which can be divided into thrust-vectoring systems and pulse jet thrusters. The aerodynamic control is utilized by means of control surfaces that can be placed in the front, back or middle of the missile. The main advantage of this type of control is the ability to change both the direction and magnitude of the generated force. The disadvantage is the need to use complicated control mechanisms using electro-hydraulic or pneumatic actuators. Similar pros and cons can be said to the thrust vector control systems that are further limited by the main motor burning time and that the control mechanisms need to be placed in the very harsh environment near the motor's nozzle.

The pulse jet thruster based control is characterized by much simpler construction of the missile, not requiring any moving parts or actuators, but the control is limited to only direction of the generated force and not the magnitude, which is determined by the thruster's propellant burn characteristic. This type of system requires the missile to spin around its longitudinal axis with high angular velocity to be able to fire the thruster in the specified direction with high resolution. The control algorithms used in that systems require the precise knowledge of the missile's roll angle which is obtained as a part of navigation solution from the information gathered by the gyroscopes. Even small errors in this information can cause the thruster to fire in the wrong direction, which is directly affecting the overall accuracy of the missile's hit. Example of the analysis of the navigation sensors' errors on the impact point accuracy for a gasdynamically controlled projectile is presented in [9].

One of the major errors of the MEMS gyroscopes that greatly affect the accuracy of the missile is the saturation. Standard MEMS gyroscopes are operating in the range of 2000 deg/s which is about 5.5 rotations per second (Hz), whereas the guided STS missiles can rotate with angular velocities reaching more than 40 Hz. After the saturation effect occurs during the flight the information about the missile's attitude in roll is irretrievable, which makes the missile lose any hit capability. To keep the accurate information about missile's attitude additional types of sensors need to be used. The ones proposed in this article are the light, infrared and ultraviolet detectors that can be placed on the missile's circumference and keep track of the missile's rotations after the saturation of the gyros.

This system, by providing additional navigation information, should allow to keep accurate track of the missile's attitude angles, therefore allowing precise hit onto the target. The major requirement for this system is to carry on with the information about missiles rotational velocity, after the saturation of the gyros, in all whether conditions.

3 Results and Discussion

A preliminary attempt to investigate the phenomenon of the possibility of determining the rotational speed of a bullet using visible light and infrared sensors has already been undertaken in our country and was presented in detail in the article [10]. However, it should be emphasized that the work was carried out to a very limited extent, using a research stand with limited functionalities. As a part of this work, a dedicated measurement station was developed based on industrial automation components such as a SIEMENS controller, on which the appropriate control software was implemented. The appearance of the control cabinet of the test stand during startup is shown in Fig. 1.

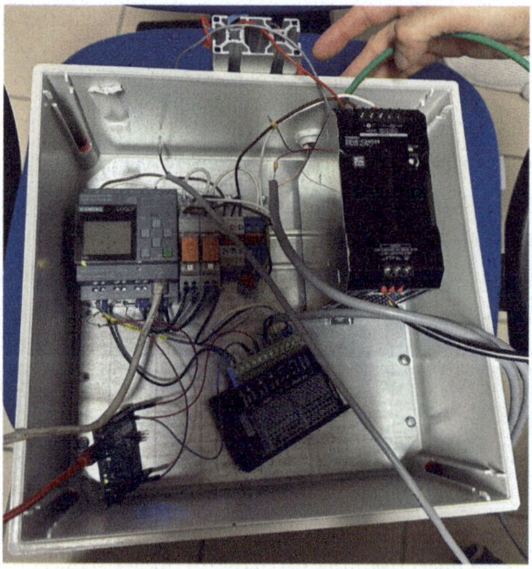

Fig. 1. Stand control cabinet during commissioning tests.

In order to reflect the possible real operating parameters of the system in relation to the potential operation of the system on an effector for use in missile artillery systems, an appropriate research stand was designed. The basic assumptions of the test stand design were: allowing the mounting of the recording electronics on a rotary disk rotating up to 1800°/s (5 Hz) (position of the rotating disk, arrow no. 1) and tilting the entire rotating system using an electric actuator in the range of 0–90° (arrow no. 2). In the following months of work, it is planned to expand the functionality of the test stand in terms of rotational speeds to the order of magnitude of 14,400°/s (40 Hz).

The diagram of the operating concept of the inertial navigation system equipped with correction resulting from the measurement of rotational speed using various electromagnetic wave sensors is presented in Fig. 2. In order to understand the concept of operation, a schematic diagram was developed showing the hypothetical operation of a navigation system containing an additional rotational speed calibration system. The proposed system should be integrated with the rest of the sensors (INS, GNSS) using advance filtering techniques such as Kalman filter and be the integral part of the overall navigation system. A view of the test stand during the first launches and tests is presented in Fig. 3.

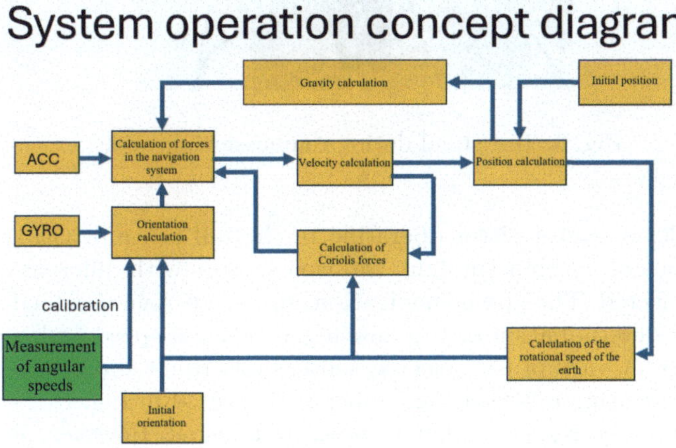

Fig. 2. System operation concept diagram.

It is assumed that the system being developed will cover a wide range of electromagnetic waves, including those visible and invisible to the human eye. For this reason, electromagnetic wave sensors were selected, with an emphasis on infrared sensors, which led to the selection of three types of sensors that were subjected to preliminary tests. The selected sensors are: L-53P3C 5mm phototransistor (visible light sensor), HMS M11 L3.0 F8.0 infrared sensors (with a narrow measurement angle) and a UV-C ultraviolet radiation sensor (GVGR-T11GD). Particularly important is the UV-C sensor, which detects electromagnetic radiation in the range from 300nm to 510nm, i.e. in the ultraviolet. Ultraviolet light, invisible to the human eye, affects the skin aging process because its radiation penetrates the deep layers of the skin and stops only in the bone tissue. Ultraviolet is also effective at breaking through natural barriers such as fog, clouds, water vapor and atmospheric pollution. UV sensors seem ideal for distinguishing Earth's radiation from the sky during the day. Additionally, the

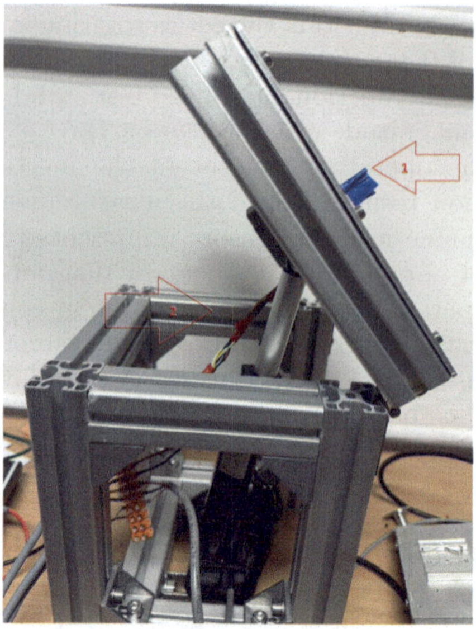

Fig. 3. Test stand during commissioning works.

system includes a photoresistor operating in the full range of visible light. For night conditions or without sunlight, infrared sensors with different observation angles were selected. The missile navigation support system designed in this way should enable determination of the missile's rotational speed and correct gyroscope readings. A view of the selected sensors described above as well as their size scale shown using a measuring caliper is presented in Fig. 4.

The preliminary research plan is presented below. However, it should be emphasized that the evolution of the research plan is assumed depending on the advancement of the stage of work being carried out. Preliminary Research Plan:

1. Launching the station in laboratory conditions
2. Checking the correct operation of the station based on illumination with IR lamps and ultraviolet lamps
3. Research in changing environmental conditions
   (a) Research during the day
      (i) Cloudy sky
      (ii) Clear sky
   (b) Research during the night
      (i) Cloudy sky
      (ii) Clear sky
   (c) Research during the Rainy/foggy day
   (d) Research during the Rainy/foggy night
4. Analysis of conclusions from the obtained results
5. Repeat steps 2–4 for the next set of sensors
6. Drawing final conclusions and selecting the final set of sensors

Fig. 4. View of the sensors selected for testing (left) and representation of the size of the GVGR-T11GD sensor relative to the caliper scale (right).

The preliminary results of running this system for the working conditions of a clear night and a cloudy day are presented in Fig. 5. In the case of the visible light sensor, in accordance with the expected results, the sensor did not work at night or on a cloudy day. The results obtained for the infrared sensor and the ultraviolet sensor allow to clearly determine the rotation speed of the system.

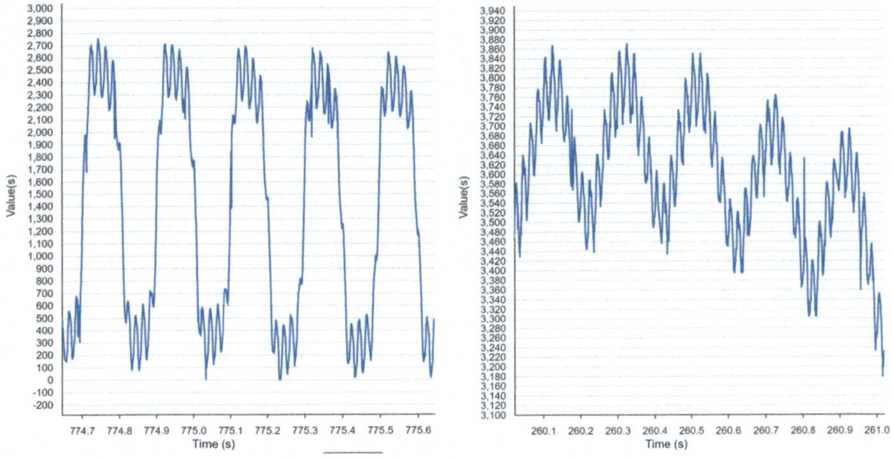

Fig. 5. HMS M11 L3.0 F8.0 infrared sensors working at night 1800°/s (left) and UV-C ultraviolet radiation sensor (GVGR-T11GD) working at cloudy day 1800°/s (right).

4 Conclusions

The preliminary results obtained during the launch of the test stand allow for drawing clear conclusions regarding the positive possibility of applying this concept of the rotating missile navigation support system by measuring changes in external factors. In the following months of work, it is necessary to carry out a series of tests in various lighting conditions (items from the research plan) in order to select the final set of sensors. It should be emphasized that this publication presents only a small fragment of the results obtained for a given set of sensors. In order to properly investigate the above phenomenon, it is necessary to analyze a wide range of sensors and select the set that gives the most realistic measurements of rotational speeds.

During the research, it was observed that it was necessary to design a dedicated algorithm controlling the sensor gain in order to avoid saturation of the system and to obtain self-calibration properties of the systems depending on the prevailing weather conditions.

The proposed system was tested on the 70 mm rocket test platform and the gathered data are being currently under processing. In the following months several tests are planned for a 122 mm rocket platform which will be equipped with the system under discussion. The results obtained from the field tests, combined with detailed results obtained in the following months at the research station, will allow for a thorough understanding of the possibilities of calibrating inertial navigation using external sensors operating in diverse electromagnetic waves.

Acknowledgments. This work was supported by The National Centre for Research and Development Grant number DOB-SZAFIR/03/B/018/01/2021.

References

1. Christian, J.A.: StarNAV: autonomous optical navigation of a spacecraft by the relativistic perturbation of starlight. Sensors **19**(19) (2019). https://doi.org/10.3390/s19194064
2. Chernodarov, A.V., Patrikeev, A.P., Ivanov, S.A.: An integrated inertialodometric navigation system with satellite calibration, pp. 1–6 (2019). https://doi.org/10.23919/ICINS.2019.8769401
3. Pachwicewicz, M., Weremczuk, J.: Accuracy estimation of the sounding rocket navigation system. In: 2018 XV International Scientific Conference on Optoelectronic and Electronic Sensors (COE), pp. 1–4 (2018). https://doi.org/10.1109/COE.2018.8435180
4. Rupniewski, M.W.: PCA-aided calibration of systems comprising multiple unbiased sensors. In: 2023 IEEE Statistical Signal Processing Workshop (SSP), pp. 502–506 (2023). https://doi.org/10.1109/SSP53291.2023.10207946
5. Rupniewski, M.W.: Statistical reconstruction of pulse shapes from pulse streams. In: 2023 IEEE Statistical Signal Processing Workshop (SSP), pp. 235–239 (2023). https://doi.org/10.1109/SSP53291.2023.10207947

6. Braun, B., Grillenberger, A., Markgraf, M.: Performance analysis of an IMUaugmented GNSS tracking system on board the MAIUS-1 sounding rocket. CEAS Space J. **10** (2018). https://doi.org/10.1007/s12567-018-0206-8
7. Wang, K., Hu, T., Zhao, Y., Ren, W., Wang, Y.: Design of an intelligent MEMS safety and arming device with a condition feedback function. Micromachines **14**(6) (2023). https://doi.org/10.3390/mi14061130
8. Scaysbrook, I.W., Cooper, S.J., Whitley, E.T.: A miniature, gun-hard MEMS IMU for guided projectiles, rockets and missiles. In: PLANS 2004. Position Location and Navigation Symposium (IEEE Cat. No. 04CH37556), pp. 26–34 (2004). https://doi.org/10.1109/PLANS.2004.1308970
9. Jacewicz, M., Lichota, P., Miedziński, D., Głębocki, R.: Study of model uncertainties influence on the impact point dispersion for a gasodynamicaly controlled projectile. Sensors **22**(9), 1–20 (2022). https://doi.org/10.3390/s22093257
10. Florczak, D., Głębocki, R., Kopyt, A.: System wsparcia nawigacji wirującego pocisku rakietowego. Mechanika w Lotnictwie ML-XX 2022 **1**(1), 65–75 (2023). https://doi.org/10.15632/ML2022/65-75

Research of Microrobot Movement Accuracy

Viktor Masalskyi(✉), Jurga Subačiūtė-Žemaitienė, Rokas Bagdonas, Vytautas Bučinskas, and Andrius Dzedzickis

Vilnius Gediminas Technical University, 10223 Vilnius, Lithuania
vikror.masalskyi@vilniustech.lt

Abstract. The range of applications of micropositioning systems is rapidly expanding in various fields such as manufacturing processes, medical diagnostics, scientific research and many others. As a result, the demand for high accuracy, reliability and high precision control of micropositioning systems is growing. The main problems requiring research in this area are the search for solutions related to vibrations, inertial effects, mechanical, thermal and assembly errors. This paper describes a micromanipulation solution consisting of drivers of 4-axis (X, Y, Z1, Z2) micropositioning system and determining errors of system (drives with stepper motors, ball screws and slider) from the movement distance. Experimental studies have been conducted on high accuracy micropositioning based on position data received in almost real time with low latency. The experiments included series of measurements the accuracy of the micropositioning system and analyzing the results. The system is control by a computer with user interface and generates commands of movement and send it to motion controller. The results obtained allow an assessment of the influence of the accumulated error, friction of the actuator and etc. The experimental results show that the positioning errors of the microrobot positioning drive are within acceptable limits and the high accuracy is sufficient for micromanipulation purpose.

Keywords: microrobotics · micromanipulation · position errors · error compensation

1 Introduction

Micropositioning devices are used in various industries and technological applications to solve a wide range of issues. First of all, micron positioning accuracy with high precision is important for bacterial studies. Issues of positioning accuracy are solved by different methods and algorithms, but all of them require improvement in connection with the global trend. To solve complex problems, different technologies are used to compensate for the shortcomings of different methods and ways of operating systems. The main problems that these systems have are the accumulation of errors and insufficient positioning accuracy.

Types for micropositioning systems with servo drives and stepper motors based on ball screws are the most popular due to their high accuracy and availability. The need

R. Szewczyk et al. (Eds.): AUTOMATION 2024, LNNS 1219, pp. 282–288, 2024.
https://doi.org/10.1007/978-3-031-78266-4_24

to determine microscope positioning errors is caused by the need to verify the characteristics of the system and its suitability for specific studies. It was possible to achieve the maximum possible parameters of relative and absolute control values. Microrobots have great potential in the biomedical field, where their use is possible in surgery, drug delivery, and bio-sensory [1]. The paper [2] presents three methods of evaluating the accuracy of the movement of robotic systems for magnetic resonance guided focused ultrasound, determining their high accuracy in all axes of movement, which is important for certification in clinical applications.

There are various control strategies for piezoelectric actuators, which are widely used in micro-systems to achieve high-precision motion with fractional force and torque. The paper [3] focuses on solving the major issues of piezoelectric actuators, such as hysteresis and creep, to improve the overall performance of micro-systems. A control scheme based on deep reinforcement learning and iterative learning control can be used to compensate for the hysteresis nonlinearity in piezoelectric micro-positioning platforms with high accuracy [4]. Piezoelectric micropositioners offer high precision and wide applicability in various industries [5]. A novel method for the identification of embedded piezoelectric stack actuator parameters in combination with a real-time capable hysteresis compensation measure is presented in [6]. A new nonlinear hysteresis, linear creep, and linear vibration model can accurately describe the main characteristics of a piezoelectric actuator in [7]. Also for improve the performance of piezo-actuated stick-slip micro-drives used hybrid dynamic modeling and waveform optimization can [8].

A machine learning, one of today's most rapidly growing technical fields. A combination of eye tracking and computer vision can be used to automate the approach of a robot to its targeted point by acquiring its 3D location with an average error of 5.5 cm [9]. A method combining machine vision and machine learning algorithms can be used to estimate adjusting displacement and to determine the correctness of the reed positioning operation in the manufacturing process of free reed aerophones with an accuracy of 100% and a measuring precision of $\pm$ 0.025 mm [10].

In paper [11] presents a comparative dynamic analysis of flexible manipulators and categorizes and studies control strategies. It also introduces fuzzy logic, neural network, and genetic algorithm approaches and presents a range of contributions of such methods in flexible robot control. The development of new learning algorithms and theory, increasing in the availability of online data and low-cost computation, and the adoption of data-intensive machine-learning methods across science, technology, and commerce was review in [12]. Reinforced machine learning improves the positioning accuracy of industrial robots [13]. Online deep Q-learning-based compensation significantly enhances positioning accuracy at key points in robotics [14].

Recent progress in the field of microrobotics, focusing on the ability to grab an object firmly without damaging its surface and structure. In article [15] discuss the limitations and restrictions of microgrippers in various applications, as well as their successful implementation history. A deep reinforcement learning-based control scheme can be used to control the X-Y position of an object in a two-dimensional positioning system with electromagnetic actuators [16].

New methods and algorithms for improving micropositioning are currently being actively researched. A new approach [17] to thermal error control in machine tool feed systems based on presents a novel high-speed/high-precision nut/screw air cooling ball screw system to avoid thermal errors and improve the positioning accuracy of ball screw feed drive systems. A novel 3-DOF flexure-based XYθ nanopositioner achieves a large stroke and sub-nanometer resolution [18]. A new method for modeling and theoretical research of micropositioning systems with stepper motors was proposed in [19]. A novel quasi-finite element and analytical model can be used for accurately predicts the static pull-in parameters of hybrid levitation micro-actuators and researched in [20].

Before conducting biological experiments with bacteria, it is necessary to determine the positioning accuracy of the microscope along the X and Y axes with multiple measurements for each axis and calculate the range of positioning error. It is also necessary to determine the possible accumulation of positioning error along each axis. To carry out these measurements, it is possible to use high-precision sensors and instruments. This approach not only increases the accuracy of the research, but also ensures the reliability of the results necessary for the correct analysis of the results of biological experiments.

Aim of this paper is to provide experimental research of micro-positioning device in order to define actual resolution and evaluate positioning error due to repeatable movement of the robot.

2 Materials and Methods

Experimental research was intended to define accuracy and repeatability of entire micro-robot kinematical system. Experiments were performed using straight line trajectory along X and Y axis. This research tested microrobot drives for non-systematic clearances, fluctuations of the crew pitch and clearances in slides and joints. Resolution of the drives was of interest as well.

During each test drive moved along X and Y axis a line in length of 1, 10 and 50 mm, in order to obtain pitch error in 4 revolutions of the shaft, moderate length of the screw and full range of the crew drive.

Each test was repeated 10 times for each distance. Short distances of the stroke were placed along all screw in order to cover broader area of the screw.

Real displacement of the microrobot was measured using laser distance sensor (Fig. 2), placed on the frame, therefore distance was driven in respect to fixing frame. In other hand, displacement of the microrobot drives was measures using optical microscope and calibrated grid with a cell size of 2 μm was used (Fig. 1).

Results of the measurements was processed using mean values and processed using statistical methods. Error level was evaluated and presented in the result sections.

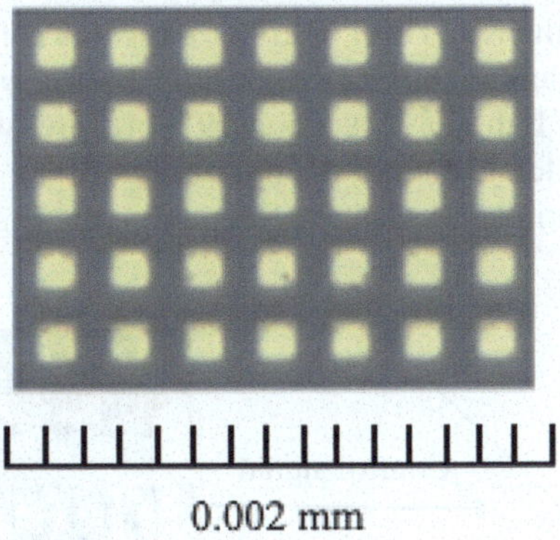

Fig. 1. Calibration grid of cells

3 Equipment

This experimental research was performed using the standard and original equipment. All measurement sensors, microrobot was placed on antivibration table with pneumatic isolation "Standa". Microrobot system includes a universal platform for micromanipulations with optical microscope on Z-axis and separately controlled microtool on Z'-axis. Microrobot is originally designed and produced in Vilnius Gediminas technical university. Microrobot control system is also originally designed and produced in the same laboratory. Externally mounted high-precision laser distance sensor LAT series brand (Germany, Di-Soric GmbH & Co.) was fixed on the antivibration table and used for movement measurement and presented in the Fig. 2. Optical microscope on microrobot was (50X-2000X 4000X Sony IMX385 V2 60FPS HDMI, Japan), with signal connection through USB 3.0 port to PC. Microscope has 500X optical zoom and a 4K color camera sensor.

Fig. 2. Laser distance sensor

Drive control was arranged by stepper motor drivers and data delivered by Orange Pi IV, running separate operating system ARMBian. Directly controlling software was

LinuxCNC package MachineKit. This controller allows to calibrate drives and remove regular clearance using internal software algorithm. Entire system of microrobot controlled by master PC with LabVIEW (version 19) software, which enables to perform optical view analysis and define path of the microrobot.

Structure of used experimental system presented in the Fig. 3.

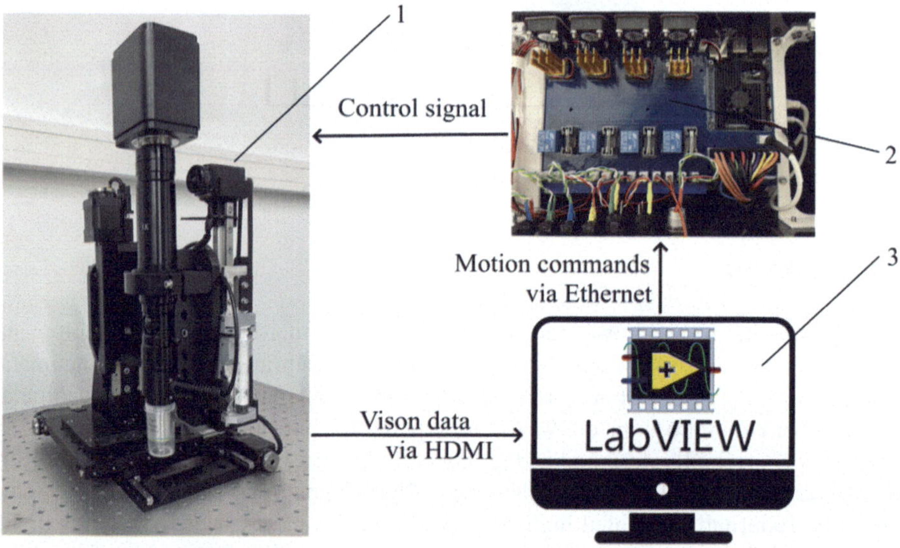

Fig. 3. Composition of the microrobot system: 1 – microrobot, 2 – LinuxCNC based on motion controller, 3 – LabVIEW based on host computer

All experimental research performed in the laboratory of Department of Mechatronics and robotics.

4 Results of Experimental Research

Checking the accuracy of the microrobot was carried out experimentally. Results are presented for the X and Y axes and are averaged measurements. The obtained experimental data show almost the same accuracies of the actuators. The test results allow us to draw a conclusion about the effect of drive friction, accumulated error, and the size of the motor micro step on the stroke error. Errors in the number of micro steps are not frequent, the inertial effect is small and the load on the drive is quite stable. The average measurement value is shown in Table 1.

More significant pitch errors are influenced by accumulated errors, while minor errors are a consequence of roll friction. The kinematic error decreases as the drive warms up.

Table 1. Measuring system accuracy

Distance and axis		Average of measurements
1 mm, X	Measurement error	0.001
10 mm, X	Measurement error	0.005
50 mm, X	Measurement error	0.003
1 mm, Y	Measurement error	0.001
10 mm, Y	Measurement error	0.001
50 mm, Y	Measurement error	0.001

5 Conclusions

After performed experimental research was obtained, that designed microrobot has sufficient accuracy for many microscale operations. Accuracy of movement reached 1 μm, however there was some erroneous measurement with error of 7 μm.

Resolution of drive was 0.5 μm, but clearances and friction in the system makes real accuracy of movement of 1 μm. It was noticed that:

- Clearances in the microrobot is well compensated by software;
- Stability of measurements was not dependent of number of trials;
- Thermal stability and error compensations still need to be researched.

References

1. Jiang, J., Yang, Z., Ferreira, A., Zhang, L.: Control and autonomy of microrobots: recent progress and perspective (2022). https://doi.org/10.1002/aisy.202100279
2. Antoniou, A., et al.: Simple methods to test the accuracy of MRgFUS robotic systems. Int. J. Med. Robot. Comput. Assist. Surg. **17**, e2287 (2021). https://doi.org/10.1002/RCS.2287
3. Sabarianand, D.V., Karthikeyan, P., Muthuramalingam, T.: A review on control strategies for compensation of hysteresis and creep on piezoelectric actuators based micro systems. Mech. Syst. Signal Process. **140**, 106634 (2020). https://doi.org/10.1016/J.YMSSP.2020.106634
4. Zhou, M., Yu, Y., Zhang, J., Gao, W.: Iterative learning and fractional order PID hybrid control for a piezoelectric micro-positioning platform. IEEE Access **8**, 144654–144664 (2020). https://doi.org/10.1109/ACCESS.2020.3014725
5. Juuti, J., Leinonen, M., Jantunen, H.: Micropositioning. Piezoelectric Acoust. Mater. Transducer Appl., 319–340 (2008). https://doi.org/10.1007/978-0-387-76540-2_16/COVER
6. Juhász, L., Maas, J., Borovac, B.: Parameter identification and hysteresis compensation of embedded piezoelectric stack actuators. Mechatronics **21**, 329–338 (2011). https://doi.org/10.1016/J.MECHATRONICS.2010.12.006
7. Tayouri, S.; Izadi, I.; Ghaisari, J.: Modeling and parameter identification of piezoelectric actuator in micropositioning systems. In: ICEE 2019 - 27th Conference on Electrical Engineerin, pp. 1193–1198 (2019). https://doi.org/10.1109/IRANIANCEE.2019.8786431
8. Nguyen, X.-H., Mau, T.-H., Meyer, I., Dang, B.-L., Pham, H.-P.: Improvements of piezo-actuated stick-slip micro-drives: modeling and driving waveform. Coatings **8**, 62 (2018). https://doi.org/10.3390/coatings8020062

9. Leroux, M., Raison, M., Adadja, T., Achiche, S.: Combination of eyetracking and computer vision for robotics control. In: IEEE Conference on Technologies for Practical Robot Applications (TePRA 2015) (2015). https://doi.org/10.1109/TEPRA.2015.7219692
10. Xu, J., Qin, K., Xu, Y., Ji, W.: Method combining machine vision and machine learning for reed positioning in automatic aerophone manufacturing. In: 2019 4th International Conference on Robotics and Automation Engineering (ICRAE 2019), pp. 140–147 (2019).https://doi.org/10.1109/ICRAE48301.2019.9043784
11. Rahimi, H.N., Nazemizadeh, M.: Dynamic analysis and intelligent control techniques for flexible manipulators: a review. Adv. Robot. **28**, 63–76 (2014). https://doi.org/10.1080/01691864.2013.839079
12. Jordan, M.I., Mitchell, T.M.: Machine learning: trends, perspectives, and prospects. Science (80-.). **349**, 255–260 (2015)
13. Sumanas, M., Petronis, A., Bucinskas, V., Dzedzickis, A., Virzonis, D., Morkvenaite-Vilkonciene, I.: Deep Q-learning in robotics: improvement of accuracy and repeatability. Sensors **22**, 3911 (2022). https://doi.org/10.3390/s22103911
14. Bucinskas, V., et al.: Improving industrial robot positioning accuracy to the microscale using machine learning method. Machines **10**, 940 (2022). https://doi.org/10.3390/machines10100940
15. Bučinskas, V., Subačiūtė-Žemaitienė, J., Dzedzickis, A., Šutinys, E., Morkvėnaitė-Vilkončienė, I.: Robotic micromanipulation: b) grippers for biological objects. Robot. Syst. Appl. **2**, 1–14 (2022). https://doi.org/10.21595/RSA.2022.22324
16. Bejar, E., Moran, A.: Deep reinforcement learning based neuro-control for a two-dimensional magnetic positioning system. In: Proceedings - 2018 4th International Conference on Control, Automation and Robotics (ICCAR 2018), pp. 268–273 (2018).https://doi.org/10.1109/ICCAR.2018.8384682
17. Xu, Z.Z., Liu, X.J., Lyu, S.K.: Study on positioning accuracy of nut/shaft air cooling ball screw for high-precision feed drive. Int. J. Precis. Eng. Manuf. **15**, 111–116 (2014). https://doi.org/10.1007/S12541-013-0312-7/METRICS
18. He, S., et al.: A novel flexure-based XYθ motion compensator: towards high-precision wafer-level chip detection. In: Proceedings 2016 IEEE 18th Electronics Packaging Technology Conference. EPTC 2016, pp. 381–387 (2017). https://doi.org/10.1109/EPTC.2016.7861509
19. Subačiūtė - Žemaitienė, J., et al.: Mathematical modelling and theoretical research of micropositioning system. In: IOP Conference Series: Materials Science and Engineering, vol. 1239, p. 012011 (2022). https://doi.org/10.1088/1757-899X/1239/1/012011
20. Poletkin, K.: Static pull-in behavior of hybrid levitation micro-actuators: simulation, modelling and experimental study. IEEE/ASME Trans. Mechatronics **26**, 753–764 (2020). https://doi.org/10.1109/TMECH.2020.2999516

Experimental Research of Carrageenan-Based Compression Force Sensors

Jurate Petroniene[1(✉)], Uldis Žaimis[2], Gediminas Braziulis[1], Andrius Dzedzickis[1], Egle Pastarnokiene[1], and Vytautas Bučinskas[1]

[1] Faculty of Mechanics, Department of Mechatronics, Robotics, and Digital Manufacturing, Vilnius TECH, Plytines str. 25, 10105 Vilnius, Lithuania
jurate-jolanta.petroniene@vilniustech.lt

[2] Institute of Science and Innovative Technology, Liepaja University, Liela str. 14, Liepaja, Latvia

Abstract. Carrageenan is natural polysaccharide derived from Baltic seaweed *Furcellaria lumbricalis*. This natural polysaccharide has wide application in sensor industry. Such products as the humidity sensors, stretch sensors are already proposed with various solutions. Since the abundance of these seawead on the coast poses ecological problems, it is necessary to carefully evaluate the circumstances of the more active use of this product of natural origin in production. Our scientific group conducted research on the production of a stretch sensor from carrageenan collected on the shores of Riga. Carrageenan itself is electrically conductive until it has some moisture in its structure. The formed carrageenan biopolymer film was enriched with various metal oxides. The most suitable and cheap additive is iron oxides. We chose Fe_2O_3 oxide because of its unique properties already confirmed in previous studies. The performed stretching tests of the product confirmed the possibility of using this carrageenan, which is not fully chemically purified, as the main material of the stretching sensor after protecting it from moisture loss with an additional layer of ecologically friendly polymer. Such a sensor can be used as a disposable product suitable for recording small deformations. The advantage of such a sensor is sustainability, since carrageenan is environmentally friendly and non-toxic in contact with the human body. The resulting characteristic of the force sensor is close to quadratic. Such a sensor can be used for loads up to 1 N per square centimeter unit. Elaborated carrageenan-based sensor has quadratic dependence of resistance on applied force.

Keywords: Stretch sensor · Resistance · Seaweed application · *Furcellaria lumbricalis* · Carrageenan · Iron (III) oxide

1 Introduction

Force sensors as a group of devices are designed using different materials to measure the applied force on a surface [1, 2]. The most popular current force sensors have the ability to adapt and adapt to various shapes and surfaces, allowing them to be integrated into a wide range of applications [3]. Recently, it has been emphasized that the long-lived

R. Szewczyk et al. (Eds.): AUTOMATION 2024, LNNS 1219, pp. 289–298, 2024.
https://doi.org/10.1007/978-3-031-78266-4_25

population of the world is increasing and with the overall increase in the number of older people, the need to control the parameters of aging-related diseases is also increasing, and there is a great need for health monitoring and care adapted to personal needs [4]. Consumers are concerned with sensitive wearable sensors that can be safely applied to human skin. Sensors integrated with textile products are no less attractive. Such sensors are designed to monitor the physiological parameters of the human body or the surrounding environment [5]. Nanomaterial-based sensors, which are convenient wearable sensor units, are most attractive to manufacturers [4]. Currently, there are many publications discussing the latest nanomaterials adapted for the advancement of wearable sensors. There are sensors providing simultaneous information on temperature, electrophysiological, deformation, tactile, electrochemical changes [3]. When developing complex structures based on nanomaterials for multimodal sensing and incorporation into portable systems, it is not always taken into account that the product is in some sense disposable. Most of the new product development highlights the typical healthcare application capabilities of nanomaterial-based wearable sensors, including continuous health monitoring, daily and sports activity monitoring, and multi-functional electronic skin, where the processed information is delivered to the user on their device. However, the reuse of raw materials used for the sensor is much less valued, especially since successfully extracting relevant chemicals or components from nanostructures may not be a financially viable solution. Thus, the search for nature-friendly materials that correspond to the idea of sustainability of the age is a relevant enough topic, no less important than the development of new complex nanostructures. The application of sensor technologies has been developed over the past decades, ushering in the stages of better understanding and monitoring of structural and environmental health [6, 7]. Global digitization poses certain challenges for nature conservation and sustainability. One of the attractive substances of natural origin is carrageenan, which is extracted from seaweed [8]. There are many proposals for force sensors using the natural polymer carrageenan using various technical solutions, ionic liquids and conductive nanomaterials [9]. However, the commonly used sensors have certain advantages as well as disadvantages. Due to their good biocompatibility, hydrogel conductors are considered as sustainable bioelectronic materials. They are used for use in wearable sensors and implantable devices [10]. The carrageenan extracted from red algae *Furcellaria lumbricalis* from Baltic seacoast already has wide applications in various industries, including food production[11–13]. Red seaweed contains this naturally occurring polysaccharides that fill the voids in the plant's cellulose structure. This family of polysaccharides includes carrageenan and furcellaran [14]. The molecular composition of carrageenan as a biopolymer can have variations depending on the change of environmental conditions, and at the same time this can affect the quality of the biopolymer. Thus, its neutrality in contact with the human body has been studied in detail, its safety has been confirmed for use in devices that come into contact with human skin [15]. Historically, a galactan mixture from Furcellaria lumbricalis was one of the first hydrocolloids produced industrially from red algae. Thus, the first attempts to produce furcellarans were made in Denmark more than a hundred years ago. Thus, the long practical multifaceted application of this material seems to leave no room for new ideas, but with the expansion of scientific knowledge, there is an opportunity to take a fresh look at the well-studied material [11, 16]. The preparation of the gel by heating and cooling the solutions of these polysaccharides and using additional materials depends on whether soft, elastic gels with iota-carrageenan and firm, brittle gels with kappa-carrageenan and furcellaran will be produced. Lambda-carrageenan is also obtained and gives viscous solutions. As an interesting fact, we should mention that carrageenan interacts with proteins to stabilize, thicken and solidify various products, including milk drinks and shakes, ice creams and dairy desserts [17], also, carrageenan has been successfully applied to vaccine production due to its known antiviral activity [18, 19]. But these red algae are not just an attractive raw material. Their excess causes problems in the Baltic Sea [11]. Gel polymers have already found applications for flexible electronic devices due to their unique properties. However, it is still a big challenge to create multi-environmentally resistant gel polymer-based soft electronics that remain non-freezing, non-drying, and resistant to environmental solvents [20]. As an interesting fact, the property of this organo-gel is the change behavior of permeability in water and based on this function, which could realize functional information recording and erasing applications. The ability of this organo-gel to change its permeability when in contact with water is considered by some authors not as a drawback, but as a potential opportunity to use it for the function of creating and erasing a record, but more detailed works on this topic could not be found [20, 21].

2 Experimental

2.1 Sample Preparation

All chemicals and materials were purchased according laboratory protocols. Biopolymer sample preparation was performed in following steps. Extracted carrageenan was cleared from large plant part, heated in water bath to 70 °C. After, the polymerization initiator (isopropanol) was poured into hot liquid of carrageenan and mixed until visible biopolymer "threads" are formed. The stiff gel was separated from liquid. After, the carrageenan was mixed with glycerol and iron (III) oxide and left to dry in natural conditions. After three days, the prepared mixture turned into a conductive film that retains its flexible shape. Such freshly prepared films were stored at room temperature in plastic Petri dishes.

2.2 Devices

Mechanical properties of samples were investigated by computer-controlled tension-compression test system Mecmesim *MultiTest 2.5-i* (Mecmesim Limited, United Kingdom, maximum load: 2500 N; maximum sample diameter: 134 mm; load sensor measurement error: ± 0.1%; speed range: 1–1000 mm/min). The testing machine was controlled by the software "Emperor" (Mecmesin Limited, United Kingdom). 25 N and 1000 N Load cells were used for testing machine holders. For resistance measurement the UNI-T UT55 Multimeter was applied.

The prepared carrageenan film sample was embedded into a flexible plastic plate performing the function of a protective layer, the aluminum foil was used as contact wires and mounted to multimeter for resistance measurements during compression tests. The micrograps of samples were performed by original micro-positioning system operated by LinuxCNC.

3 Results

Before load experiments, the samples were investigated as an appropriate material for force sensor. In our experiments, incompletely purified carrageenan was purposely sampled. Since the main guideline of our research was sustainability and cheapness, the complete purification of carrageenan would not meet the objectives. Since incompletely purified carrageenan is significantly cheaper and likely easier to utilize, our research was directed to evaluate the suitability of such a product for the development of force sensors. The compositions with different additional salts and oxides were investigated by measuring resistance and visual evaluation of produced films using various optical equipment. In Fig. 1 are presented micrographs of carrageenan film.

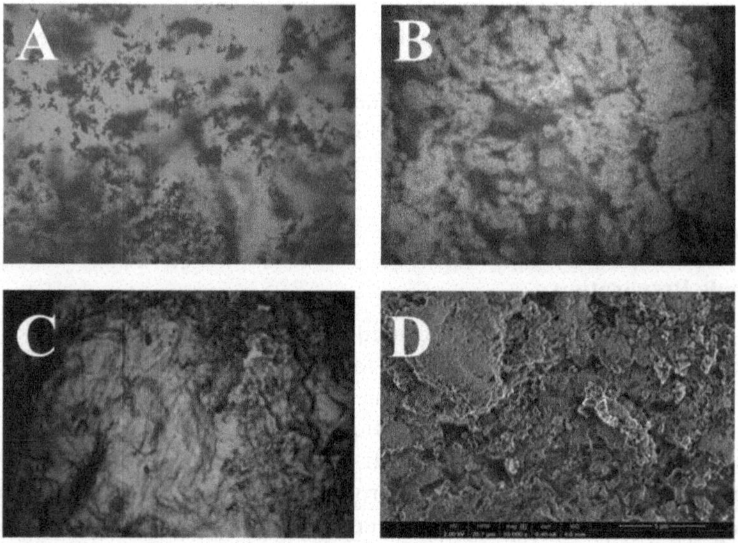

Fig. 1. Optical investigation of carrageenan **1A**. Freshly prepared liquid carrageenan biopolymer, optical magnification: 20 × 0.40 × 1.5; **1B.** Carrageenan film by optical microscope, optical magnification: 20 × 0.40 × 1.5; **1C.** Micrograph of pure carragenan film by scanning optical system, optical magnification: 20 × 0.40 × 1.5; **1D.** Surface micrograph by Helios Nanolab 650 scanning electron microscope (FEI, Eindhoven, The Netherlands) with a voltage of 2 kV and a current of 0.4 nA. Magn 10000, Figure was made by Antanas Zinovicius.

Prepared carrageenan biopolymer mass, mixed with selected concentration of iron (III) oxide and glycerol, was investigated as the electric charge carrier of the compression sensor. Samples with different concentrations of iron (III) oxide were prepared and resistance of resistance was measured (Fig. 2).

Before load the experiment, the carrageenan sample was placed on a polyethylene film. Copper foil and silver threads and adhesive aluminum foil were examined as an electrode for sensor. After careful evaluations, aluminum foil wires were chosen as electrodes. To ensure the transfer of electric charge through the biopolymer film, the electrodes were anchored on different sides of the film and at the maximum possible

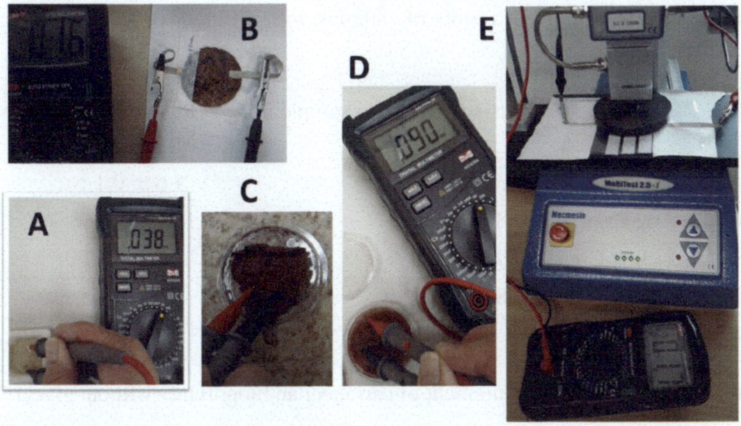

Fig. 2. Resistance measurements of carrageenan samples from prepared biopolymer in liquid to different stages of drying. A) Resistance of liquid with carrageenan biopolymer; B) resistance of pure carrageenan film; C) measurement of dried sample; D) Resistance of prepared mass for drying procedure; E) Stage of applied force measurement with sample and multimeter

distance. Distance between the two electrodes was set at 3 cm. A multimeter, megohm scale was used to measure resistance. The resistance of the fresh prepared carrageenan sample prepared for force measurements with connected electrodes was 0.16 ± 0.1 MΩ before the test.

Over time, signs of aging of the carrageenan films were observed, then the initial resistance increased. A time limit of five months was reached when the non-environmentally insulated film lost its ability to conduct electrical current uniformly and have a stable measurable contact resistance. From the visual assessment and resistance measurements, we can say that the optimal amount of iron oxide is within quite wide limits and only 10% concentration already affects the resistance of the product. It was found that without glycerol the product dries quickly and the film loses its flexibility and electrical conductivity (Table 1).

As we can see from the Table 1, the measured resistance values on the surface of the film are influenced by the optimal amount of iron (III) oxide. An excessive amount of iron oxide damages the electrical conductivity of the carrageenan film.

In order to evaluate the necessity of applying glycerol in the production of this film, a series of tests were conducted to determine how long the product loses its ability to pass an electric current. Sample placed to dry in ambient temperature at relative humidity 82% at the first and second measurement. After 120 h humidity was 86%. The influence natural drying conditions on product quality was explored by registration humidity and resistance of sample surface and visual assessment of quality. After 24, 48, 120, 168 ours the resistance was measured. Results are presented in Table 2.

Then confirming the necessity of glycerol as a moisture-retaining additive in the biopolymer film, further tests were conducted to determine the minimum amounts of glycerol to maintain the carrageenan film with sufficient water molecules and maintain flexibility. In this paper, studies with glycerol concentration searches are not presented.

Table 1. Resistance measurement results of samples with different concentration of iron (III) oxide

composition	Resistance of wet sample, MΩ	Resistance of dried sample, MΩ
Pure carrageenan (Crgn)	0.38 ± 0.1	1.6 ± 0.1
0,36g Fe_2O_3 + 3,6g Crgn	0.81 ± 0.3	12.99 ± 0.6
0,0036g Fe_2O_3 + 3,6g Crgn	0.91 ± 0.3	1.36 ± 0.25
0,002g Fe_2O_3 + 3,6g Crgn	0.82 ± 0.2	0.6 ± 0.2

Table 2. Resistance measurement of carrageenan biopolymer without glycerol

Fe_2O_3 amount ins sample	Resistance, MΩ After 24h	Resistance, MΩ After 48h	Resistance, MΩ After 120h	Resistance, MΩ After 168h
0.002g	1.26 ± 0.1	1.58 ± 0.1	128.6 ± 2.1	∞
0.01g	1.571 ± 0.1	1.7 ± 0.1	80.7 ± 4.5	∞
0.02g	1.65 ± 0.2	1.5 ± 0.1	129.6 ± 6.7	∞
0.04g	1.5 ± 0.1	1.8 ± 0.2	34.6 ± 1.2	∞
0.2	1.762 ± 0.1	1.9 ± 0.1	34.2 ± 0.9	∞

All experiments were carried out consecutively without time breaks. As it is presented in Fig. 3, the visible reproducibility of the results of "just prepared" samples are reasonably good. So, in the measurement zone from 50 N to 150 N, the dependence of the resistance on the applied force is visible.

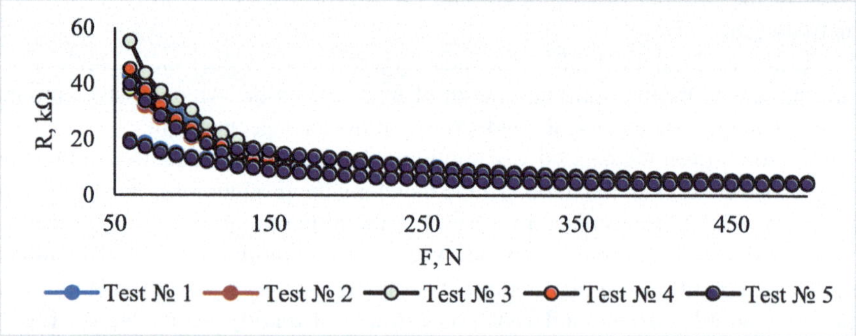

Fig. 3. Dependence of resistance on applied force of sample with 15% glycerol,2% Fe_2O_3 in carrageenan biopolymer.

The difference between the measured initial resistance and the final resistance is about 23 kΩ in the first experiment, and about 36 kΩ in the fifth measurement. Thus, material fatigue is visible even in a new product, but there are quite a lot of solutions presented in the literature to reduce this natural carrageenan fatigue.

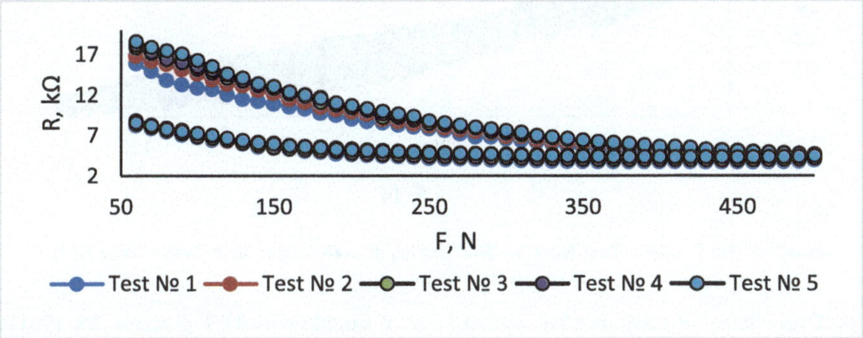

Fig. 4. Dependence of resistance on applied force of sample with 18% glycerol, 2% Fe_2O_3 in carrageenan biopolymer

Carrageenan biopolymer enriched with 18% of glycerol on first day of experiment demonstrate more repeatable results. The mathematical difference between the resistance measured at the beginning and at the end of the experiment is from 7 to 10 kΩ. This is a much better result (Fig. 4) than the measurement data of the example discussed earlier in Fig. 3. However, this composition of carrageenan biopolymer (Fig. 4) is questionable in order to select suitable operating limits for the sensor, because no clear qualitative dependence of the resistance on the applied force is seen.

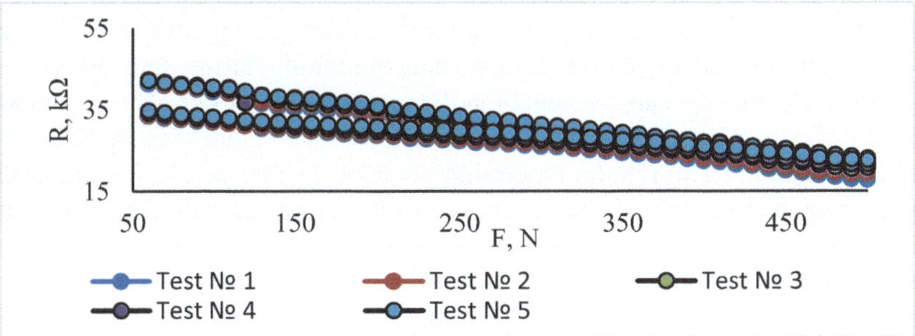

Fig. 5. Dependence of resistance on applied force of sample with 15% glycerol, 2% Fe_2O_3 in carrageenan biopolymer, after 39 days from first measurement.

The same recipe samples were left for 39 days in ambient conditions, not sealed. Such a decision was made in order to find out how long the biopolymer loses its relevant technical characteristics.

As we can see from Fig. 6, the dispersion of the data after performing one after the other consecutive measurements is obvious compared to another sample that was also aged (Fig. 5). As we can see from the graphs (Fig. 6), at the applied force of 200 N, the

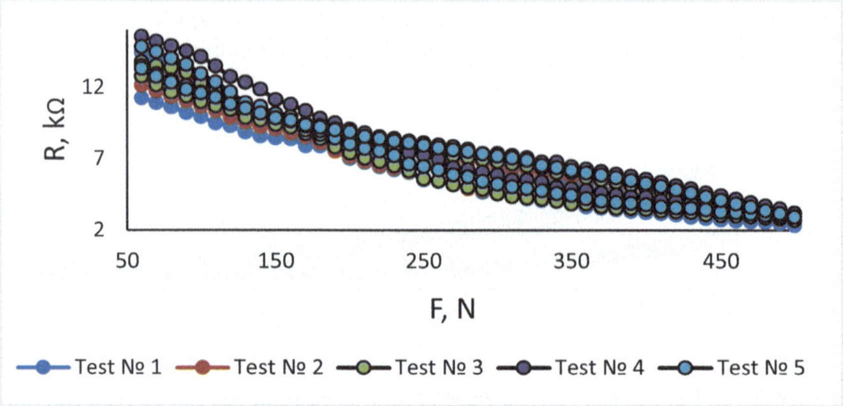

Fig. 6. Dependence of resistance on applied force of sample with 18% glycerol, 2% Fe_2O_3 in carrageenan biopolymer after 38 days from the first measurement.

shape of the curves changes slightly. This may be due to excess glycerol in the film. This may indicate the displacement of iron oxide particles in the film structure. Thus, glycerol in the carrageenan biopolymer structure not only plays a softening role in the product and helps to retain water molecules inside it, but also gives the components a certain looseness, which can be seen as a negative factor or should be controlled by choosing the concentrations of the components more carefully.

4 Conclusions

Not isolated from the from the effects of environmental conditions, such a carrageenan force sensor partially loses its properties after more than thirty days from the time of first testing or more than five weeks from the date of manufacturing. Overall, we can say that incompletely purified carrageenan biopolymer with moieties of plant parts and non-sterile can basically be used to produce low-cost eco-sustainable sensors. Such sensors would be disposable. The reliable operating zone of the sensor can be adjusted with various compounds that help to change the film's flexibility, electrical conductivity and other relevant parameters. If accidentally lost in nature, such a sensor will not cause an ecological problem, and it be easy to recycle. The sensor prototype tested in laboratory conditions showed reliable sensitivity in designed resistance region. The resistance of the sensor has a dependence on the squared of the applied force.

Funding. This research was supported by Lithuanian Research Council Project No T22T091692 "Development and improvement of a structural parameter sensor using a scanning electrochemical microscope".

References

1. Bhaskar, A., et al.: Flexible force sensors for e-textiles. IOP Conf. Ser. Mater. Sci. Eng. **254**(7), 072007 (2017). https://doi.org/10.1088/1757-899X/254/7/072007

2. Wei, Y., Xu, Q.: An overview of micro-force sensing techniques. Sens. Actuators A Phys. **234**, 359–374 (2015). https://doi.org/10.1016/J.SNA.2015.09.028
3. Cheng, M., et al.: A review of flexible force sensors for human health monitoring. J. Adv. Res. **26**, 53–68 (2020). https://doi.org/10.1016/J.JARE.2020.07.001
4. Yao, S., Swetha, P., Zhu, Y.: Nanomaterial-enabled wearable sensors for healthcare. Adv. Healthc. Mater. **7**(1), 1700889 (2018). https://doi.org/10.1002/ADHM.201700889
5. Ng, P.C., Murray, S.S., Levy, S., Venter, J.C.: An agenda for personalized medicine. Nat. **461**, 7265, 724–726 (2009). https://doi.org/10.1038/461724a
6. Hashim, K.S., Kot, P., Muradov, M., Al-Khaddar, R.: How can sensors be used for sustainability improvement? Methods Sustain. Sci. Assess. Prioritization, Improv. Des. Optim., 321–344 (2021). https://doi.org/10.1016/B978-0-12-823987-2.00011-8
7. Ramírez-Moreno, M.A., et al.: Sensors for sustainable smart cities: a review. Appl. Sci. **11**(17), 8198 (2021). https://doi.org/10.3390/APP11178198
8. Serra, J.P., et al.: Carrageenan-based hybrid materials with ionic liquids for sustainable and recyclable printable pressure sensors. ACS Sustain. Chem. Eng. **10**(26), 8631–8640 (2022). https://doi.org/10.1021/ACSSUSCHEMENG.2C02374/ASSET/IMAGES/LARGE/SC2C02374_0008.JPEG
9. Macedo, V.M., Pereira, N., Tubio, C.R., Martins, P., Lanceros-Mendez, S., Costa, C.M.: 3D-printed carrageenan-based nanocomposites for sustainable resistive sensing devices. Polymer (Guildf). **262**, 125456 (2022). https://doi.org/10.1016/J.POLYMER.2022.125456
10. Zeng, L., Liu, B., Duan, L., Gao, G.: Tough, recyclable and biocompatible carrageenan-modified polyvinyl alcohol ionic hydrogel with physical cross-linked for multimodal sensing. Int. J. Biol. Macromol. **253**, 126954 (2023). https://doi.org/10.1016/J.IJBIOMAC.2023.126954
11. Tuvikene, R., et al.: The extraction, structure, and gelling properties of hybrid galactan from the red alga Furcellaria lumbricalis (Baltic Sea, Estonia). J. Appl. Phycol. **22**(1), 51–63 (2010). https://doi.org/10.1007/S10811-009-9425-X/FIGURES/8
12. Ozolina, S., Zaimis, U.: Some aspects of extraction and application of seaweed furcellaria lumbricalis carrageenan in production of recycled paper. https://doi.org/10.22616/ERDev.2022.21.TF200
13. Zaimis, U., Ozolina, S., Jurmalietis, R.: Production of seaweed derived bioplastics. https://doi.org/10.22616/ERDev.2021.20.TF370
14. Elmarhoum, S., Mathieu, S., Ako, K., Helbert, W.: Sulfate groups position determines the ionic selectivity and syneresis properties of carrageenan systems. Carbohydr. Polym. **299**, 120166 (2023). https://doi.org/10.1016/J.CARBPOL.2022.120166
15. Tennakoon, P., Chandika, P., Yi, M., Jung, W.-K.: Marine-derived biopolymers as potential bioplastics, an eco-friendly alternative (2023). https://doi.org/10.1016/j.isci
16. Abad, L.V., Nasimova, I.R., Relleve, L.S., Aranilla, C.T., De La Rosa, A.M., Shibayama, M.: Dynamic light scattering studies of irradiated kappa carrageenan. Int. J. Biol. Macromol. **34**(1–2), 81–88 (2004). https://doi.org/10.1016/J.IJBIOMAC.2004.03.012
17. Imeson, A.P.: Carrageenan and furcellaran. Handbook of Hydrocolloids. Second Ed., pp. 164–185 (2009). https://doi.org/10.1533/9781845695873.164
18. Frediansyah, A.: The antiviral activity of iota-, kappa-, and lambda-carrageenan against COVID-19: a critical review. Clin. Epidemiol. Glob. Heal. **12**, 100826 (2021). https://doi.org/10.1016/J.CEGH.2021.100826
19. De Ruiter, G.A., Rudolph, B.: Carrageenan biotechnology. Trends Food Sci. Technol. **8**(12), 389–395 (1997). https://doi.org/10.1016/S0924-2244(97)01091-1
20. Zheng, G., et al.: A κ-carrageenan-containing Organohydrogel with adjustable transmittance for an Antifreezing, Nondrying, and solvent-resistant strain sensor. Biomacromolecules **23**(11), 4872–4882 (2022). https://doi.org/10.1021/ACS.BIOMAC.2C01044/ASSET/IMAGES/LARGE/BM2C01044_0006.JPEG

21. Macedo, V.M., Pereira, N., Tubio, C.R., Martins, P., Costa, C.M., Lanceros-Mendez, S.: Carrageenan based printable magnetic nanocomposites for actuator applications. Compos. Sci. Technol. **224**, 109485 (2022). https://doi.org/10.1016/J.COMPSCITECH.2022.109485

Understanding Properties of DriveBay Electromagnetic Tracker for Use in Soft Robot Experiments

Tomasz Smaż and Igor Zubrycki(✉)

Institute of Automatic Control, Lodz University of Technology, Stefanowskiego 18, 90-537 Lodz, Poland
igor.zubrycki@p.lodz.pl

Abstract. Soft Robotics experiments necessitate advanced measurement techniques to track robot pose and deformations accurately. Electromagnetic tracking systems provide substantial benefits over vision-based systems, especially in capturing both the position and orientation of a robot's surface without occlusion issues. However, these systems are typically designed for applications in controlled environments, with multiple guidelines stated by the producers, necessitating an evaluation of their effectiveness in different contexts. This paper presents a series of foundational experiments to assess the capabilities and limitations of such a system in Soft Robotics. We investigate the impact of distance from the base station, resistance to environmental interference, effects of metallic elements, and the accuracy of rotational measurements. The results demonstrate a significant influence of positional accuracy relative to the distance from the transmitter, with a tenfold decrease in accuracy observed. Additionally, the presence of metallic objects in close proximity (less than 20 mm) to the probes adversely affects accuracy, as do large metallic elements such as construction materials. This effect is particularly pronounced when the sampling frequency of the system aligns with multiples of the electrical network frequency, resulting in more than a tenfold reduction in accuracy.

Keywords: Soft Robotics · Electromagnetic Tracking · Sensor Accuracy · Robotics Measurement Systems · Environmental Interference in Robotics

1 Introduction and Motivation

Soft robotics presents a set of challenges in pose estimation. Generally, such devices can deform both as a mode of operation (e.g., PneuNets, which work by inflating internal channels, causing the structure to bend or expand) and while

Supported by [National Centre for Research and Development], grant No. LIDER/50/0203/L-11/19/NCBR/2020.

R. Szewczyk et al. (Eds.): AUTOMATION 2024, LNNS 1219, pp. 299–308, 2024.
https://doi.org/10.1007/978-3-031-78266-4_26

interacting with the environment. Moreover, they are capable of complex movement even with a single actuator, as exemplified by Octopus Arm-Inspired soft actuators that can coil around an object. While these built-in capabilities make them attractive for use in various applications, such as medical procedures or interactions in unstructured environments, tracking their movement, particularly during experiments, poses a challenge.

Proprioceptive sensors (such as motor encoders, using forward kinematics) cannot be used to accurately determine the position and orientation of a particular robot's surface, as the robots are not rigid and potential deformations of links, especially due to contact with the environment, can introduce errors. Solutions include direct surface measurement (such as skin-like sensors [7]) and sets of sensors like flexion sensors coupled with sophisticated measurement algorithms [9]. Here, the key challenges are the possible introduction of errors during the calibration procedure and limited resolution. Other systems such as using IMU's also exist with their limitations (mainly difficulty in tracking position change and affinity to position drift) [4]. Vision systems are commonly used for external position tracking. Advances in computer vision allow for accurately determining the position of robot parts and measuring surface deformations through 3D vision systems. The primary limitation is the line of sight problem, i.e., the inability to observe the robot's pose when occlusion occurs. This occlusion can occur during interaction with the environment (such as a soft robot grasping an object or a wearable soft robot) or due to self-occlusion.

Electromagnetic tracking systems offer significant benefits in soft robot experiments. Operating based on a central station that emits an electromagnetic signal, these systems use antenna-like probes attached to the robot to calculate each probe's position and rotation relative to the central station through signal processing. Compared to vision systems, electromagnetic tracking has two main advantages for soft robots. First, since these devices are typically made from materials like silicone rubber, which minimally affect electromagnetic wave propagation, the line of sight problem due to self-occlusion is reduced. Second, each probe's position and rotation are directly accessible, yielding comprehensive data from experiments. However, employing these systems in soft robotics deviates from their original intended use, which is tracking medical instruments during surgery. The manufacturer of the driveBay Electromagnetic Tracker, Ascension Technology Corporation, provides specific guidelines for the device's use, as outlined in their documentation [1]. Producers state that orientation and position could be tracked with an accuracy of 0.5° RMS and 1.4 mm) RMS:

1. Sensors should be placed at least 33 mm apart from each other.
2. The distance between the computer and the transmitter should be at least 66 cm.
3. Avoiding the presence of highly conductive metals, such as aluminum and copper, in the system's working field is recommended.
4. The presence of magnetic metals should also be avoided in the system's working field.
5. Electronic devices should be kept away from the system.

These guidelines, while thorough, could limit the system's applicability in soft robot experiments. It is worth noting that electromagnetic sensors were already used in conjunction with soft robots. In a review of soft robotics in minimally invasive surgery, Ruciman et al. found seven applications of electromagnetic tracking devices in such applications [6]. Tiziani et al. used trakSTAR electromagnetic tracking system to compare the results to optical sensor for muscle position estimation [8]. Chen et al. used Northern Digital EM tracker to evaluate experimentally their contact detection algorithm [3].

1. **Spatial Constraints**: Soft robots, especially those with smaller or compact structures, may not meet the requirement of maintaining a 33mm distance between sensors, posing a challenge for sensor placement.
2. **Proximity to Electronics**: Soft robots inherently require control systems, often incorporating built-in electronics or having them near actuators, which could interfere with electromagnetic tracking.
3. **Metal Restrictions**: The need to avoid highly conductive and magnetic materials in the vicinity of the tracking system can limit the choice of materials used in the construction of soft robots, as well as the objects they are designed to interact with.
4. **Environmental Limitations**: The operational environment of soft robots might contain elements, such as certain metals or electronic devices, to which the electromagnetic tracking system is sensitive, potentially impacting its accuracy and reliability.

While studies exist evaluating the behavior of the sensors in a medical environment [10,11], there is a lack of basic metrological studies of such sensor setups. This study experimentally investigates the basic properties of the probes to understand how each factor might influence the accuracy and other tracking properties and to assess the potential role of such sensors in soft robot experiments or operations.

2 Methods

2.1 System Configuration

The 3DGuidance system consists of a control module, a transmitter, and a receiver. The control module, particularly in the driveBAY version, is designed for integration with desktop computers. The system's primary function is to track the orientation and position of sensors within the coordinate system relative to the transmitter. In this experiment, 3dG trakSTAR/ driveBAY Mid-Range Transmitter with a set of four 6DOF Sensors Model 130 (130dB-tS) was used.

The system communicates with the computer via a USB interface (Serial emulator). C API is available as well as documentation on the Serial commands for data acquisition and control. In this experiment, we acquired data via the Linux operating system and custom-written Python software based on serial commands.

2.2 Experimental Setup

A testing frame was constructed to mount four sensors, as shown in Fig. 1. The frame was laser cut from plywood with holes cut 20 mm apart. The sensors were mounted in holes on a square 20 cm × 20 cm. Due to the high precision of the laser plotter used for its construction, the distances between mounted sensors were considered exact. The position values of the sensors were then translated into distances between each sensor and compared with expected values (200 mm).

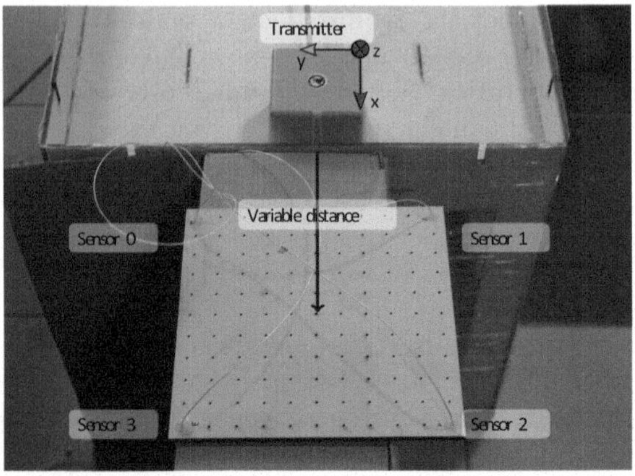

Fig. 1. Laser cut test frame used to verify electromagnetic sensors. Sensors are placed on a square 20 cm × 20 cm. Top left: D0, Top right: D1, Bottom left: D2, Bottom right: D3.

2.3 Data Collection and Analysis

A custom program written in Python for the Linux Operating System, using a specially developed driver, was utilized to calculate the distances between sensors. The error values Δ were defined as the difference between the average of the last 160 samples. The error for a specific sensor n is calculated as follows:

$$\Delta_n = \frac{1}{160} \sum_{i=0}^{159} \|D_0 - D_i\| - \text{dist}_i \tag{1}$$

Here, $\|D_0 - D_i\|$ represents the Euclidean (L2) norm of the distance between the reference sensor position D_0 and the ith sensor as stated by the system, and dist_i is the actual distance.

The uncertainty measure σ was calculated as the standard deviation of the calculated distances, corresponding to 2 s of data collection. The standard deviation is given by:

$$\sigma = \sqrt{\frac{1}{N-1}\sum_{i=1}^{N}(x_i - \overline{x})^2} \tag{2}$$

where x_i represents each individual distance measurement, $\overline{x}$ is the mean of all distance measurements, and N is the total number of measurements (in this case, $N = 160$).

System settings were largely kept default, with an adjustment to the power network frequency parameter from the standard 60 Hz to 50 Hz.

2.4 Influence of Sensor Distance from Transmitter

The first set of tests determined how the distance between the sensors and the transmitter affects data quality. The testing frame and transmitter were placed away from metallic objects and electronic devices. The distance was defined as the space between the center of the testing frame and the edge of the transmitter, with measurements labeled as "D01", "D02", and "D03" representing distances between sensor "0" and sensors "1" to "3", respectively. The distance was measured with a ruler.

2.5 System's Susceptibility to Interference

The system's resilience to environmental interference was another aspect of the evaluation. Particularly, the influence of sampling frequency on the accuracy and reliability of measurements was examined. The system allows for a range of sampling frequencies from 20 Hz to 255 Hz, with a default setting of 80 Hz. Adjustments to the sampling frequency were explored as a means to counteract specific interfering frequencies present in the environment. The test was conducted in a KUKA cube which is a metal frame with industrial robots with a shape of 2,5 × 2,5 × 2,5 m [5].

2.6 Effect of Mounting Elements on Measurement Quality

The manufacturer strongly advises against the use of metallic elements near the sensors in their documentation. However, in practice, it's not always possible to avoid this. To evaluate the impact of metallic mounting screws on the quality of measurements, tests were conducted in three different configurations:

1. Without screws
2. With two M5×20 screws placed at a distance of 28 mm from sensor "0"
3. With two M5×20 screws placed at a distance of 20 mm from sensor "0"

These tests were aimed at determining how significantly metallic mounting elements, specifically screws, affect the accuracy and reliability of the measurements. The configurations included setups with and without screws placed at different distances from the sensor to gauge their impact.

2.7 Verification of Rotational Measurements

The sensor reports orientation in as yaw, pitch, roll angles (global X, Y, Z) (see Fig. 1), a rotation matrix or a quaternion. We calculate a relative orientation of two sensors using formula 3:

$$R_{rel} = RR_0^T = R_z(\theta_z)R_y(\theta_y)R_x(\theta_x) \tag{3}$$

$$R_{\text{rel}} = \begin{bmatrix} \cos(\theta_y)\cos(\theta_z) & \sin(\theta_x)\sin(\theta_y)\cos(\theta_z) - \sin(\theta_z)\cos(\theta_x) & \sin(\theta_x)\sin(\theta_z) + \sin(\theta_y)\cos(\theta_x)\cos(\theta_z) \\ \sin(\theta_z)\cos(\theta_y) & \sin(\theta_x)\sin(\theta_y)\sin(\theta_z) + \cos(\theta_x)\cos(\theta_z) & -\sin(\theta_x)\cos(\theta_z) + \sin(\theta_y)\sin(\theta_z)\cos(\theta_x) \\ -\sin(\theta_y) & \sin(\theta_x)\cos(\theta_y) & \cos(\theta_x)\cos(\theta_y) \end{bmatrix}$$

and extract yaw, pitch, and roll angles using the formula 4:

$$\begin{aligned} \theta_x &= \arctan(R_{rel}(3,2), R_{rel}(3,3)) \\ \theta_y &= \arcsin(-R_{rel}(3,1)) \\ \theta_z &= \arctan(R_{rel}(2,1), R_{rel}(1,1)) \end{aligned} \tag{4}$$

The correctness of the sensor data interpretation concerning rotation was verified using a test frame the frame was then rotated around the X, Y, and Z axes, and the frame's orientation was compared with the orientation from the readouts. Focus was mainly on understanding the noise of the readout, assessed through measurement of standard deviation of the read angles.

3 Results

3.1 Accuracy of Sensor Positioning and the Role of Distance

The initial tests focused on the accuracy of relative sensor positioning. The impact of sensor distance from the transmitter on measurement quality was investigated. Measurements were conducted at varying distances, with relative distances "D01", "D02", and "D03", corresponding to the distances between sensor "0" and sensors "1" to "3".

The table below summarizes the relative errors and standard deviations for sensors D01, D02, and D03 at distances of 10 cm, 30 cm, 50 cm, 70 cm, and 90 cm from the board. The results are presented in graphical form as a graph of absolute errors to a distance in Fig. 2 and uncertainty in relation to distance in Fig. 3

These results indicate a trend of increasing relative errors and standard deviations as the distance from the board increases.

3.2 System's Susceptibility to Interference

The results of the test are shown in Table 2. It was observed that certain frequencies, especially around 25 Hz, chosen to demonstrate the impact of electrical network interference, led to substantial deviations in measurement accuracy.

Table 1. Relative Errors and Standard Deviations at Various Distances

Distance	Sensors	Relative Error (mm)	Standard Deviation (mm)
10 cm	D01	1.065	0.108
	D02	−0.372	0.000226
	D03	0.197	0.118
30 cm	D01	2.571	0.493
	D02	−0.337	0.386
	D03	0.332	0.824
50 cm	D01	5.556	1.678
	D02	−1.791	2.413
	D03	−4.486	3.348
70 cm	D01	15.311	5.986
	D02	−5.321	7.919
	D03	−6.505	9.121
90 cm	D01	55.097	16.844
	D02	−17.437	18.523
	D03	−12.566	23.267

Table 2. Impact of Sampling Frequency on Measurement Accuracy

f_s [Hz]	σ [mm]			Δ [mm]		
	D01	D02	D03	D01	D02	D03
160.0	1.02	1.05	1.63	−15.24	−8.44	−13.24
80.0	1.41	1.40	1.65	−8.90	−8.34	−11.54
25.0	133.63	5.11	35.62	397.47	−48.30	48.65
20.0	5.08	2.84	5.72	2.98	−13.02	−15.27
Control Measurement Outside the Cage						
160.0	0.49	0.39	0.82	2.57	−0.34	0.33

This result underscores the importance of selecting an appropriate sampling frequency and the need for an interference-free environment for optimal system performance (Table 1).

The results indicated that changing the sampling frequency within the system's operational range did not significantly alter the error values under controlled conditions. However, in environments with potential electronic interference, the choice of sampling frequency became crucial. A frequency of 25 Hz, deliberately chosen to demonstrate the effects of electrical network interference, exhibited a marked increase in measurement errors and uncertainties. This finding highlights the importance of selecting an optimal sampling frequency to minimize the impact of environmental noise, especially in settings with potential electronic disturbances.

3.3 Effect of Mounting Elements on Measurement Quality

Results of the test are shown in Table 3. The findings suggest that while metallic screws can be used for mounting, their proximity to the sensors should be carefully managed to minimize measurement errors.

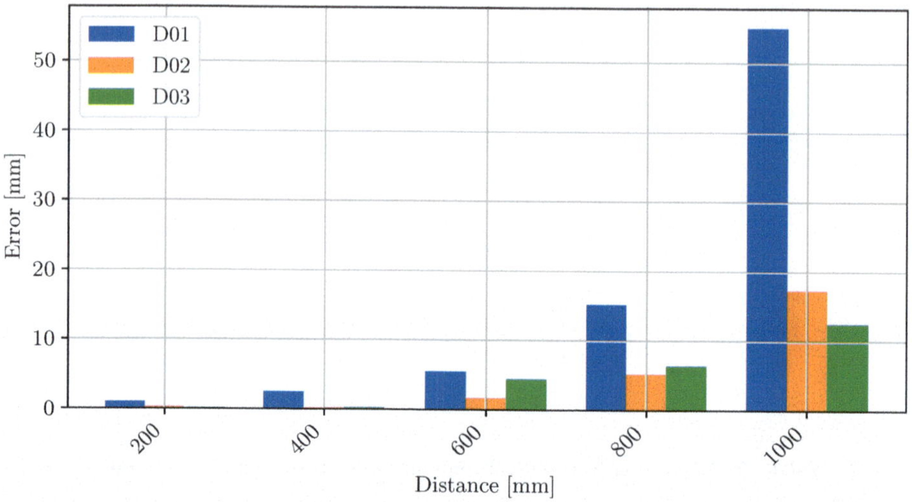

Fig. 2. Absolute error in calculating distances between sensors.

Table 3. Comparison of the Impact of Mounting Screws

Variant	σ [mm]			Δ [mm]		
	D01	D02	D03	D01	D02	D03
1 (no screws)	0.49	0.39	0.82	2.57	−0.34	0.33
2 (screws 28mm apart from sensor)	0.63	0.43	0.71	2.23	−0.20	0.04
3 (screws 20 mm from sensor)	0.67	0.49	0.79	10.51	13.12	13.80

3.4 Verification of Rotational Measurements

The system's ability to accurately measure rotation was assessed. The results are shown in Table 4. Comparisons between these reported orientations and the actual orientation of a test frame were conducted, revealing some discrepancies in rotational measurements, particularly along specific axes.

Table 4. Errors and Uncertainties in Sensor Orientation Measurement

Axis	σ [°]			Δ [°]		
	θ_x	θ_y	θ_z	θ_x	θ_y	θ_z
R2	0.018	0.024	0.026	—		
R3	0.073	0.108	0.148	—		
R23	0.079	0.110	0.145	−5.733	−0.082	−115.729

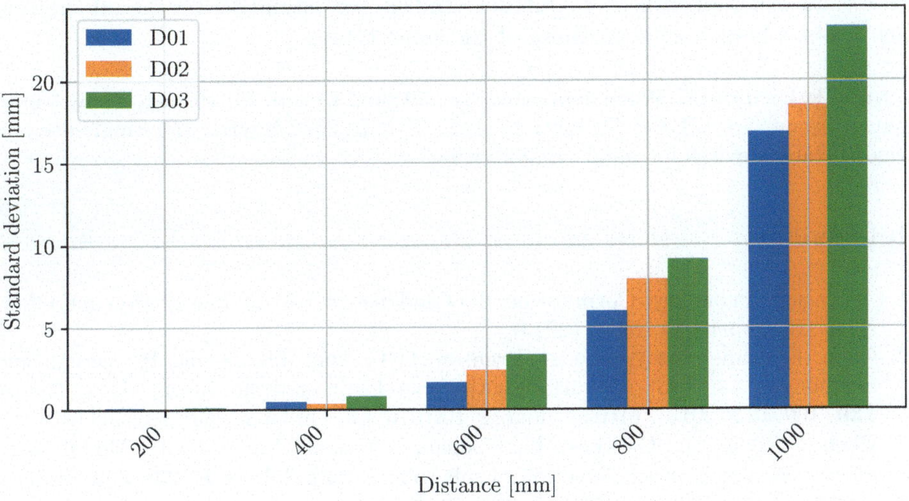

Fig. 3. Uncertainty in calculating distances between sensors.

4 Discussion and Conclusions

Overall, the system's reported errors and uncertainties in various conditions were larger than those claimed by the manufacturer, even inside the "Performance Motion Box," which is a workspace starting 20 cm from the base and shaped like a box measuring $46 \times 56 \times 60$ cm [1]. Despite this, the system still offers attractive millimeter-level accuracy when sensors are positioned close to the base (10–30 cm). Our findings suggest that measurement errors increase with the square of the distance from the transmitter, highlighting the importance of optimal sensor placement in experimental setups.

We observed a drop in performance within the large metal cage, particularly where the sampling frequency was 25 Hz (half of the electric mains frequency in Poland). There is a noticeable effect of mounting elements on system performance, especially when metal objects (such as screws) are placed less than 20 mm from the probe. However, the results for screws placed more than 28 mm away from the sensor are minimal.

In general, we believe that such sensors hold value in soft robot studies. While the environment does influence the results, these sensors enable studies with limited line of sight (self-occlusion) or inside non-transparent media. However, it is important to note that such sensors are usually used in conjunction with other tracking systems, particularly vision systems ([2]), as changes in the curvature of soft robots can be traced more accurately when actuators are visible.

In our future studies, we aim to understand better the influence of distance between the sensors, a capability our current setup lacks. We also plan to investigate further the impact of sensor rotation on accuracy. Finally, we noticed that the distance between probes 0 and 1 was consistently measured with a larger error than other distances. In further studies, we intend to better understand each probe's behavior, even those of the same type.

Acknowledgements. Work supported by National Centre for Research and Development, grant No. LIDER/50/0203/L-11/19/NCBR/2020 "myHAND: Single-use and personalizable devices for hand rehabilitation".

References

1. Ascension Technology Corporation: 3D Guidance driveBAY Installation and Operation Guide, 940035 Rev 5 (2013)
2. Chen, L., Yang, C., Wang, H., Branson, D.T., Dai, J.S., Kang, R.: Design and modeling of a soft robotic surface with hyperelastic material. Mech. Mach. Theory **130**, 109–122 (2018). https://doi.org/10.1016/j.mechmachtheory.2018.08.010
3. Chen, Y., Wang, L., Galloway, K., Godage, I., Simaan, N., Barth, E.: Modal-based kinematics and contact detection of soft robots. Soft Rob. **8**(3), 298–309 (2021)
4. Gastinger, J., Müller, D., Hildebrandt, A., Sawodny, O.: Pose estimation and tracking control of a pneumatic soft robotic hand. IFAC-PapersOnLine **53**(2), 9962–9967 (2020)
5. Granosik, G., Zubrycki, I., Soghbatyan, T., Zarychta, D., Gawryszewski, M.: Kube - platforma robotyczna dla badań naukowych i prac wdrożeniowych. Prace Naukowe Politechniki Warszawskiej. Elektronika **195**(1), 223–234 (2016), bibliogr. 35 poz., rys
6. Runciman, M., Darzi, A., Mylonas, G.P.: Soft robotics in minimally invasive surgery. Soft Rob. **6**(4), 423–443 (2019). https://doi.org/10.1089/soro.2018.0136
7. Shih, B., et al..: Electronic skins and machine learning for intelligent soft robots. Sci. Robot. **5**(41), eaaz9239 (2020). https://doi.org/10.1126/scirobotics.aaz9239. https://www.science.org/doi/abs/10.1126/scirobotics.aaz9239
8. Tiziani, L.O., Hammond, F.L.: Optical sensor-embedded pneumatic artificial muscle for position and force estimation. Soft Rob. **7**(4), 462–477 (2020)
9. Xavier, M.S., et al.: Soft pneumatic actuators: a review of design, fabrication, modeling, sensing, control and applications. IEEE Access **10**, 59442–59485 (2022). https://doi.org/10.1109/ACCESS.2022.3179589
10. Yaniv, Z., Wilson, E., Lindisch, D., Cleary, K.: Electromagnetic tracking in the clinical environment. Med. Phys. **36**(3), 876–892 (2009). https://doi.org/10.1118/1.3075829
11. Zhou, J., Sebastian, E., Mangona, V., Yan, D.: Real-time catheter tracking for high-dose-rate prostate brachytherapy using an electromagnetic 3D-guidance device: a preliminary performance study. Med. Phys. **40**(2), 021716 (2013). https://doi.org/10.1118/1.4788641

Methods of Reduction of Fuel Consumption Measurement Error by Thermo-Anemometric Flow Meter

Andrii Ilchenko[1] and Igor Korobiichuk[2](✉)

[1] Agricultural Engineering and Technical Service Department of Polissya National University, Zhytomyr, Ukraine
avi_77@ukr.net

[2] Łukasiewicz Research Network – Industrial Research Institute for Automation and Measurements PIAP, Warsaw, Poland
igor.korobiichuk@piap.lukasiewicz.gov.pl

Abstract. It was analyzed the ways of influence on the indexes of radial heat flow in thermo-anemometric flow meter that will influence the indexes of axial heat flow, will make them more informative and this, in its turn, will affect the consumption measurement error. It was formulated the condition of selecting the temperatures distribution with the minimal measurement error for multi-sectional flow meter, it can be obtained by the indexes of thermal transformers of the very pipe of multi-sectional thermo-anemometric flow meter, where the maximal total temperatures difference between all the neighboring thermal transformers was measured. Expanding the range of fuel consumption measurement by the thermo-anemometric flow meter was suggested due to the application of sequentially axially installed sections with different diameters.

Keywords: measurement error · internal combustion engine · fuel consumption · thermo-anemometric flow meter · temperature distribution · hot wire

1 Introduction

The fuel consumption of a vehicle's internal combustion engine (ICE) is a com-plex indicator of its technical condition. The fuel consumption of a vehicle's internal combustion engine is influenced by all the design and operational parameters of the stated system. Measuring the fuel consumption of the vehicle's internal combustion engine is obligatory to identify malfunctions, prevent failures of all units, systems and components of the vehicle. Such measurements can be unequivocally considered an integral part of the process of general vehicles diagnostics in general, and their systems as well. In most cases, fuel flow meters are not installed on vehicles. Software registration of the nozzles (injectors) open state time of the of the internal combustion engine power system is used, and based on the known pressure in front of the nozzles (injectors) and considering their geometric parameters (cross-section of the "saddle-needle" pair or similar), it is calculated the amount of fuel entering through the nozzle to the chamber of the internal

R. Szewczyk et al. (Eds.): AUTOMATION 2024, LNNS 1219, pp. 309–317, 2024.
https://doi.org/10.1007/978-3-031-78266-4_27

combustion engine. When operating a vehicle, specialists are more interested in the of absolute fuel consumption numerical value (for a certain period, on a certain route, in certain driving modes (loading), etc.). At the same time, the absolute fuel consumption can be significantly different and differ significantly from the one measured by the method of accounting for the time of the nozzles opening.

Fuel flow meters of various types, designs and principles of operation can be used on road transport. However, the specifics of its operation impose many fundamental demands on flowmeters, as it influences the measurement error [1]. However, it is important to realize that the accuracy of measuring fuel consumption with such a flow meter is affected by friction in the rotor supports, pressure pulsations in front of it, and possibly the location of the flow meter relative to the horizon and the longitudinal axis of the car.

The industry offers a really large number of fuel flow meters; however, they are in one way or another not adjusted to the vehicles operating conditions [2]. Therefore, the most attractive in this case are flowmeters whose operation is based on the principle of recording the heat transfer by the body, the flow of which is recorded – thermal flowmeters [3–5]. They have significant advantages over flowmeters of other operating principles. The most important of them are: measurement accuracy does not depend on the orientation of the flowmeter in space, enabling you to install it on the car in a convenient place under the hood, cabin, interior, etc.; the measurement error is not affected by vibrations, shock loads, pressure and pulsations of the fuel at the entrance to the flow meter; installation of a two-circuit heat flow meter allows to reduce the fuel consumption measurement error in internal combustion engine power systems with reverse fuel flow into the vehicle tank [6].

The essential drawback of the heat flow meter is the relatively narrow range of measuring fuel consumption. As a rule, they are concentrated on relatively stationary costs, and hence heat flow meters are not able to cover the entire possible range of vehicle's internal combustion engine fuel consumption of the in all possible modes of its operation. Another important drawback is that thermal flowmeters have a non-constant error in the flow measurement at different fuel flow rates (different recorded flows) [7]. This must be taken into account in the measurement process, and this fact also requires the improvement of the heat flow meter design, which can be solve in different ways and approaches [8].

Therefore, main aim of this research to develop ways to reduce the measuring error of the vehicle's diesel fuel consumption using a thermo-anemometric flow meter by improving its design.

2 Materials and Methods

The classic design of a thermo-anemometric flow meter is shown in Fig. 1. It consists of a body-tube 1, through which fuel flows, an electric heating element 2 and heat exchangers 3. The dimensions of the body-tube 1 can be different and are chosen depending on the maximum possible flow that is measured and the required strength, since the fuel can be supplied under different pressures. For the vehicle's internal combustion engine, when choosing the design of a thermo-anemometric flow meter for use, it is necessary to consider the wall thickness of the tube body for the specified pressure values created by the fuel pump of the internal combustion engine power system, and sufficient throughput taking into account the maximum fuel consumption of the vehicle's given specific engine (maximum possible fuel consumption). These parameters are taken into account at the stage of the flowmeter development and they are, as a rule, indicated in the technical data sheet for the finished product. The body-tube 1 can be made of different materials, however, in the process of developing and manufacturing the flowmeter, the thermal conductivity of the material, the convenience of its mechanical processing for further connection to the internal combustion engine power system should be considered. The material of the casing-tube 1 will affect the process of heat transfer from the fuel to the atmosphere, that is, it will influence the radial heat flow value in the process the fuel consumption measuring. Since the amount of heat at different flow modes, which will be discharged into the atmosphere, will be different, consequently, the error of fuel consumption measurement will also be different. To reduce the heat exchange between the casing-tube 1 and the environment, it is desirable to opt for the material of the casing-tube 1 with a lower thermal conductivity, that will reduce the radial heat flow variations and heat loss to the atmosphere for different amounts of fuel consumption, and this, in turn, will decrease the error of measuring the consumption data.

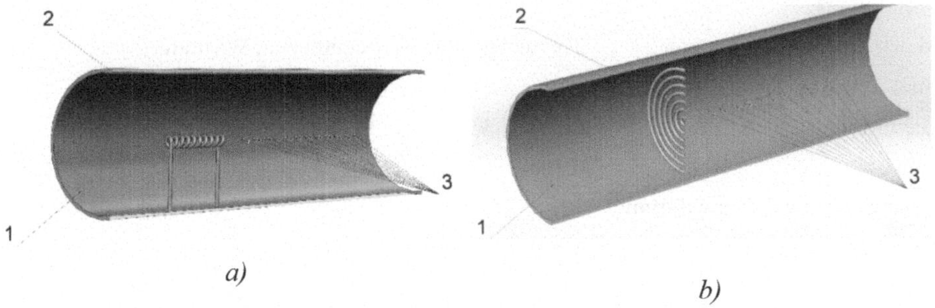

Fig. 1. Thermo-anemometric flow meter with different shape and location of the heating element relative to the body-tube axis: a). Small radius, along the axis; b). Large radius, in a section perpendicular to the axis; 1 – body-tube; 2 – heating element; 3 – thermal converters

It was already mentioned above that the casing-tube 1 should have a capacity to ensure the maximum possible internal combustion engine fuel consumption, for which the flow meter is selected, at all possible speed and load modes of its operation. Consequently, the inner diameter of the housing-tube 1 must be greater than a certain value from the point of view of the flowmeter's throughput capacity as a whole for this specific internal combustion engine. The wall thickness of the housing-tube 1 of the flowmeter and its material must provide the necessary strength, the tube should withstand a certain pressure in the engine power system without destruction, gouging and/or loss of tightness. In the process housing-tube 1 material choice, it is necessary to consider its linear expansion coefficient. Transport means operate in a fairly wide range of temperatures, for example, the KrAZ car is designed for operation in a temperature range of −45… + 50 °C [9], while the body-tube should not change its geometric dimensions, or this change should be minimal, which can be considered by software in the process of using a heat flow meter with a computer connection [8].

To reduce the radial heat flow in the housing-tube 1 of the thermo-anemometric flow meter, it can be recommended its thermal insulation, that can be provided by various materials. However, they should also meet certain requirements, including resistance to the aggressive influence of the fuel itself. The authors suggest using fluoroplastics as the heat-insulating material of the body-tube 1.

Fluoroplast F-4 is a crystalline polymer. It has a melting temperature of crystallites of 327 °C, a glass transition temperature of amorphous areas −100…−120°C. At the temperature higher than the decomposition temperature (415 °C), F-4 fluoroplastic does not go into a viscous state, and at 370 °C, the viscosity of its melt is approximately 1011 P, e.g. 106 times more than the viscosity required for casting under pressure. To consider the change in heat flow for the body-tube 1 different materials of the heat flow meter and to measure different fuels consumption, it is necessary to analyze their heat transfer properties, e.g. thermal conductivity coefficients (Table 1).

Table 1. Thermal conductivity of some petroleum fuels, the body-tube materials and insulating materials of the thermal flow meter, t = 20 °C [10, 11]

Material/substance	Thermal conductivity coefficient, W/(m•K)
Fuel	
Diesel	0,108
Petrol	0,106
Casing-tube and/or insulation	
Duralumin	160
Brass	110
Fluoroplast F-5	0,25

It can be seen from the table, that the coefficients of fuels thermal conductivity have significantly lower values than the materials of the casing-tube 1 and/or its insulation. To maximally exclude the influence of the thermal conductivity of the body-tube 1 material of the thermo-anemometric flowmeter on the radial heat flow, and accordingly, on the error of fuel flow measurements, it is necessary to find materials with such a coefficient of thermal conductivity that completely excludes heat transfer in the radial direction, that is possible only theoretically. Thus, complete elimination of heat transfers in the radial direction in the thermo-anemometric flow meter by choosing the material of the body-tube 1 and/or insulation is practically impossible.

Heating element 2 (Fig. 1) heats the fuel flow to a certain temperature, that, under safety conditions, should not exceed its flash point. The flash point is the lowest fuel temperature at which, under the established conditions, a fuel-air mixture is formed above its surface, which is capable of igniting in the air under the influence of an ignition source. At the same time, this mixture is formed slowly and is not able to burn stably. The higher the boiling point of petroleum products is, the higher the flash point is. Petrol fractions have a flash point of up to −40 °C, kerosene – over 28 °C, oil fractions from 130–350 °C. The flash point gives an idea of volatile fractions presence in the fuel and indicates the level of fire hazard and petroleum products explosive hazard. In fact, the flash point is the temperature at which combustion begins. Modern Ukrainian standards set a flash point limit of 61 ºC, below which fuels are classified as flammable, and above – as flammable liquids.

It should also be considered that there is no air in the housing-tube 1 of the thermo-anemometric flow meter and the formation of a fuel-air mixture is practically impossible. The fuel is under a certain pressure, which is created by the engine power system fuel pump, and when the engine is stopped, the pressure in the system should remain above atmospheric to prevent fuel leakage (air entering the system). Therefore, it is possible to consider the ignition temperature in a closed crucible and/or boiling, including for fire safety reasons. The boiling point for diesel fuel is in the range of 170–380 °C [12], since these values are determined for normal atmospheric conditions, it is considered possible to heat diesel fuel in a thermo-anemometric flow meter up to a maximum of 170 °C.

The heating element 2 can have a different geometric shape. As a rule, it is made of a high specific resistance wire (nichrome) and is installed axially with the axis of the housing-tube 1. If the heating element 2 is made of wire, it often has the spiral shape of a relatively small (Fig. 1) or large radii (Fig. 1b), which also affects the magnitude of radial and axial heat flows and requires additional research.

The heating element 2 can be installed on the outer surface of the housing-tube 1, which facilitates the issues solution of ensuring the supply of electric current to it (the thermal conductivity of the material of the housing-tube 1 must be taken into account) and creates a different temperature distribution, changing the magnitude and direction of the radial heat flow in the fuel, the consumption of which is measured.

Thermal converters 3 are also in mostly located along the axis of the body-tube 1 of the thermo-anemometric flow meter. They can be of different operating principles and types and record the temperature of the fuel in the process of its flow along the axis of the housing-tube 1 at certain distances from the heating element 2. The specified temperature distribution is the basis for calculating the speed of fuel movement and, accordingly, its volume consumption, which are connected by a known dependence, m^3/s:

$$Q = VS \tag{1}$$

where: V – the fuel flow rate, m/s; S – the cross-sectional area of the casing-tube by the inner diameter, m^2.

It follows from (1) that for a given volumetric fuel flow, with a decrease in the cross-sectional area S, the fuel flow rate V increases proportionally, and vice versa, with an increase in the cross-sectional area S, the flow rate decreases proportionally. This enables to adjust the temperature distribution along the body-tube axis for different values of flow rates (fuel consumption).

The number of thermoconverters 3 and the distance between them, the distance from the heating element 2 to the first thermoconverters 3 influence the general picture of temperature distribution for a given specific fuel consumption. They depend on the range of velocities (the range of possible flows that can be measured), that should be registered by the thermo-anemometric flow meter. This range can be quite large for a car diesel engine – from zero (no fuel supply to the diesel engine) in forced idling mode to the maximum value corresponding to the maximum instantaneous of this diesel engine fuel consumption. From (1), it is difficult to calculate that, for example, for fuel consumption from 0.05–0.25 m^3/h, the speed of fuel movement in tubes with diameters of 0.01–0.1 m will be 1.77–884.6 mm/s. Such a wide range of speeds requires the development of constructive solutions with regards to the possibility of registering instantaneous fuel temperatures values at a given time.

It is also not hard to draw the conclusion that the parameters of the heat transfer process in the radial and axial directions in the housing-tube 1 of the thermo-anemometric flow meter at different fuel speeds will differ substantially and this will not always make it possible to determine the temperatures at the locations of the heat converters 3 and calculate the fuel flow rate, its consumption [5, 8]. For example, when the fuel speed (expenditure) increases, a temperature will be registered by the thermal converters 3 for a certain speed, which is little or practically no different from the temperature registered by two consecutively installed thermal converters 3, then this can mean that the fuel flow rate has reached a critical value, above which its definition is impossible. The algorithmic task appears: to reduce the error of fuel consumption measuring in a given mode of operation of the internal combustion engine, consider the temperatures of two neighboring thermotransducers with a larger difference in readings (for all serially installed thermotransducers 3). Then, in the measurement process, it is necessary to influence (change in the required range) the fuel flow rate for a specific value (range) of its consumption. These reflections lead to the idea of using a body-tube of a thermo-anemometric flow meter with a different diameter. Hence, the thermo-anemometric flow meter (Fig. 1.) has limitations regarding the range of measurement of the vehicle's internal combustion engine fuel consumption.

3 Results and Discussion

Expansion of internal combustion engines fuel consumption measurement range (reduction of measurement error) can be achieved by using several sequentially installed tubes 1 of different diameters, the number of which depends on the range of fuel consumption being measured, that is, by creating a multi-stage thermo-anemometric flow meter (Fig. 2). It consists of tubes 1 connected serially and axially by cones 2 (four tubes 1 are considered in the example). The cones conicity value 2 is chosen for reasons of exclusion (or minimization) of swirling the fuel flow at all speeds of its movement through all tubes 1. This fact requires additional research.

The use of tubes 1 with diameters d_1–d_4 enables to vary the fuel speed flowing through the tubes 1 of the thermo-anemometric flow meter, as well as to influence its heating temperature in an increased flow range. This makes it possible to register the maximum possible temperature differences of the fuel in the stream via one thermoconverter 4 to the next, located in the stream at a distance of k.

Due to the correct choice of diameters d_1–d_4 of tubes 1, it is possible to achieve overlapping a wide range of possible fuel consumption. They are selected on the basis of simple considerations: if we consider the law of fluid flow continuity and establish, for example, that $d_4 = d_3/2 = d_2/4 = d_1/8$, it is not difficult to obtain that the speed of V_1

is 64 times smaller than the speed of V_4. Therefore, by choosing the number of thermo-anemometric flow meter tubes and the ratio of their diameters, it is possible to cover almost the entire possible range of fuel consumption of a transport internal combustion engine of any capacity.

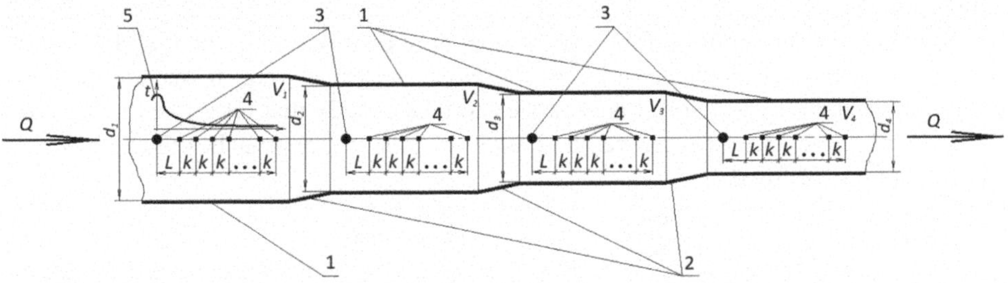

Fig. 2. Scheme of a multi-stage thermo-anemometric flow meter [13]: 1 – tube; 2 – cone; 3 – heating element; 4 – thermal converter; 5 – temperature along the tube axis, °C; Q – volumetric fuel consumption, m^3/s; d_1 - d_4 – tube diameters; V_1 -V_4 – fuel speed along the tube axis, m/s; L – distance from the heating element to the first heat exchanger, m; k – the distance between the thermoconverters, m.

The generalized formula for determining the tubes diameters ratio of the thermo-anemometric flow meter and the fuel flow rates looks in the following way:

$$\frac{V_i}{V_{i+1}} = \left(\frac{d_{i+1}}{d_i}\right)^2, \tag{2}$$

where $i = (1 \ldots n\text{-}1)$ – the serial number of the thermo-anemometric flow meter tube in the direction of fuel movement; n – the number of thermo-anemometric flow meter tubes.

Heat converters 4 and heating elements 3 are installed axially with tubes 1. The first heat converter 4 in the fuel flow is placed at a distance L from the heating element 3, the distance from which to the next heat converter 4 and further between the following heat converters is k (Fig. 2). The number of thermotransducers depends on the flow rate Q and is chosen in such a way that it is possible to reproduce the temperature distribution along the axis of each tube 1 with the maximum possible temperature difference between all neighboring thermotransducers. It has already been mentioned before that for the fuel consumption value Q, its movement speed along the axis of the tube (V_1–V_4) will be higher, the greater the value of Q and/or the smaller the diameter (d_1–d_4).

Due to the increase in the fuel flow rate in tube 1 of the thermo-anemometric flow meter, the thermal transducers 4 will register an increasingly uniform temperature, and after establishing rapid heat transfer, the indications of the thermal transducers 4 will not vary significantly and will approach the fuel temperature near the heating element 3. Absence (reduction) of the difference in readings between by neighboring thermotransducers 4 increases the temperature determination error (fuel consumption measurement error) or makes it impossible to determine the temperature difference (consumption) at all. Then it can be suggested to reduce the speed of the fuel, which in our case can be observed in tubes 1 of a larger diameter. Conversely, as the speed of fuel movement decreases, heat transfer slows down, heat converters 4 can register a non-sufficient change in the temperature of the flow, which will also lead to an increase in the fuel flow measurement error. In this case, it is necessary to increase the speed of the fuel, which in the proposed design of the flow meter can be observed in tubes 1 of a smaller diameter. In this way, the measurement range of fuel consumption is not only expanded, but also the error of measuring liquid fuel consumption with a thermo-anemometric flow meter is reduced.

To obtain the maximum informative temperature distribution in the fuel flow (with the largest difference between adjacent thermoconverters 4), e.g. to reduce the fuel consumption measuring error, it is necessary to take exactly that tube 1 of the thermo-anemometric flow meter for temperature analysis, where it approaches the view when the condition is fulfilled:

$$\sum\nolimits_{y=1}^{x} (|t_y - t_{y+1}) \to max, \tag{3}$$

where x – the number of thermal converters in tube 1 of the thermo-anemometric flow meter; y – serial number of the heat converter in the fuel flow, y = 1…x–1; t – the temperature registered by the converter under number y.

The vehicle's internal combustion engine fuel consumption at different speed and load modes of its operation can change both in the direction of its increase and in the direction of its decrease. The modulus sign in (3) makes it possible to analyze specifically the magnitude of the temperature difference between adjacent heat exchangers. It is in that tube that the condition (3) is fulfilled, e.g. the distribution of measured temperatures with the maximum difference between all neighboring heat exchangers, greater informativeness regarding the change in flow rates (fuel consumption) is obtained.

Besides, it should be noted that it is necessary to check the movement of the fuel flow in the tubes for turbulence, since it can significantly influence the difference in indications between adjacent heat exchangers (measurement error). The turbulent motion will contribute to the equalization of the thermotransducers indications and depends on the flow turbulence intensity. Turbulent movement can manifest in separate modes of fuel movement and be characteristic for some modes and/or for some tubes of the thermo-anemometric flow meter. Such a movement can introduce a non-constant error into the process of temperature measurement and subsequent calculation of fuel velocities (consumption).

4 Conclusion

It was formulated the condition regarding the informativeness of the flowmeter's axial heat flow indicators: the minimum error can be obtained from the thermotransducers indication of that particular tube of the multi-section thermo-anemometric flow meter, where the maximum total temperature difference between all neighbouring thermotransducers is obtained.

It was offered the ways and constructive solutions to reduce fuel flow measurement errors with a thermo-anemometric flow meter for internal combustion engines of vehicles, which comprise expanding the flow measurement range, determining a better distribution of temperatures (velocities) in the fuel flow, using materials for thermo-anemometric flow meter tubes with certain physical properties. It was outlined the ways to thermo-anemometric flow meter design further improvement to reduce the errors in measuring the fuel consumption of the vehicle's internal combustion engine.

References

1. Papathanasiou, P., Kissling, B., Berberig, O., Kumar, V., Rohner, A., Bezděk, M.: Flow disturbance compensation calculated with flow simulations for ultrasonic clamp-on flowmeters with optimized path arrangement. Flow Measur. Instrum., **85**, 102167 (2022). https://doi.org/10.1016/j.flowmeasinst.2022.102167
2. Büker, O., et al.: Characterisation of a Coriolis flow meter for fuel consumption measurements in realistic drive cycle tests. Flow Measur. Instrum. **93**, 102424 (2023). https://doi.org/10.1016/j.flowmeasinst.2023.102424
3. Massoud, M.: Engineering thermofluids: thermodynamics, fluid mechanics, and heat transfer, pp. 1–1119 (2005). https://doi.org/10.1007/b138870
4. Yoder, J.: New-technology flowmeters: Volume I, pp. 1-162 (2022). https://doi.org/10.1201/9781003130017
5. Korobiichuk, I., Bezvesilna, O., Ilchenko, A., Shadura, V., Nowicki, M., Szewczyk, R.: A mathematical model of the thermo-anemometric flowmeter. Sensors **15**, 22899–22913 (2015). https://doi.org/10.3390/s150922899
6. Korobiichuk, I., Kachniarz, M., Bezvesilna, O., Ilchenko, A., Szewczyk, R.: Calorimetric flow meter of motor fuel with inlet temperature regulation. In: 4th International Conference on Control, Decision and Information Technologies (CoDIT), 2017, IEEE Conference Publications, 0975–0979 (2017). https://doi.org/10.1109/CoDIT.2017.8102725
7. Sherwin, D., Dees, J.: Fuel and engine management system for gaseous-fueled spark-ignition non-road engines. In: ASME 2020 Internal Combustion Engine Division Fall Technical Conference, ICEF 2020, Art. no. V001T07A003 (2021). https://doi.org/10.1115/ICEF2020-3045
8. Korobiichuk, I., Ilchenko, A.: Optimal design parameters of thermal flowmeter for fuel flow measurement. Sensors **22**(8882), 2022 (2022). https://doi.org/10.3390/s22228882
9. Manoylo, V., et al.: Estimating dynamic and flow characteristics of electromagnetic dispenser for the Kraz truck converted to gas. Int. J. Integr. Eng. **15** (4), 146–156 (2023). https://doi.org/10.30880/ijie.2023.15.04.013
10. Hansen, A.C., Kyritsis, D.C., Lee, C.F.: Characteristics of biofuels and renewable fuel standards. Biomass Biofuels: Strat. Global Industr. 1–26 (2010). https://doi.org/10.1002/9780470750025.ch1
11. Modified fluoroplast. Available online:https://blagos.com.ua/ua/index.php?route=blog/article&blog_category_id=74&article_id=131.Accessed 29 Dec 2023
12. Biletskyi, V.S. (ed.) Small mining encyclopedia. Eastern Publishing House, Donetsk, vol. 3, p. 644 (2013)
13. Ilchenko A.V.: Improvement of the design of the heat flow meter of liquid motor fuels. In: Abstracts of the 11th International Scientific and Technical Internet Conference "Problems and Prospects of Road Transport Development", pp. 136–138. VNTU, Vinnytsia (2023)

Improving the Operational Reliability of Ground-Based Navigation Systems

Igor Korobiichuk[1(✉)], Viktorij Mel'nick[2], Vera Kosova[2], Zhanna Ostapenko[2], Nonna Gnateiko[2], Galyna Boiko[2], and Iana Piskivets[3]

[1] Łukasiewicz Research Network – Industrial Research Institute for Automation and Measurements PIAP, Warsaw, Poland
igor.korobiichuk@piap.lukasiewicz.gov.pl
[2] National Technical University of Ukraine "Igor Sikorsky Kyiv Polytechnic Institute", Kyiv, Ukraine
vera_62@ukr.net, g.boyko@kpi.ua
[3] Independent Researcher, Kyiv, Poland

Abstract. The most commonly used suspension in primary measuring instruments and vibration reproduction systems is a suspension on flat springs. In the paper analyses one of the methods for reducing the natural frequency of an elastic suspension on flat springs. A theoretical substantiation of the possibility of increasing the vibration isolation efficiency of navigation equipment by reducing the natural frequency of the elastic suspension by using a system of additional loading of the moving part is presented. The correlations between the natural frequency of the suspension and the system parameters are determined.

Keywords: vibration isolator · elastic suspension · spring material · gimbal natural frequency · mathematical pendulum

1 Introduction

It is well known that one of the main elements of primary measuring instruments, vibration protection, and vibration reproduction systems is an elastic suspension. The most commonly used suspension in the above-mentioned objects is a suspension on flat springs, which is explained, first of all, by the technical simplicity of such a suspension, the simple technology of manufacturing springs, and the comparative simplicity of their calculation.

In [1] paper the mathematical methods used and the resulted criteria for designing a flat ride suspension system which will perform in different speeds is presented. In [2] was investigated the benefits of forming side-by-side oppositely buckling blades from a single monolithic sheet of spring material and was shown that optimal shaping typically improves the inconveniently small spring working range by over 60% compared to an equivalent rectangular blade. New suspension technique uses flat springs in Euler

I. Piskivets—Independent Researcher, Poland.

R. Szewczyk et al. (Eds.): AUTOMATION 2024, LNNS 1219, pp. 318–324, 2024.
https://doi.org/10.1007/978-3-031-78266-4_28

buckling mode was presented by authors in [3], there was shown numerical analysis of the basic lever constrained configuration identifies the effect of the main geometrical parameters.

A characteristic feature of the flat-spring suspension is the practical absence of relative angular rotation and relative translational movement of the connected parts of the suspension in all directions except the selected one, as well as the practical absence of dry friction. This is springs was used for design of the double-ring dynamically tuned gravimeter is given [4]. This design it possible to increase the measurement accuracy of Earth's gravitational field anomalies [5]. Research [6] provides a frame of reference and guidance for improving the stable cyclic operating life of flat springs and show perspective of use them in MEMS devices [7]. When using a resilient suspension in the form of an inverted pendulum on vertically located and cantilevered flat springs, it is possible to obtain a natural frequency of the suspension in the horizontal plane of about 1 Hz [8]. The high kinematic selectivity of such a suspension concerning the direction of the measured or reproduced vibration and the possibility of achieving relatively low natural frequencies make it promising in terms of use for many practical tasks, in particular, vibration isolation of navigation equipment in the region of low disturbance frequencies characteristic of ground-based carriers of such equipment [9]. Further reduction of the natural frequency of the elastic suspension by changing the geometric dimensions of the springs is associated with a significant increase in their length and, consequently, the dimensions of the suspension [10, 11].

This work analyzes one of the methods for reducing the natural frequency of an elastic suspension on flat springs, the essence of which is to use an additional system for loading the moving part.

2 Materials and Methods

The design scheme of the suspension on vertically mounted flat springs, which contains an electromagnetic system for additional loading of the moving part, is shown in Fig. 1.

This figure denotes: 1 – magnetic system core; 2 – electric winding; 3 – magnetic core; 4 – base; m – mass of the moving part; l – working length of the springs; δ and δ_n – initial and current value of the gap between the magnetic system and the moving part; x and Δ – horizontal and vertical movement of the moving part relative to the base, respectively; P – the force acting on the moving part in the horizontal plane (caused by the bending of flat springs when the moving part deviates from the base along the X axis by x, the pendulum moment of forces, the magnetic system, and an external disturbance or control force). With the optimal choice of the design parameters of the magnetic system related to the reduction of scattering fluxes, the magnetic energy W_M, will be mainly concentrated in the air gap and will be proportional to the magnetic conductivity of the gap [12, 13]. Due to the kinematic connection between the movements of the moving part in the vertical and horizontal directions, the current value of the gap δ_n between the core of the magnetic system and the moving part will decrease when the shape of the springs is close to the sinusoidal one in accordance with the formula [14]:

$$\delta_n = \delta - \frac{\pi^2}{16} \cdot \frac{x^2}{l}. \quad (1)$$

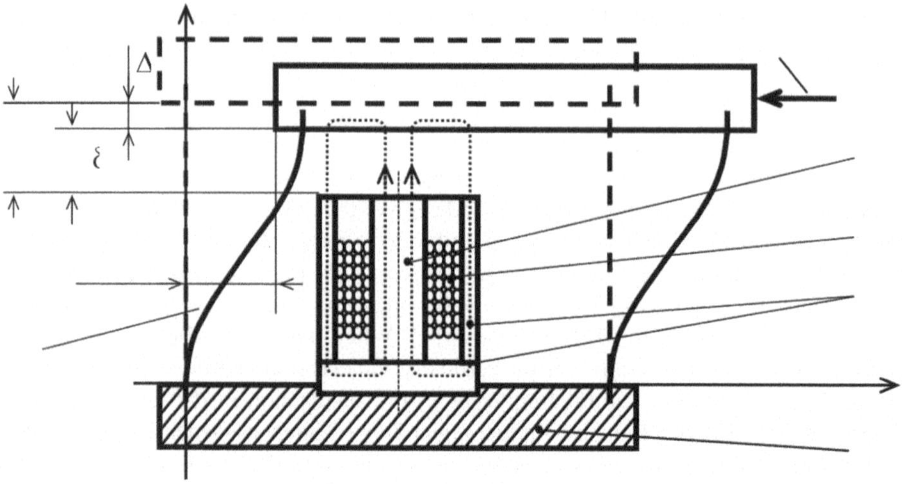

Fig. 1. The design scheme of the suspension

The largest possible values of displacement x and air gap δ are of the same order of smallness, with $x < 0{,}1\ l$. Then the magnetic energy can be expanded into a power series

$$W_M = \mu_o \frac{(iw)^2}{2}\frac{S}{\delta}\left(1 + \frac{\pi^2}{16}\frac{x^2}{l\delta} + \frac{\pi^4}{256}\frac{x^4}{l^2\delta^2}\right), \tag{2}$$

where iw – is the ampere-wattage of the electric winding, S – is the cross-sectional area of the air gap. The equation of oscillations of the moving part in the horizontal plane, taking into account the nonlinear components of the force, will take the form

$$m\ddot{x} + \frac{\pi^2}{8l}\left(\pi^2\frac{EI}{l^2}n - mg - \frac{k}{\delta^2}\right)x + \frac{\pi^2}{8l}\left(\frac{\pi^4}{16}\cdot n\cdot\frac{EI}{l^4} - \frac{\pi^2}{8l}\frac{k}{\delta^3}\right)x^3 = P_3, \tag{3}$$

where n – is the number of flat springs; E – s the modulus of elasticity of the material of flat springs; I – is the moment of inertia of the cross-section of flat springs; P_3 – is the external disturbance or control force; $k = \frac{1}{2}\mu_o S(iw)^2$.

As can be seen from the above equation, the presence of the magnetic system simultaneously reduces the linear and nonlinear components of the force P, caused by the bending of the springs and the magnetic system. Natural angular frequency of a linear conservative system with a magnetic system of additional load – ω_o

$$\omega_o^2 = \frac{\pi^2}{8}\frac{1}{l}\left(\pi^2\frac{EI}{l^2}n - mg - \frac{k}{\delta^2}\right)\Big/ m. \tag{4}$$

will decrease with an increase in the coefficient k and a decrease in the size of the initial air gap between the magnetic system and the moving part. The condition for the stability of the linear system is the ratio

$$\frac{k}{\delta^2} < \frac{\pi^2}{8l}\left(\pi^2\frac{EI}{l^2}n - mg\right). \tag{5}$$

at

$$\frac{k}{\delta^3} > \frac{\pi^2}{2} \cdot n \cdot \frac{EI}{l^3}, \tag{6}$$

the presence of a magnetic system causes the nonlinear component of the restoring force to become negative, i.e., the rigid overall characteristic of the suspension is transformed into a soft one. In this case, an increase in the modulus of the nonlinear component of the restoring force may cause a loss of system stability due to the fact that the overall stiffness coefficient of the suspension may become negative. The value of the oscillation amplitude x_a, which, if exceeded, will lead to a loss of stability in this case, is determined by the formula

$$x_a < \sqrt{\left(\frac{k}{\delta^2} + mg - \pi^2 \frac{EI}{l^2} n\right) \Big/ 3\frac{\pi^2}{8l}\left(\frac{\pi^2}{2} \cdot n \cdot \frac{EI}{l^3} - \frac{k}{\delta^3}\right)}. \tag{7}$$

Reducing the natural frequency of the suspension with the help of a magnetic system of additional loading is also achieved in the case when the plane of the flat springs – the XY plane– coincides with the horizontal plane. In this case, the natural frequency of the linear system is determined from Eq. (4) at $mg = 0$

$$\omega_o^2 = \frac{\pi^2}{8}\frac{1}{l}\left(\pi^2 \frac{EI}{l^2} n - \frac{k}{\delta^2}\right) \Big/ m. \tag{8}$$

Compared to the suspension on vertically mounted flat springs, the horizontal arrangement of the springs leads (with the same k/δ2) to a significant increase in the overall stiffness of the suspension and its natural frequency. In this case, the condition for the linear system stability is the ratio

$$\frac{k}{\delta^2} < \pi^2 \frac{EI}{l^2} n, \tag{9}$$

i.e., from the standpoint of ensuring the linear system stability, the possible value of the ratio k/δ^2, which characterizes the parameters of the magnetic system and its location relative to the moving part, increases. With a horizontal arrangement of flat springs, the possibility of obtaining a general negative elastic characteristic of the suspension is achieved at higher values of the oscillation amplitude of the moving part. With a vertical arrangement of flat springs according to the usual, not inverted, pendulum scheme, formula (4) takes the form

$$\omega^2 = \frac{\pi^2}{8}\frac{1}{l}\left(\pi^2 \frac{EI}{l^2} n + mg - \frac{k}{\delta^2}\right) \Big/ m, \tag{10}$$

and the stability condition of the linear system is

$$\frac{k}{\delta^2} < \frac{\pi^2}{8l}\left(\pi^2 \frac{EI}{l^2} n + mg\right). \tag{11}$$

As can be seen from formulas (4), (8), and (10), the additional magnetic system, which works to mutually attract the moving and stationary parts of the suspension along the flat springs, significantly reduces the natural frequency of the elastic suspension in all variants of the spring arrangement.

3 Results and Discussion

Let's illustrate the possibility of a real reduction in the natural frequency of the suspension with the following numerical example. Let $m = 4$ kg; $n = 3$; $E = 1.25{\cdot}10^{11}$ N/m^2 (Bronze BrB2); l = 0,12 m; spring width – b = 0.02 m; spring thickness – h = 0.0005 m; natural frequency of oscillations of the moving part in the horizontal plane without an additional load system is 0.97 Hz. Let's assume that the natural frequency of the suspension should be reduced by a factor of 4. Then $k/\delta^2 = 13.4$ N and the dependence $k = f(\delta)$ will take the form shown in Fig. 2 (curve 1). The same figure shows the dependence $k/\delta^3 = f(\delta)$ (curve 2), which gives an idea of the possible change in the nature of the suspension stiffness when using a magnetic system of additional loading (at $k/\delta^3 = 223.1$ the nonlinear component of stiffness $c_н = 0$). At $k/\delta^3 < 223.1$ the elastic response of the suspension will be stiff, and at $k/\delta^3 > 223.1$ – it will be soft. From the viewpoint of ensuring a small nonlinear component of the suspension stiffness, the initial gap should be chosen quite large (60 mm for the example given). However, a large value of δ leads to a significant increase in the power of the magnetic system, such that its practical use becomes impractical. The existing contradiction in the choice of the initial gap is resolved by analyzing the stiffness characteristics of the suspension with a magnet for several real values of k and δ (see Fig. 3) and ensuring such possible oscillation amplitudes of the moving part that would not violate the suspension stability. At $x_{max} \leq 5,\ldots,6$ mm the choice of δ = 12 mm can be considered satisfactory. Then $k = 0.00193$ N/m^2.

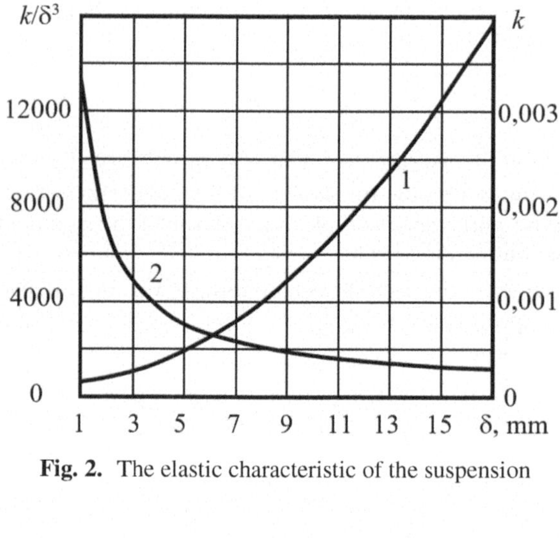

Fig. 2. The elastic characteristic of the suspension

Further calculation of the interrelated geometric and electrical parameters of the magnetic system itself is determined by the suspension geometry. When choosing the value of S = 1600 mm^2 and the permissible current density of 10 A/mm^2, we get that the number of ampere turns should be 1380 A [15].

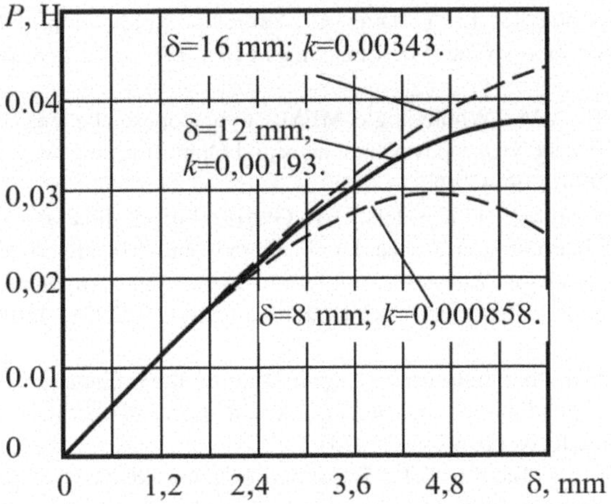

Fig. 3. Stiffness characteristics of the suspension

4 Conclusion

Thus, in the construction of vibration isolation systems for navigation equipment based on elastic suspension on flat springs, the use of additional loading of the moving part along the flat springs leads to a significant reduction in the natural frequency of the suspension, an increase in the vibration isolation efficiency and an increase in the operational reliability of navigation equipment by reducing errors caused by low-frequency horizontal vibration. When such a suspension is used in measuring devices or for vibration reproduction, a certain choice of parameters of the additional load system can significantly reduce (theoretically to zero) the coefficient of nonlinear distortions of the moving part oscillations.

References

1. Marzbani, H., Simic, M., Fard, M., Jazar, R.N.: Sustainable flat ride suspension design. Smart Innov. Syst. Technol. **40**, 251–264 (2015). https://doi.org/10.1007/978-3-319-19830-9_23
2. van Heijningen, J.V., Winterflood, J., Wu, B., Ju, L.: Multi-blade monolithic Euler springs with optimised stress distribution. J. Sound Vibr. **552**, 117614 (2023). https://doi.org/10.1016/j.jsv.2023.117614
3. Winterflood, J., Barber, T.A., Blair, D.G.: Mathematical analysis of an Euler spring vibration isolator. Phys. Lett. Sect. A: Gener. Atomic Solid State Phys. **300**(2-3), 131–139 (2002). https://doi.org/10.1016/S0375-9601(02)00259-1
4. Korobiichuk, I., Nowicki, M., Szewczyk, R.: Design of the novel double-ring dynamical gravimeter. J. Autom. Mob. Robot. Intell. Syst. **9**(3), 47–52 (2015). https://doi.org/10.14313/JAMRIS_3-2015/23
5. Korobiichuk, I., Bezvesilna, O., Tkachuk, A., Chilchenko, T., Nowicki, M., Szewczyk, R.: Design of piezoelectric gravimeter for automated aviation gravimetric system. J. Autom. Mob. Robot. Intell. Syst. **10**(1), 43–47 (2016). https://doi.org/10.14313/JAMRIS_1-2016/6

6. Lai, L., Yu, G., Wang, G., Li, Y., Ding, G., Yang, Z.: Application of Ni/SiCw composite material in MEMS microspring. Micromachines **14**(9), 1767 (2023). https://doi.org/10.3390/mi14091767
7. Senkal, D., Shkel, A.M.: Whole-angle MEMS gyroscopes: challenges and opportunities. Whole-Angle MEMS Gyroscopes: Challenges and Opportunities, pp. 1–153 (2020). https://doi.org/10.1002/9781119441908
8. Karachun, V.V., Melnik, V.N., Kovalets, O.Y.: Gyro-stabilized three-axis platform. Automatic compensation of external interference, Kyiv: "Korneychuk" 192 p. (2010)
9. Ge, L.-F.: Solid Acoustic Waves and Vibration: Theory and Applications. Solid Acoustic Waves and Vibration: Theory and Applications, pp. 1–425 (2021). https://doi.org/10.1142/12231
10. Odintsov, A.A.: Method of automatic compensation for the influence of external noise acting on gyroscopes and pendulum accelerometers. Automation and instrument making, Collection of scientific works, Kyiv, pp. 87–94 (1973)
11. Mehta, Y., Gehlot, S., Sakthivel, P.: Static and dynamic analysis of parabolic leaf spring with design optimization for light commercial vehicle. In: 2019 International Conference on Nascent Technologies in Engineering, ICNTE 2019 - Proceedings, art. no. 8945899 (2019). https://doi.org/10.1109/ICNTE44896.2019.8945899
12. Frolov, K.V.: Reducing the amplitude of oscillations of resonant systems by controlled changes in parameters. Mech. Sci. **3**, 38–42 (1965)
13. Galeazzi, R., Pettersen, K.Y.: Controlling parametric resonance: Induction and stabilization of unstable motions. Parametric Resonance in Dynamical Systems, pp. 305–327 (2012). https://doi.org/10.1007/978-1-4614-1043-0_15
14. Pisarenko, G.S., Yakovlev, A.P., Matveev, V.V.: Handbook on strength of materials. Nauk. Dumka, Kyiv, 736 p. (1988)
15. Bublik, G.F., Kochmar, B.F., Larionchik, V.N.: Analysis of the elastic characteristics of the shock absorber. West. Kyiv. Polytechnic in–t. Instrumentation, vol. 12, pp. 12–15 (1982)

Method of Fitting a Nonlinear Function to Measurement Data, Its Uncertainty and as Example the Application in Thermistor Circuits

Jacek Puchalski[1] and Zygmunt L. Warsza[2](✉)

[1] Central Office of Measures, 00-139 Warsaw, Poland
jacek.puchalski@gum.gov.pl
[2] Polish Metrological Society, Warsaw, Poland
zlw1936@gmail.com

Abstract. The paper presents a method of determining parameters and uncertainty bands of a specific non-linear function fitted to given examined measured data. One or both of the variables of this non-linear function are changed so as to linearize it. Using the linear regression method, fined are the most favorable parameters of this straight line for its adjustment to the measured values of the coordinates of points tested according to the weighted total mean square WTLS criterion. Their autocorrelation and cross-correlation coefficients as well as uncertainties estimated according to the rules of the GUM guide [1] are considered. The parameters and the uncertainty band of the non-linear function result from the parameters of this straight line and its uncertainty band. In the part 2 numerical examples of determining the parameters and uncertainty bands for the characteristic of thermistor (two methods) alone and with resistance twoport sensor are given.

Keywords: Linear regression of non-linear functions · fit · uncertainty and its band · correlation · weighted total least squares method · characteristic function of thermistor

1 Part 1. Theoretical Basis of the Linear Regression Method for Nonlinear Functions

1.1 Introduction

The nominal characteristics of many sensors, transducers and measuring instruments, as well as the dependencies of the parameters of studied phenomena, material properties, characteristics of devices and their components are described by nonlinear functions. In the metrological practice (e.g., calibration and periodic inspection of instruments and sensors), the actual values of the parameters of the tested characteristics are determined. Their characteristics and uncertainties are also compared with the nominal ones whether they meet specific requirements.

R. Szewczyk et al. (Eds.): AUTOMATION 2024, LNNS 1219, pp. 325–341, 2024.
https://doi.org/10.1007/978-3-031-78266-4_29

The basic mathematical tool used in fitting measured parameters of various linear and nonlinear functions is the linear regression. Measured are values of one or more coordinates of the studied points. For two variables, it concerns the determination of function y $=f(P_i, x)$ with a linear dependence on m parameters. P_i for $i = 1, \ldots, m$. As example, for the parabola $y = ax^2 + bx + c$, the parameters a, b, c have to be determined. For characteristic of thermistor $r(T) = r_{\infty}exp(B/T)$, parameters r_{∞}, B have to be find. As a criterion various variants of the least squares method are used, including its most general form with the English acronym WTLS (Weighted Total Least Squares). It considers all the data of the measured points, i.e., coordinate values, distributions, and uncertainties, as well as autocorrelations and mutual correlations of the results of their measurements. The equation of the tested function and the limits of its extended uncertainty band (also known as the uncertainty corridor) as the interval of occurrence of errors with a certain probability are determined. Linear regression is used in scientific, natural, and social scientific disciplines, in medicine and health care, in the economy and technical fields, including instrumentation branch and the service of measures [23].

There are many linear regression methods for fitting linear and different nonlinear functions to measurement data [2–8]. They are numerically complex and usually named after their creators. The oldest is the Gauss–Newton method published at 1809, then the Quasi-Newtonian methods include the Davidson [9], Fletcher and Powell methods [10], the BFGS [11], Levenberg-Marquardt methods [12–14], and the general numerical method for solving Raphson-Newton nonlinear equations. The Monte Carlo method is also used. But in the basic mathematical papers on regression methods, the uncertainty of measurements is usually not assessed strictly according to the principles of the internationally used Guide to the to the expression of Uncertainty in Measurement with the acronym GUM [1], commonly now used in measurement practice. This is also due to the fact that the first version of GUM, initiating the concept of measurement uncertainty, was published in 1993, but many regression methods were developed earlier.

The many recent works usually do not consider the type B uncertainty (u_B), which is estimated by the combined randomized effect of a set of unknown biases in the instruments used in a particular measurement experiment. This is done on the basis of the maximum permissible instrument error given by the manufacturers. If the measurements of the coordinates of the studied points are repeated, then their random complex components u_A should be also determined.

The parameters of searched function $y = f(x)$ are determined by linear regression according to the least squares criterion for the coordinates of the studied points, considering their standard uncertainty $u = \sqrt{u_A^2 + u_B^2}$, autocorrelations, and mutual correlations [20]. The expanded uncertainty band $U_P(y) = k_P\ u(y)$ with a given probability p is also estimated.

1.2 Backgrounds of Proposed Method

To determine the parameters of mathematical models describing the characteristics of devices and their components, including nonlinear functions, the computer calculations and the method of least squares are now commonly used. Previously, this method was avoided as it was very computationally cumbersome without computer. A visual adjustment of curve to measured points marked on graph paper was very popular. The nonlinear curves were linearized by selecting an appropriate scale on this paper for the function sought, e.g., square, logarithmic, or exponential scale. Such simply method was not enough precise now in the most measurements.

A similar numerical method for determining the parameters of a non-linear function $y = f(x)$ fitted to experimental data, but implemented by computer, is proposed below. In this method, the variables x, y determining the coordinates of the measurement points are changed on ψ and ξ to obtain a straight line in the new Cartesian system.

Let us assume case that they are parameterized by only one common parameter β. The equation of a straight line in new coordinates is then of the linear form:

$$\psi(y, \beta) = \theta_1 \xi(x, \beta) + \theta_0 \tag{1}$$

In this method the parameters of Eq. (1) will be adjusted by the linear regression to the measurement data according to the generalized weighted least squares criterion WTLS. From the measurement data of the nonlinear function, the values of the parameters $\theta_1, \theta_0, \beta$ will be determined and from them the parameters f nonlinear function. Due to the inaccuracy of measurements, both the slope coefficient θ_1 and the free term θ_0 will have uncertainties $u(\theta_1) \neq 0$, $u(\theta_0) \neq 0$.

In solving problems of the measurement accuracy also in regression methods, a linear approximation of the values of errors and uncertainty by the term with the first derivative is used. This is permissible when their values do not exceed a few percent of the range and the deviations of the position of the points from the determined function (residual variable) are not too large. That is why the method we proposed here we called as approximate.

After such transformation, the nonlinear function $y = f(x)$ is expressed in Eq. (1) by the new variables ψ and ξ, and there is also a β parameter. For each set of measurement data, it has the most favorable constant value of β with uncertainty $u(\beta) \approx 0$. Limiting the dispersion to the term with the first derivative allows the criterion function to be minimized $\phi_{\psi\xi}$ as two-dimensional θ_1 and β. The solution requires the use of complex numerical algorithms. A larger number of selectable parameters would greatly complicate the procedure of minimizing the function $\phi_{\xi\psi}$. There could be also local minimums or even a criterion function without a global minimum. The time of numerical calculations would also increase significantly.

The accuracy of the result obtained by the method discussed here is assessed by the width of the uncertainty band of the determined function, and in the case of a known rated waveform, by the discrepancy of both curves.

As computational examples of the proposed method, two variants of determining the parameters of the second-degree parabola and the complex exponential function, considered in the classic monograph of N. Draper and H. Smith [23] as impossible to linearize by changing variables, will be considered. The parameters of both nonlinear functions will be adjusted to the measurement data of the examined points for different uncertainties and correlations of random components of their coordinates. The nonlinear function $y = f(x)$, after converting the variables y and x to ξ and ψ, becomes a straight line $\psi(y, \beta) = \theta_1 \xi(x, \beta) + \theta_0$ fitted to points $M_i(\xi_i, \psi_i)$ by the linear regression method, as is shown in Fig. 1. We will use for that our considerations of linear straight-line regression for measurements of both variables and different cases of coordinate dispersion, their uncertainty, autocorrelation, and mutual correlation of variables. They are contained in our earlier works [15–24], including ones in Polish [15–22].

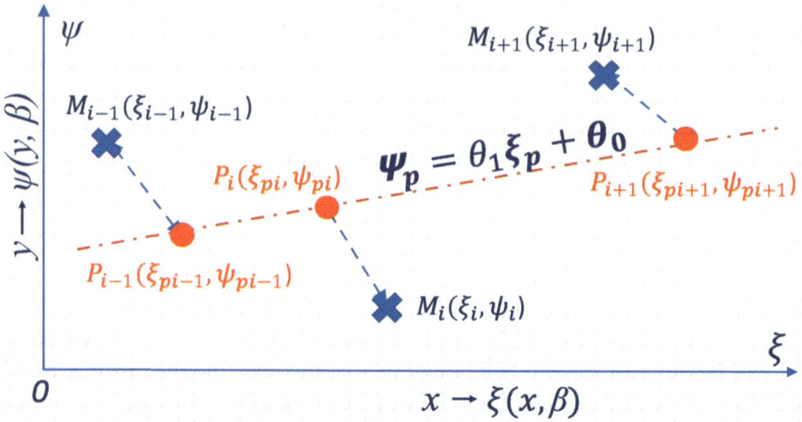

Fig. 1. The result of transforming the non-linear function $y = f(x)$ into a straight line.

The uncertainties of the parameters θ_1 and θ_0 of the line (1a) depend on the coordinates and uncertainties of the coordinates of measured points $M_i(x_i, y_i)$ for $i = (1,\ldots, n)$ and on all autocorrelations and correlations of the mutual sets of coordinates that may occur in these measurements. These influences are generally described by the covariance matrix $\boldsymbol{U_Z} = \begin{bmatrix} \boldsymbol{U_X} & \boldsymbol{U_{XY}} \\ \boldsymbol{U_{XY}^T} & \boldsymbol{U_Y} \end{bmatrix}$. It consists of four matrices as its elements and is used in the criterion function$\phi(\boldsymbol{\Delta X}, \boldsymbol{\Delta Y})$:

$$\phi(\boldsymbol{\Delta X}, \boldsymbol{\Delta Y}) = [\boldsymbol{\Delta X}, \boldsymbol{\Delta Y}]\begin{bmatrix} \boldsymbol{U_X} & \boldsymbol{U_{XY}} \\ \boldsymbol{U_{XY}^T} & \boldsymbol{U_Y} \end{bmatrix}^{-1}\begin{bmatrix} \boldsymbol{\Delta X} \\ \boldsymbol{\Delta Y} \end{bmatrix} \to min \tag{2}$$

where: $\boldsymbol{\Delta X} = \boldsymbol{X} - \boldsymbol{X_p}$, $\boldsymbol{\Delta Y} = \boldsymbol{Y} - \boldsymbol{Y_p}$ are the deviation vectors as differences between the coordinate vectors of measured points $\boldsymbol{X} = [x_1, \ldots, x_n]^T$, $\boldsymbol{Y}=[y_1, \ldots, y_n]^T$ and vectors $\boldsymbol{X_p}=[x_{p1}, \ldots, x_{pn}]^T$,$\boldsymbol{Y_p}=[y_{p1}, \ldots, y_{pn}]^T$ of the corresponding coordinates of points on the curve; $\boldsymbol{U_X}$, $\boldsymbol{U_Y}$ and $\boldsymbol{U_{XY}}$ – are covariance matrices of coordinates x and y of the measurement points and their cross-correlations.

A nonlinear function $y = f(x)$ fitted to the results of measured coordinates (x_i, y_i) of the test points is find from the straight-line $\psi = \theta_1\xi + \theta_0$ on Fig. 1 fitted to the new point coordinates M_i (ξ_i, ψ_i). These points correspond to the points of the P_i (ξ_{pi}, ψ_{pi}) for $i =$ 1, …, n, lying on this straight line. The segments connecting the two types of points, in the general case, have different lengths and slopes. They are respectively dependent on the slope of the simple regression and on the elements of the covariance matrix described the measured coordinates. In the absence of correlation, this matrix becomes diagonal and the slopes of the segments connecting the points depend only on the slope of the simple regression and the ratio of the squares of the uncertainty of the two coordinates at each measurement point. In the special case, for equal uncertainty of both coordinates of the points, these segments are perpendicular to the simple regression. On this line, the measuring points M_i correspond to the P_i points with a position dependent on the M_i coordinates and their uncertainties and correlation coefficients. If there are only y_i uncertainties, then the P_i points are projections of M_i points in the vertical direction y_0.

Relations describing the propagation of deviations when changing variables can be presented in the form of Taylor series expansion. Small deviations of the residual variable together with the measurement errors of the coordinates of the points are approximated with sufficient accuracy by the first terms of this expansion. The first derivative is used to determine the deviations ψ and ξ as $\Delta\xi \approx \xi\prime(x, \beta)\Delta x$ and $\Delta\psi \approx \psi\prime(y, \beta)\Delta y$. They are stored together in a diagonal matrix $\boldsymbol{L}$ about the size $[2n \times 2n]$:

$$\boldsymbol{L} = \begin{bmatrix} \xi\prime(x_1, \beta) & 0 & \dots & & 0 \dots 0 & \\ \dots & \dots & \dots & & \dots\dots\dots & \\ 0 & & \dots \xi\prime(x_n, \beta) & & 0 \dots 0 & \\ & 0 \dots 0 & & \psi\prime(y_1, \beta) & 0 & 0 \\ & \dots\dots\dots & & \dots & \dots & \dots \\ & 0 \dots 0 & & 0 & & \dots \psi\prime(y_n, \beta) \end{bmatrix} \tag{3}$$

for which $\boldsymbol{L}\boldsymbol{L}^{-1} = \boldsymbol{L}^{-1}\boldsymbol{L} = \mathbf{I}$, where $\mathbf{I}$ – is an identity matrix.

The uncertainties in the new Cartesian coordinates result indirectly from this linear approximation. For Gaussian density distributions: $u(\psi) \approx |\psi\prime(y, \beta)|u(y)$ and $u(\xi) \approx |\xi\prime(x, \beta)|u(x)$, and the Eq. (2) takes the form:

$$\phi(\boldsymbol{\Delta X}, \boldsymbol{\Delta Y}) = [\boldsymbol{\Delta X}, \boldsymbol{\Delta Y}]\boldsymbol{L}\boldsymbol{L}^{-1}\begin{bmatrix} \boldsymbol{U_X} & \boldsymbol{U_{XY}} \\ \boldsymbol{U_{XY}^T} & \boldsymbol{U_Y} \end{bmatrix}^{-1}\boldsymbol{L}^{-1}\boldsymbol{L}\begin{bmatrix} \boldsymbol{\Delta X} \\ \boldsymbol{\Delta Y} \end{bmatrix}$$

$$\approx [\boldsymbol{\Delta\xi}, \boldsymbol{\Delta\Psi}]\begin{bmatrix} \boldsymbol{U_\xi} & \boldsymbol{U_{\xi\Psi}} \\ \boldsymbol{U_{\xi\Psi}^T} & \boldsymbol{U_\Psi} \end{bmatrix}^{-1}\begin{bmatrix} \boldsymbol{\Delta\Xi} \\ \boldsymbol{\Delta\Psi} \end{bmatrix} = \phi_{\xi\psi}(\boldsymbol{\Delta\xi}, \boldsymbol{\Delta\Psi}, \beta) \tag{4}$$

In the new coordinates the measurement points are given by vectors respectively $\boldsymbol{\xi}$ and $\boldsymbol{\Psi}$. Their covariance matrices are $\boldsymbol{U_\xi}$, $\boldsymbol{U_\Psi}$ i $U_{\xi\psi}$. Only the β is unknown parameter. Its constant value should be chosen in order to minimize the criterion function $\phi_{\xi\psi}$. It is assumed that the uncertainty of this operation is negligibly small, i.e., $u(\beta) \approx 0$. When transforming vectors $\boldsymbol{X}$, $\boldsymbol{X_p}$ and $\boldsymbol{Y}$, $\boldsymbol{Y_p}$ respectively into vectors $\boldsymbol{\xi}$, $\boldsymbol{\xi_p}$ and $\boldsymbol{\Psi}$, $\boldsymbol{\Psi_p}$, with (2) and (4) in order to fit a nonlinear function, a system of two equations is obtained: describing the minimization of the criterial function $\phi_{\xi\psi}$ and the equation of a straight line. These are given below as n-dimensional vector:

$$\begin{cases} \phi_{\xi\psi}(\boldsymbol{\Delta\xi}, \boldsymbol{\Delta\Psi}, \beta) \rightarrow min \\ \boldsymbol{\Psi_p} = \theta_1\boldsymbol{\xi_p} + \boldsymbol{\theta_0} \end{cases} \tag{5a,b}$$

The general case with all possible correlations is described in [14, 19, 22]. The solution of this system in the new Cartesian coordinates ξ and ψ is the same as for simple regression using the WTLS together with specific cases [4–14, 17]. Criterion function

$\phi_{\xi\psi}$ shall be determined numerically. It depends on the slope of the straight line θ_1 and its offsets linked unambiguously as a relation $\theta_0 = (S_y - \theta_1 S_x)/S$ with auxiliary parameters S, S_x and S_y, which are different functions of coordinates x_i, y_i and uncertainties defined and described in the [17, 19, 21]. Using the characteristics calculated at several points of the variable θ_1, parameter is selected β and finds a minimum of features $\phi_{\xi\psi}(\theta_1)$, to obtain a straight line aligned with the measuring points. In the new coordinates ψ and ξ uncertainty are related only to the parameters of the θ_1 and θ_0. When they are correlated with each other, the correlation coefficient is also an important parameter $\rho_{\theta_0\theta_1}$. These elements can be determined from the matrix specified in (4) using the uncertainty propagation and approximate formulas for sensitivity coefficients described in [17, 19]. The boundaries of the uncertainty band, as the area around the function to be sought, are determined numerically in the software environment, e.g., *R*, MATLAB, and Monte Carlo Method.

In the generation of samples from the covariance matrix (4), the Choleski decomposition method is used [24]. This method allows you to generate a multidimensional variable $\boldsymbol{Z}$ with correlated components defined by the covariance matrix $\boldsymbol{U_Z}$. Cholesky's distribution is to represent the covariance matrix $\boldsymbol{U_Z}$ as the product of two triangular matrices: $\boldsymbol{H}$ and the upper one as its transposition $\boldsymbol{H}^{\boldsymbol{T}}$. Then $\boldsymbol{U_Z} = \boldsymbol{HH^T}$. In this way, random multidimensional vectors are obtained $\boldsymbol{Z}$ with components correlated by matrix $\boldsymbol{U_Z}$. This is obtained by generating vectors $\boldsymbol{R}$ as uncorrelated samples from the normal distribution $N(0,1)$, and then multiplying the matrix $\boldsymbol{H^T}$ by these vectors, it is $\boldsymbol{Z} = \boldsymbol{H^T R} + \boldsymbol{E}$. Vector $\boldsymbol{E} = [\boldsymbol{\xi}, \boldsymbol{\Psi}]$ is the multivariate expected value for the original distribution. The generated correlated random vectors are used to determine from the Eq. (5b) directional coefficient distributions θ_1 and intercept θ_0, to find the extension interval U_ψ about probability 0.95 for any selected value of the variable ξ. The values of parameters of the line are determined numerically and marked with indices '*min*'. They are: θ_{1min}, θ_{0min}, β_{min}, $u(\theta_{1min})$, $u(\theta_{0min})$ and coefficient of correlation $\rho_{\theta_0\theta_1}$. An inverse function is assumed to exist ψ^{-1}. The fitted nonlinear function can then be written as:

$$y = \psi^{-1}(\theta_{1\text{min}}\xi(x, \beta_{min}) + \theta_{0\text{min}}) \tag{6}$$

The limits of the uncertainty band of the extended output quantity are described by:

$$U_{y+/-} = \left|\psi^{-1}(\theta_{1min}\xi(x, \beta_{min}) + \theta_{0min} \pm U) - \psi^{-1}(\theta_{1min}\xi(x, \beta_{min}) + \theta_{0min})\right|, \tag{7}$$

or approximate

$$U_y \approx U/\psi\prime \tag{8}$$

where: $U = t_{1-\frac{\alpha}{2},n-2}\sqrt{u^2(\theta_{1min})\xi^2 + 2\rho_{\theta_0\theta_1}\xi u(\theta_{1min})u(\theta_{0min}) + u^2(\theta_{0min})}$

- is the extended uncertainty of the argument of the inverse function for correlated distributions θ_{1min} and θ_{0min}, and the coefficient $t_{1-\frac{\alpha}{2},n-2}$ corresponds to the inverted distribution of the *t*-Student's distribution with *n* - 2 degrees of freedom and coverage with probability of $1 - \alpha = 0.95$ (where : $\alpha = 0.05$).

Then, the width of the uncertainty bands of the nonlinear function $y = f(x)$ results from the uncertainties of the linear function (1a).

2 Part 2. Example of Application: Matching Characteristics of the Thermistor and Its Circuit to the Measurement Data

The characteristics of many types of measuring instruments, sensors, and transducers, as well as the properties of the tested materials, are determined from the measurements of the coordinates of the appropriate number of test points. When they are described by nonlinear functions, the parameters of the equation of this function can be determined using the linear regression method. When calibrating instruments, the results are also compared with the rating data and their accuracy is estimated. As the example will be considering the fitting of a characteristic of thermistor alone and with its resistor circuits.

Measurements were made of the temperature and resistance of a thermistor, i.e. a resistor made of semiconductor material, the resistance of which depends much more on temperature than in classic metal conductors. The form of the function describing the dependence of the thermistor resistance on the absolute temperature expressed in Kelvin is known and described by an exponential curve depending on three parameters r∞, *B* and *C* related by the formula:

$$r(T) = r_{\infty} exp\left(\frac{B}{T}\right) + C \tag{2.1}$$

The parameter *C* is in practice negligible as it is close to zero ($C \approx 0$).

Measurements of the exponential characteristics of the thermistor were made with a meter with a standard uncertainty of $\delta = 2\ \%$. Table 1 shows the measured values of the coordinates of the tested points defined as the absolute temperature *T* [K] and the thermistor resistance $r(T)$ [kΩ].

Table 1. Coordinates of measured points obtained and their estimated deviations.

No	Measured coordinates			Deviations from ideal characteristics (2.1) /Errors Δr [kΩ]/
	t [°C]	T_i[K]	r_i[kΩ]	
1	1	274	844.38	8.44
2	10	283	542.26	−5.43
3	15	288	430.40	−2.15
4	25	298	275.60	4.14
5	35	308	183.30	3.10
6	45	318	124.11	2.10
7	60	333	73.28	−1.23
8	90	363	26.45	−0.25
9	100	373	19.36	−0.38
10	120	393	14.06	0.16

The nominal transmitter characteristic has the following parameter values:

- $r_\infty = 0.8\Omega$ – thermistor resistance at $T \rightarrow \infty$,
- material constant of the thermistor $B = 3800$ K, $C = 0$.

The relative uncertainties of the thermistor resistance measurements were assumed to be 2%, and the absolute uncertainties of the temperature measurements were 0.1°C. Fitting the parameters of Eq. (2.1) was performed using the weighted least squares (WTLS) method. This was done in two ways: using a prepared EXCEL workbook and the R and MATLAB environments.

2.1 Method I of Fitting the Exponential Function of Thermistor

In method I is using a new variable described as $\xi(T) = \exp(\beta/T)$ and coefficient $\beta = B$. The exponential function in Cartesian coordinates y and ξ has the equation:

$$y = \theta_1 \xi + \theta_0, \tag{2.2}$$

because $\psi(y,\beta) = y$ and the substitutions should be applied: $\theta_1 = r_\infty$, $\theta_0 = C$. The laws of error propagation and uncertainty propagation result in:

$$\Delta\xi \approx \xi'(T)\Delta T = -\frac{\beta}{T^2} exp\left(\frac{\beta}{T}\right)\Delta T,$$

$$u(\xi) = |\xi'(x)|u(x) = \frac{\beta}{T^2} exp\left(\frac{\beta}{T}\right) u(T). \tag{2.2a,b}$$

In this example, it was assumed that the uncertainties of the parameters θ_1 and θ_o, i.e., $u(\theta_1) \neq 0$ and $u(\theta_0) \neq 0$ result from the relative uncertainty of measurement points ($\delta = 2\%$). It is also assumed that the parameter β is already fitted, constant and has negligible errors and standard uncertainty $u(\beta)\approx 0$. It is also assumed that there is no correlation of measured data, i.e., then their covariance matrices $\boldsymbol{U}_\Xi$, $\boldsymbol{U}_\Psi$ are diagonal and $\boldsymbol{U}_{\Xi\Psi} = 0$.

Figure 2 shows the measurement points and the exponential curve fitted to them, together with its uncertainty band. The nominal characteristic given in Fig. 2 is within the uncertainty band of the non-linear curve fitted to the measurement observations.

For comparison, a parabola was also determined adapted to the same test points with coordinates as given in Table 1, but with correlated measurement data.

Equal autocorrelation coefficients of +0.2 were assumed for the elements of the $\boldsymbol{U}_X$, $\boldsymbol{U}_Y$ matrix and negative cross-correlation coefficients of –0.2 for the $\boldsymbol{U}_{XY}$ matrix.

Comparison of the uncertainty band for correlated and uncorrelated measurement data – Fig. 3.

2.2 Fitting the Thermistor Exponential Characteristic by Method II

Calculations were also made using method II, which uses Eq. (2.1) after transferring the constant C and taking logarithms on both sides, i.e.:

$$\ln(r(T) - C) = \frac{B}{T} + \ln r_\infty \tag{2.3}$$

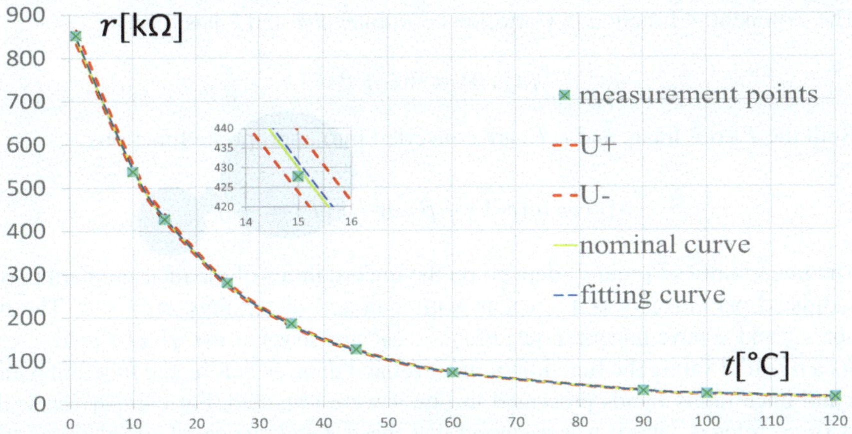

Fig. 2. Measuring points (turquoise squares), fitted exponential characteristic (blue dashed line) for the following parameters: fitting parameters $r_\infty = 0.75928\ \Omega$, $B \approx 3816$ K, $C \approx 148.89\ \Omega$, $\phi_{\psi\xi\min} \approx 4.146$ along with the uncertainty band (red dashed lines) – uncorrelated measurements, solid line in yellow - nominal characteristic $r(T) = 0.8 \cdot \exp(3800/T)$.

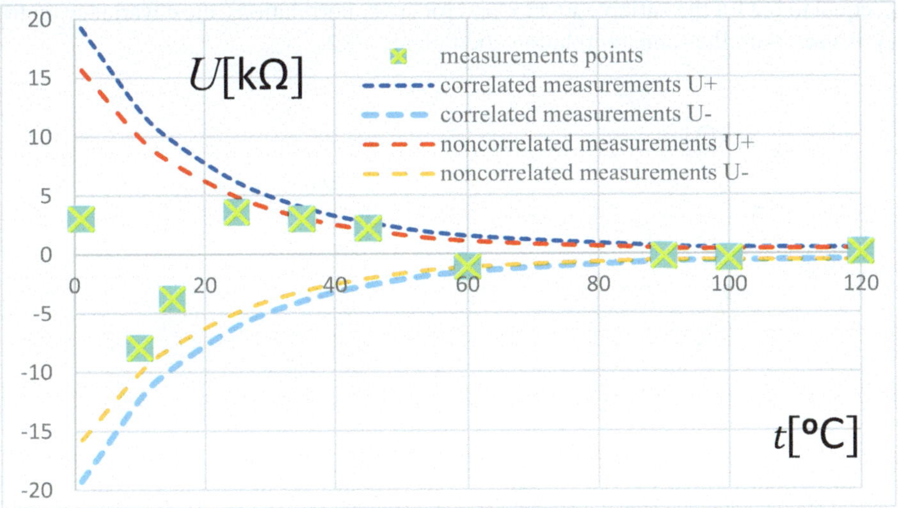

Fig. 3. Comparison of estimated expanded uncertainty bands for correlated measurements (coefficients: +0.2 for the $\boldsymbol{U_X}$, $\boldsymbol{U_Y}$ matrix elements and negative – 0.2 cross-correlations for $\boldsymbol{U_{XY}}$ matrix) and for uncorrelated measurements.

Equation (2.3) becomes linear after applying the following substitutions:

$$\psi = \ln(r(T) - \beta),\ \xi(T) = \frac{1}{T},\ \theta_1 = B,\ \theta_0 = \ln r_\infty,\ \beta = C \tag{2.4}$$

The exponential function in Cartesian coordinates ψ and ξ has form:

$$\psi(y, \beta) = \theta_1 \xi + \theta_0.$$

Both the T coordinate and $r(T)$ are converted into appropriate functions

$$\psi(r) = \ln(r(T) - \beta) \text{ and } \xi(T) = \frac{1}{T}.$$

The uncertainties θ_1 and θ_o depend on the uncertainties of the measurement points. The adjusted parameter β is a constant with standard uncertainty $u(\beta) \approx 0$. The new variables ξ and ψ have uncertainties: $u(\xi) = \frac{u(T)}{T^2}$ and $u(\psi) = u(r)/|r(T) - \beta|$.

As a result of fitting the function of uncorrelated data, as before, the function parameters and uncertainty bands presented in Fig. 4 were obtained. For comparison with a series of correlated data, it was assumed that these data have equal correlation coefficients for the variable x and y in the covariance matrix $\boldsymbol{U_X}$ and $\boldsymbol{U_Y}$ equal to 0.2. The cross-correlation coefficients in the covariance matrix $\boldsymbol{U_{XY}}$ are also the same and amount to –0.2.

For comparison with a series of correlated data, it was assumed that these data are characterized by the same correlation coefficients for the covariance matrices $\boldsymbol{U_X}$ and $\boldsymbol{U_Y}$, equal to 0.2 for the variable x and y, and for cross-correlations, the covariance matrix $\boldsymbol{U_{XY}}$ is used with the same correlation coefficients –0.2.

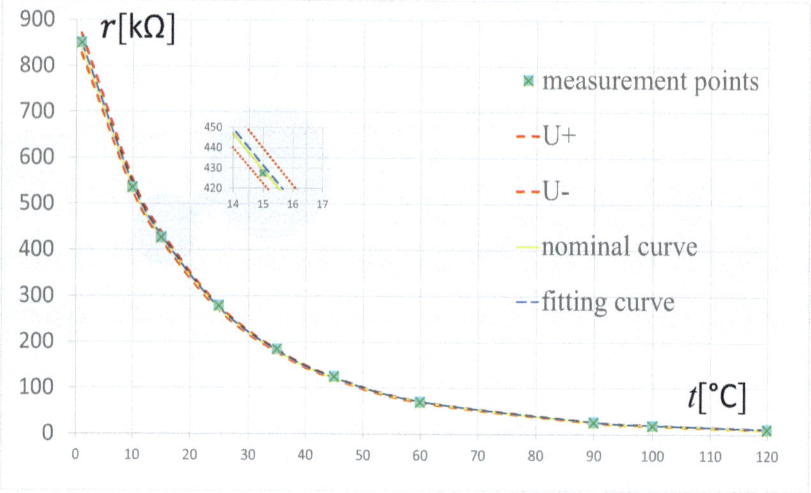

Fig. 4. Measuring points (turquoise squares), fitted exponential characteristic (blue dashed line) for the following parameters: fitting parameters $r_\infty = 0.74642\ \Omega$, $\phi_{\psi\xi\min} \approx 4.113$ with uncertainty band (red dashed lines) - uncorrelated measurements, solid yellow line - nominal function $r(T) = 0.8 \cdot \exp(3800/T)$, $B \approx 3820.8$ K, $C \approx 207\ \Omega$.

2.3 Fitting the Thermistor Exponential Curve by Method II for Parameter $C = 0$

Calculations were also made using method II, which uses Eq. (1) for $C = 0$ and two-sided logarithm, i.e.:

$$\ln(r(T)) = \frac{B}{T} + \ln r_\infty \tag{2.5}$$

Equation (3) becomes linear after applying the following substitutions:

$$\psi = \ln(r(T)),\ \xi(T) = \frac{1}{T},\ \theta_1 = B,\ \theta_0 = \ln r_\infty, \tag{2.6}$$

The exponential function in Cartesian coordinates ψ and ξ has the Eq. (1), i.e.:

$$\psi(y, \beta) = \theta_1\xi + \theta_0.$$

The coordinates T and $r(T)$ turn into functions $\psi(r) = \ln(r(T))$ and $\xi(T) = \frac{1}{T}$, respectively. The uncertainties θ_1 and θ_o depend on the uncertainty of the measurement points. The new variable ξ has uncertainty $u(\xi) = \frac{u(T)}{T^2}$ and the new variable ψ has uncertainty $u(\psi) = u(r)/|r(T)|$. As a result of fitting the function for uncorrelated data, as before, the function parameters and uncertainty bands were obtained, shown below in Fig. 5. For the comparison with a series of correlated data, it was assumed that these data have the same correlation coefficients for the covariance matrices $\boldsymbol{U_X}$ and $\boldsymbol{U_Y}$ equal to +0.2 for the variable x and y, and for cross-correlations the covariance matrix $\boldsymbol{U_{XY}}$ with the same correlation coefficients - 0.2. In this case the following parameters are obtained: $r_\infty = 0.78\Omega$, $B \approx 3808.44°\text{K}$, $C = 0\Omega$, $\phi_{\psi\xi min} \approx 5.38$ along with the uncertainty band (red lines dashed) – correlated measurements, solid yellow line - nominal characteristic $r(T) = 0.8 \cdot \exp(3800/T)$.

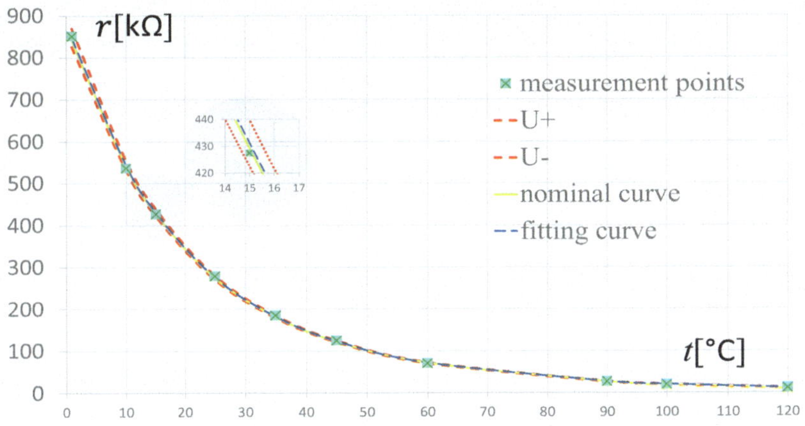

Fig. 5. Measurement points (turquoise squares) fitted exponential characteristic (blue dashed line) for the following parameters: fitting parameters $r_\infty = 0.7815\ \Omega$, $B \approx 3808.39$ K, $C = 0\ \Omega$, $\phi_{\psi\xi\text{min}} \approx 4.308$ along with the uncertainty band (red dashed lines) - uncorrelated measurements, solid yellow line – nominal characteristic $r(T) = 0.8 \cdot \exp(3800/\text{T})$.

The criterion function $\phi_{\psi\xi}$ for correlated measurement data depending on the parameters θ_1 and θ_0 is illustrated in: Figs. 6 and 7.

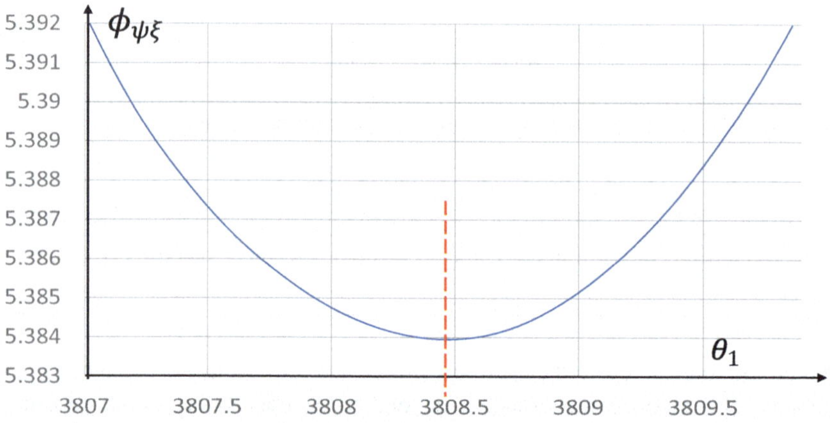

Fig. 6. Numerical dependencies of the criterion function $\phi_{\psi\xi}$ on a) parameter θ_1 obtained from the EXCEL sheet and the R environment, respectively.

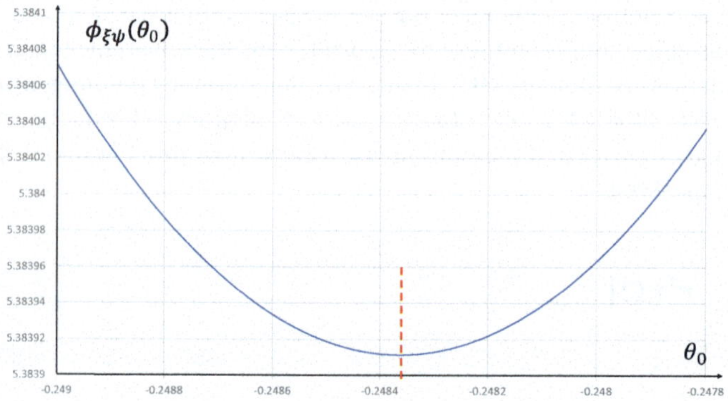

Fig. 7. Numerical dependencies of the criterion function $\phi_{\psi\xi}$ on parameter θ_0 – obtained from the EXCEL sheet and the R environment, respectively.

Figure 8 gives the absolute uncertainty bands for both methods I and II.

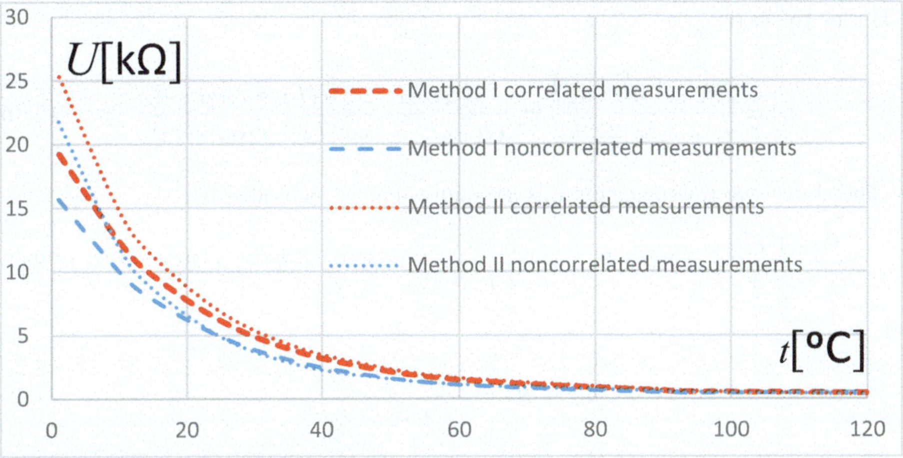

Fig. 8. Comparison of the expanded absolute uncertainty bands for both exponential curve fitting methods for correlated and uncorrelated measurements.

2.4 Matching Parameters of Thermistor with T- or Π-type Twoports

The thermistor's operation was adjusted for temperatures around 25ºC by connecting it to the terminals of a passive twoport consisting of three resistors R_1, R_2 and R_3 in a T-type arrangement or three resistors r_1, r_2 and r_3 in a π-type arrangement. Below, on Fig. 9, are given diagrams of the thermistor connection with passive quadrants and equivalent resistances, seen from the twoport terminals.

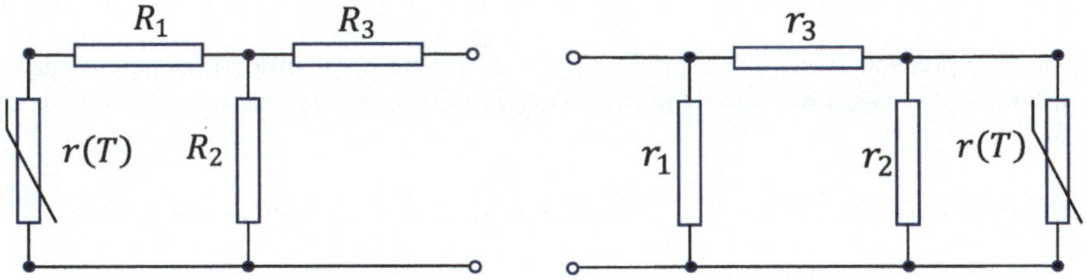

Fig. 9. Diagrams of the thermistor connection with twoport circuits type T and Π.

Their equivalent resistances seen from the input terminals are:

$$R_Z = ((r(T) + R_1)\|R_2) + R_3, \qquad r_Z = ((r(T)\|r_2) + r_3)\|r \tag{2.7}$$

The equivalent resistances of the thermistor-passive circuits are as follows:

- T-type twoport

$$R_Z = R_3 + \frac{R_2 r(T) + R_1 R_2}{r(T) + R_1 + R_2} = \frac{(R_2{+}R_3) r(T) + R_1 R_2 + R_1 R_3 + R_2 R_3}{r(T) + R_1 + R_2} \tag{2.7a}$$

- Π type twoport

$$r_Z = \frac{r_1\left(r_3 + \frac{r_2 r(T)}{r(T)+r_2}\right)}{r_1 + r_3 + \frac{r_2 r(T)}{r(T)+r_2}} = \frac{(r_1 r_3 + r_1 r_2) r(T) + r_1 r_2 r_3}{(r_1 + r_2 + r_3) r(T) + r_2(r_1 + r_3)} \tag{2.7b}$$

The above two resistance circuits are equivalent to each other if:

$$\frac{r_1 r_3 + r_1 r_2}{r_1 + r_2 + r_3} = R_2 + R_3, \quad \frac{r_1 r_2 r_3}{r_1 + r_2 + r_3} = R_1 R_2 + R_1 R_3 + R_2 R_3 \tag{2.8a,b}$$

and

$$R_1 + R_2 = \frac{r_2(r_1 + r_3)}{r_1 + r_2 + r_3}, \tag{2.9}$$

The equivalent resistance Z for parameters a, b, c has the form:

$$Z = \frac{ar(T) + b}{r(T) + c} \tag{2.10}$$

For expression $r\ (T) = r_\infty \exp \frac{B}{T}$, the first derivative respect to temperature T is:

$$r'(T) = -\frac{B}{T^2} r_\infty exp\left(\frac{B}{T}\right), \tag{2.10a}$$

and the second derivative

$$r''(T) = \frac{B}{T^4}(B2T) r_\infty exp\left(\frac{B}{T}\right) \tag{2.10b}$$

For Z expressed from the formula (2.10) at the point of inflection, i.e., when the second derivative becomes zero, the following condition is obtained:

$$r''(T) - \frac{2(r'(T))^2}{r(T) + c} = 0 \tag{2.11}$$

The parameter c is determined from:

$$c = \frac{2(r'(T))^2}{r''(T)} - r(T) = 2Br_\infty \frac{\exp\left(\frac{B}{T}\right)}{B + 2T} - r_\infty \exp\left(\frac{B}{T}\right) = \frac{B - 2T}{B + 2T} r_\infty \exp\left(\frac{B}{T}\right) \tag{2.12}$$

The inflection point, e.g. $t_0 = 25$°C ($T_0 = 298$ K), is characterized by a second derivative equal to zero. The expansion of the relationship $\boldsymbol{Z}(T)$ around this point does not have a quadratic term and the next one is a cubic term. The value of parameter c at the inflection temperature $t_0 = 25$°C) for the nominal thermistor curve with parameters $r_\infty = 0.8\ \Omega$ and $B = 3800$ K is:

$$c = \frac{3800 - 2 \cdot 298}{3800 + 2 \cdot 298} 0.8 \exp\left(\frac{3800}{298}\right) = 201237.14\ \Omega = 201.24\,\mathrm{k}\,\Omega \tag{2.13}$$

For the T-type twoport, parameter $c = R_1 + R_2$. Assuming $R_1 = 100\ \text{k}\,\Omega$, then $R_2 = c - R_1 = 101.24\ \text{k}\,\Omega$. R_3 is usually selected on level $R_3 = 100\ \text{k}\,\Omega$.

Resistance values can also be expressed in relative values related to the resistance R_1 or R_2. For R_1, the twoport matching can be given in the form of two parameters $\alpha = \frac{c}{R_1} - 1$, $\beta = \frac{R_3}{R_1}$. Hence $R_1 = \frac{c}{\alpha+1}$, $R_2 = \alpha R_1$, $R_3 = \beta R_1$. Resistance $R_1 = 100\Omega$ and dimensionless parameters $\alpha = 1.0124$, $\beta = 1$.

Similarly, for the π-type twoport parameter $c = \frac{r_2(r_1+r_3)}{r_1+r_2+r_3}$. Assuming $r_1 = 100\ \text{k}\,\Omega$ and $r_3 = 200\ \text{k}\,\Omega$, obtained is $r_2 = \frac{c(r_1+r_3)}{r_1+r_3-c} = \frac{201.24(200+200)}{200+200-201.24} = 404.98\ \text{k}\,\Omega = 404.98\ \text{k}\,\Omega$.

For dimensionless parameters: $\gamma = r_2/c$ and $\varepsilon = (r_1 + r_3)/r_2$ these values can be expressed as a function of the resistance r_2. Therefore, the resistances of the twoport are determined by the parameters γ and ε, i.e. $r_2 = c\gamma$ and $r_1 + r_3 = \varepsilon\, r_2$. The resistance values r_2 and dimensionless parameters in considered example are: $r_2 = 404.98\ \text{k}\Omega$, $\gamma = 2.012423$ and $\varepsilon = 0.9877$.

For the range 0 to 120°C, the nominal characteristic of the resultant resistance of T-type and Π-type twoports with thermistor as a function of temperature are given in Fig. 10.

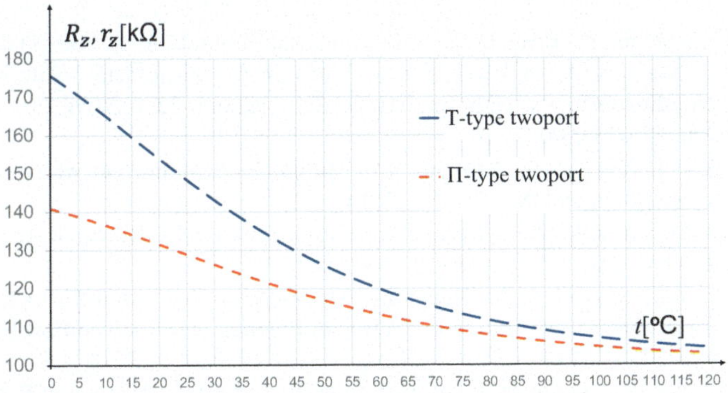

Fig. 10. Resistance of the thermistor with a T-type and Π-type two-port as a function of temperature

2.5 Summary

The exponential curve of the calibrated thermistor with nominal characteristic $r(T) = 0.8 \cdot \exp(3800/T)$ was fitted to the measurement points included in Table 1. The relative uncertainties of the resistance measurement were estimated at 2%, and absolute uncertainties of the temperature measurement were estimated at 0.1 °C.

For uncorrelated data: $r(T) = 0.7815 \cdot \exp(3808.39/T)$, minimizing $\phi_{\psi\xi min} \approx 4.308$. For correlated data – correlation coefficients for the $\boldsymbol{U}_X$ and $\boldsymbol{U}_Y$ matrices equal to 0.2 and for $\boldsymbol{U}_{XY}$ matrix – 0.2. $r(T) = 0.7812 \cdot \exp(3808.4/T)$ when minimizing $\phi_{\psi\xi min} \approx 5.4$.

The uncertainty band of determined characteristics was also compared.

The uncertainties resulting from the coverage band are a maximum of 2.8 kΩ higher for correlated data than without correlation for the lower range of the thermistor characteristics, and at high temperatures they practically disappear.

The uncertainty corridor does not exceed 3% of the thermistor resistance value matched to the measurement points.

References

1. Joint Committee for Guides in Metrology (BIPM, IEC, IFCC, ILAC, ISO, IUPAC, IUPAP and OIML) JCGM 100:2008 Evaluation of Measurement Data—Guide to the Expression of Uncertainty in Measurement. Sevres, France: International Bureau of Weights and Measures BIPM
2. Reed, B.C.: Linear least-squares fits with errors in both coordinates. II: comments on parameter variances. Am. J. Phys. **60**, 59–62 (1992)
3. Nielssen, L.: Evaluation of measurements by the method of least squares. In: Levesley, J., Anderson, I.J., Mason, J.C. (eds.) Reprint from Algorithms for Approximation IV, University of Huddersfield, pp. 170–186 (2002)
4. York, D., Evensen, M.N., Lopez Martines, M., DE Basabe Delgato, J.: Unified equations for the slope, intercept, and standard errors for the best straight line. AM. J. Phys. **72**(3), 367–375 (2004)
5. Lira, I., Grientschnig, D.: Error-in-variables models in calibration. Metrologia **54**, 133 (2017)
6. Krystek, M., Mathias, A.: A least-squares algorithm for fitting data points with mutually correlated coordinates to a straight line. Meas. Sci. Technol. **22**, 035101 (2011). https://doi.org/10.1088/0957-0233/3/035101
7. Amiri-Simkooei, A.R., et al.: Estimation of straight-line parameters with fully correlated coordinates. Measurement **48**, 378–386 (2014)
8. Davidson, W.C.: Variable Metric Method for Minimization. AEC Research and Development Report ANL 5990. https://doi.org/10.2172/4252678
9. Dennis, J.E. Jr., Schnabel, R.B.: Numerical Methods for Unconstrained Optimization and Nonlinear Equations. SIAM 1996 reproduction of Prentice-Hall 1983 edition. https://doi.org/10.1137/1.9781611971200
10. Fletcher, R.: Practical Methods of Optimization (2nd ed.). John Wiley and Sons, New York (1987). ISBN 978–0–471–91547–8
11. Mascarenhas, W.F.: The divergence of the BFGS and Gauss Newton Methods, Mathematical Programming, 2013, arXiv:1309.7922, https://doi.org/10.48550/arXiv.1309.7922
12. Levenberg, K.: A method for the solution of certain non-linear problems in least squares. Q. Appl. Math. **2**(2), 164–168 (1944). https://doi.org/10.1090/qum/10666
13. Marquardt, D.: An algorithm for least-squares estimation of nonlinear parameters. SIAM J. Appl. Math. **11**(2), 431–441 (1963). https://doi.org/10.1137/0111030
14. Kanzow, C.H., Yamashita, N., Fukushima, M.: Levenberg–Marquardt methods with strong local convergence properties for solving nonlinear equations with convex constraints. J. Comput. Appl. Math. **172**(2), 375–397 (2004). https://doi.org/10.1016/j.cam.2004.02.013
15. Warsza, Z.L., Puchalski, J.: Method of estimation uncertainties of indirect multivariable measurement including accuracy of processing function as extension of GUM S2. In: Advanced Mathematical and Computational Tools in Metrology and Testing AMCTM XII. Pavese F., Forbes A.B., Zhang N.F., Chunovkina A.G. (eds.): Series on Advances in Mathematics for Applied Sciences, vol. 90, pp. 451–63, World Scientific, New Jersey, London, Singapore (2021). ISBN 978–981–124–37–3 0I, https://doi.org/10.1142/9789811242380_0029

16. Warsza, Z.L., Puchalski, J.: Uncertainty bands of the regression line for autocorrelated data of dependent variable Y. In: Proceedings of 22nd Conference on Automation 2021, vol. 1390, 364–386 (2021). https://doi.org/10.1007/978-3-03074893-7_33
17. Puchalski, J.: A new algorithm for generalization of least square method for straight line regression in cartesian system for fully correlated both coordinates. Int. J. Autom. AI Mach. Learn. **2**(2), 20–54 (2021)
18. Puchalski, J., Warsza, Z.L.: Regression and uncertainty of a straight line for measurements of x and y variables with all correlations. (in Polish) PAR (Pomiary Automatyka Robotyka) R.26. Nr 2/2022 s.47–58 ISSN 427–9126
19. Puchalski, J., Warsza, Z.L.: Determining a straight line by the linear regression method and its uncertainty band with reference to the GUM guide from measurements of both variables x and y with their autocorrelation and cross-correlation. Monograph in Polish: METROLOGIA badania i zastosowania. Monografia M152, TU Politechnika Świętokrzyska Kielces. 235–256
20. Warsza, Z.L., Puchalski, J.: Uncertainty bands of the regression line for data with type A and type B uncertainties of dependent variable Y. In: Proceedings of 22nd Conference on Automation 2021, vol. 1390, pp. 342–363 (2021). https://doi.org/10.1007/978-3-03074893-7_32
21. Puchalski, J., Warsza, Z.L.: Linear regression method of matching the parabolic curve to tested points of both correlated coordinates Abstract 2p. Paris MathMet 2022 EMN Conference
22. Puchalski, J., Warsza, Z.L.: The method of fitting a non-linear function to data of measured points and its uncertainty band. (in Polish) Pomiary Automatyka Robotyka, R.26. Nr 3/2023 s.45–55. https://doi.org/10.14313/PAR_249/45
23. Draper, R.D., Smith, H.: Applied Regression Analysis, 3rd Edn. Willey, New York (Polish translation of ed. 1 from 1966) (1998)
24. Grafarend, E.W., Awange, J.L.: Applications of linear and nonlinear models–fixed effects, random effects, and total least squares. J. Geodetic Sci. **3**(1), 77–78 (2013). https://doi.org/10.2478/jogs-2013-0009

Application of the Polynomial Maximization Method for Estimating Nonlinear Regression Parameters with Non-gaussian Asymmetric Errors

Serhii Zabolotnii[1](✉), Oleksandr Tkachenko[2,3], Waldemar Nowakowski[4], and Zygmunt L. Warsza[5]

[1] Cherkasy State Business College, Cherkasy, Ukraine
zabolotniua@gmail.com
[2] Uzhhorod National University, Uzhhorod, Ukraine
[3] Ivan Franko National University of Lviv, Lviv, Ukraine
[4] Casimir Pulaski Radom University, Radom, Poland
w.nowakowski@uthrad.pl
[5] Polish Metrological Society, Radom, Poland

Abstract. In the article, an alternative approach to estimating parameters in nonlinear regression models under asymmetric error distributions is examined. A novel approach for adaptive estimation is proposed, which is based on the use of second-order polynomial functions. This enables a straightforward implementation to account for deviations from Gaussian idealization in the form of moments up to the fourth order. It is demonstrated that the overall problem can algorithmically be reduced to the numerical solution of a system of nonlinear stochastic equations. Analytical expressions are obtained, which facilitate the estimation of parameters and the analysis of their asymptotic variance. Statistical modeling using the Monte Carlo method was conducted, and the results indicate that the accuracy of PMM2 estimates is comparable to SLS estimates and significantly so exceeds the accuracy of OLS estimates.

Keywords: Polynomial Maximization Method · non-linear regression models · asymmetric error distributions · Monte Carlo method · higher order statistics

1 Introduction

In the contemporary realm of statistical analysis, significant attention is devoted to the study of complex interrelationships among various variables. Consequently, the development of reliable methods for parameter estimation in nonlinear regression models, particularly in the context of complex and realistic data, is of great importance. The predictability and accuracy of such models are critically vital in many scientific and applied fields, including engineering, ecology, biomedicine, economics, and finance. In these and other applied disciplines, regression analysis has made a substantial contribution to understanding and addressing numerous real-world problems [1–5].

R. Szewczyk et al. (Eds.): AUTOMATION 2024, LNNS 1219, pp. 342–356, 2024.
https://doi.org/10.1007/978-3-031-78266-4_30

However, traditional estimation methods, especially the widely used ordinary least squares (OLS) approach, which are based on the assumption of Gaussian statistical data, can be inefficient or inadequately precise in conditions where the error distribution is asymmetric or has heavy tails. This presents a challenge for researchers to develop new approaches that better accommodate these conditions.

In response to these challenges, contemporary researchers are developing and implementing novel strategies. One of the most successful is based on a modification of the least squares method, first utilized by Wang [6] to address the problem of measurement error in nonlinear regression models. This modification, known as the "Second order" Least Squares (SLS), has been successfully applied for estimating parameters in nonlinear regression with asymmetric errors [7]. Subsequently, several authors have explored the asymptotic efficiency of this approach [8], developed a series of new optimization criteria [9, 10], and applied it to solve practical problems [11–13] and is possible to use also in the automatic diagnostics of the railway tracks surface [27].

2 Purpose of the Study

In this article, we explore another approach to solving the problem of parameter estimation in nonlinear regression, based on the use of the Polynomial Maximization Method (PMM). This novel statistical estimation method is founded on the framework of Kunchenko stochastic polynomials [14]. Its application opens new possibilities for constructing sub-optimal methods for processing non-Gaussian data. Beyond parameter estimation, this can include tasks such as testing statistical hypotheses [15], pattern recognition ("matching template") [16], posterior [17] and online [18] detection of change points (problem of change point).

This study continues a series of works examining the application of the quadratic variant (at degree S = 2) of the polynomial maximization method for estimating various linear-parameter models (linear regression [19], polynomial regression [20], autoregressive [21], and moving average [22]) under the condition of asymmetric distribution of their random components and using moment-cumulant description. One of the main results of these works is the fact that in finite samples, PMM2 significantly outperforms the accuracy of OLS. Only when the non-Gaussian regression errors acquire a property of distribution symmetry (or at least with a zero of 3rd moment) do the PMM2 estimates degenerate into OLS estimates. Moreover, as the results of simulation modeling in work [7] demonstrate, a similar effect is observed for SLS estimates.

Thus, the aim of this work is to synthesize an algorithm for estimating the parameters of nonlinear regression based on the use of the polynomial maximization method at degree S = 2; to analyze the properties of the resulting PMM2 estimates and to conduct a comparative analysis of their efficiency against OLS and SLS estimates through Monte Carlo statistical modeling.

3 General Statement of the Problem

Suppose there is a regression model of observations in the form of the sum of deterministic and random components:

$$y_v = R_v(\boldsymbol{\theta}, \boldsymbol{X}) + \xi_v,\ v = \overrightarrow{1, N}, \tag{1}$$

describing the dependence of the values of the target variable $\boldsymbol{Y} = \{y_1, y_2, ..., y_N\}$ on a set of independent variables P $\boldsymbol{X}$ of independent variables (P-factor regressor).

The deterministic component $R_v(\boldsymbol{\theta}, \boldsymbol{X})$ is a nonlinear function that depends on the Q-dimensional vector of informative parameters $\boldsymbol{\theta} = \{a_0, a_1, ...a_{Q-1}\}$.

The samples of the random component (error) of the regression model ξ_v are a sequence of independent and equally distributed random variables that have the property of centeredness ($E\{\xi\} = 0$) and homoscedasticity $D\{\xi\} = const$. The probabilistic properties of the regression model error differ significantly from the Gaussian (normal) law. The general task is to find estimates of the components of the informative parameter vector $\boldsymbol{\theta}$ on the basis of statistical analysis of statistical data $\boldsymbol{Y}$ and $\boldsymbol{X}$.

4 Application of the PMM in Finding Estimates of a Vector Parameter with Unequally Distributed Data

To solve the problem of regression analysis, one can use a modification of the PMM for statistical estimation of a vector parameter with unequally distributed data [14]. It is based on the property of maximizing a functional in the form of a stochastic polynomial of the general form:

$$L_{SN} = \sum\nolimits_{v=1}^{N} \sum\nolimits_{i=1}^{S} \phi_i(y_v) \int^{a} k_{iv}(a)dz - \sum\nolimits_{i=1}^{S} \sum\nolimits_{v=1}^{N} \int^{a} \Psi_{iv} k_{iv}(a)dz, \quad (2)$$

It is assumed that random variables y_v are described by a sequence of mathematical expectations:

$$\Psi_{iv} = E\{\phi_i(y_v)\}, i = \overrightarrow{1, S}, v = \overrightarrow{1, N}, \quad (3)$$

and are twice differentiated by the parameter a that is being evaluated.

A stochastic polynomial L_{SN} of the form (2) has two main properties:

1) for any order S with asymptotic growth $v \to \infty$ of the sample size $\boldsymbol{Y}$ polynomial L_{SN} as a function of the parameter a takes a maximum in the neighborhood of the true value of this parameter,
2) with different samples $\boldsymbol{Y}$ the deviation of the maximum of the polynomial L_{SN} from the true value of the parameter a will have a minimum variance (for the corresponding order of the polynomial S).

By analogy with the maximum likelihood method, the parameter a can be found from solving the equation of the form:

$$\frac{d}{da} L_{SN}\bigg|_{a=\hat{a}} = \sum\nolimits_{i=1}^{S} \sum\nolimits_{v=0}^{N} k_{iv}\left[\phi_i(y_v) - \Psi_{iv}\right]\bigg|_{a=\hat{a}} = 0, \quad (4)$$

The optimal coefficients k_{iv} maximizing the functional (2) are found from solving the system of linear algebraic equations:

$$\sum\nolimits_{j=1}^{S} k_{jv} F_{(i,j)v} = \frac{d}{da} \Psi_{iv}, i = \overrightarrow{1, S}, v = \overrightarrow{1, N}, \quad (5)$$

where: $F_{(i,j)v} = \Psi_{(i,j)v} - \Psi_{iv}\Psi_{jv}$, $\Psi_{(i,j)v} = E\{\phi_i(y_v)\phi_j(y_v)\}$, $i, j = \overrightarrow{1, S}$.

This approach to estimation can be extended to the case of finding estimates of a vector parameter $\boldsymbol{\theta} = \{a_0, a_1, ...a_{Q-1}\}$. In this, it is necessary to use Q polynomials $L_{SN}^{(p)}, p = \overrightarrow{0, Q-1}$ of the general form (2) for each component a_p of vector parameter.

Each p -stochastic polynomial $L_{SN}^{(p)}$ as a function of the parameter a_p at known values of other components of the vector $\boldsymbol{\theta}$ at $n \to \infty$ also has a maximum in the neighborhood of the true value of the parameter a_p. Thus, the desired parameter estimates can be found as a solution to the system of equations of the form:

$$f_{SN}^{(p)}(y_v, x_v) = \sum\nolimits_{i=1}^{S} \sum\nolimits_{v=1}^{N} k_{iv}^{(p)} \left[\phi_i(y_v) - \Psi_{iv}\right]\Big|_{a_p = \hat{a}_p} = 0, p = \overrightarrow{0, Q-1}. \quad (6)$$

When using polynomials of degree $S \geq 2$ finding the PMM estimates of a vector parameter in the vast majority of cases (similar to the situation with MLE) requires the use of numerical methods for solving systems of nonlinear equations. In particular, an approach based on the iterative Newton-Raphson numerical procedure is often used. It is based on the principle of linearization by expanding the left-hand side of each nonlinear equation of system (6) into a Taylor series in the neighborhood of the true value of the vector $\boldsymbol{\theta}$. Limiting ourselves to the first two terms of the series, we can write the linear system in matrix form:

$$\hat{\boldsymbol{\theta}}_{(S)}^{(k+1)} = \hat{\boldsymbol{\theta}}_{(S)}^{(k)} - \left[\boldsymbol{Z}_S\left(\boldsymbol{Y}/\hat{\boldsymbol{\theta}}^{(k)}\right)\right]^{-1} \boldsymbol{F}_S\left(\boldsymbol{Y}/\hat{\boldsymbol{\theta}}^{(k)}\right), \quad (7)$$

which can be used to iteratively search for estimated values. To obtain system (8), it is necessary to calculate the column matrix $\boldsymbol{F}_S(\boldsymbol{Y}/\widehat{\boldsymbol{\theta}}^{(k)})$ composed of the elements of the left-hand side of each nonlinear equation of system (6) and a square matrix $\boldsymbol{Z}_S(\boldsymbol{Y}/\widehat{\boldsymbol{\theta}}^{(k)}) = \boldsymbol{H}_S(\boldsymbol{Y}/\widehat{\boldsymbol{\theta}}^{(k)}) + \boldsymbol{J}_S(\widehat{\boldsymbol{\theta}}^{(k)})$ with the components

$$H_{SN}^{(p,q)} = \sum\nolimits_{i=1}^{S} \frac{\partial}{\partial a_q} k_{iv}^{(p)} \left[\sum\nolimits_{v=1}^{N} \phi_i(y_v) - N\Psi_{iv}\right] \quad (8)$$

and elements of the matrix of the amount of information extracted $\boldsymbol{J}_S(\widehat{\boldsymbol{\theta}}^{(k)})$ type:

$$J_{SN}^{(p,q)} = \sum_{v=1}^{N} \sum\nolimits_{i=1}^{S} k_{i,v}^{(p)} \frac{\partial}{\partial a_q} \Psi_{iv}, p, q = \overrightarrow{0, Q-1}. \quad (9)$$

To start the iterative procedure, it is assumed that there is some initial approximation $\boldsymbol{\theta}^{(1)} = \left\{a_0^{(1)}, a_1^{(1)}, ..., a_{Q-1}^{(1)}\right\}^T$ which can be chosen as a "rough" estimate found by a simpler method, such as the OLS [14].

5 Finding PMM Estimates of Nonlinear Regression Parameters

5.1 Application of the PMM at Degree S = 1

To solve the problem using the method of polynomial maximization in the formation of the stochastic polynomial (2), we use the power transformations of the sample values as basic functions $\boldsymbol{Y}$, i.e.

$$\phi_i(y_v) = y_v^i. \tag{10}$$

In this case, the sequence of mathematical expectations (3) is a set of initial moments of the corresponding order:

$$\Psi_{iv} = E\left\{y_v^i\right\} = \alpha_{iv}, i = \overrightarrow{1, S}, v = \overrightarrow{1, N}, \tag{11}$$

and the functions $F_{(i,j)v} = \alpha_{(i+j)v} - \alpha_{iv}\alpha_{jv}$.

Thus, when using a stochastic polynomial of order $S = 1$ PMM, the estimates of the elements of the vector parameter $\boldsymbol{\theta}$ of the deterministic component of the nonlinear regression model (1) can be found from the solution of the system Q of equations:

$$\sum\nolimits_{v=1}^{N}\left\{k_{1,v}^{(p)}\left[y_v - R_v(\boldsymbol{\theta}, \boldsymbol{X})\right]\right\} = 0, p = \overrightarrow{0, Q-1}. \tag{12}$$

where the optimal coefficients $k_{1,v}^{(p)}$ are found as solutions to a system of linear algebraic equations, and which can be represented as

$$k_{1,v}^{(p)} = \frac{1}{\mu_2}\frac{\partial}{\partial a_p}R_v(\boldsymbol{\theta}, \boldsymbol{X}), p = \overrightarrow{0, Q-1} \tag{13}$$

We use an assumption similar to the one used in the conventional OLS for estimating the parameters of a nonlinear regression [10]. It consists in the fact that the deterministic component $R_v(\boldsymbol{\theta}, \boldsymbol{X})$ of the nonlinear model (1) can be approximated by a linear function based on the first-order Taylor series:

$$R_v\left(\boldsymbol{\theta}^{(k)}, \boldsymbol{X}\right) + \sum\nolimits_p \frac{\partial}{\partial a_p}R_v\left(\boldsymbol{\theta}^{(k)}, \boldsymbol{X}\right)\left[\boldsymbol{\theta} - \boldsymbol{\theta}^{(k)}\right] \approx R_v\left(\boldsymbol{\theta}^{(k)}, \boldsymbol{X}\right)\sum\nolimits_p H_{v,p}\Delta\boldsymbol{\theta}, \tag{14}$$

where $H_{v,p}$ – are the elements of the Jacobi matrix $\boldsymbol{H}$ that depend on the components a_p of the vector parameters $\boldsymbol{\theta}$ at fixed values of the regressors $\boldsymbol{X}$.

$$\sum\nolimits_v R_v(\boldsymbol{\theta}, \boldsymbol{X})\frac{\partial}{\partial a_p}R_v(\boldsymbol{\theta}, \boldsymbol{X}) = 0, p = \overrightarrow{1, Q}. \tag{15}$$

Then, considering (13), after certain transformations, we can represent (12) in matrix form as a system of linear equations:

$$\left(\boldsymbol{H}^T\boldsymbol{H}\right)\Delta\boldsymbol{\theta}^T \approx \frac{1}{N}\sum\nolimits_{v=1}^{N}\boldsymbol{H}^T\Delta\boldsymbol{Y}^T, \tag{16}$$

which is equivalent to the corresponding least squares system [23]. Such OLS estimates are obtained by using appropriate numerical iterative procedures (gradient descent, Newton-Hauss method, etc.). They may have a certain bias, but they can be used as a first approximation for more complex estimation methods, in particular, the OLS at higher polynomial degrees.

5.2 Application of the PMM at Degree S = 2

Using the general formula (6), we present a system of equations for finding the PMM estimates $\widehat{\boldsymbol{\theta}}_{(2)}$ of the deterministic component of the nonlinear regression model (1):

$$\sum\nolimits_{v=1}^{N}\left\{k_{1,v}^{(p)}\left[y_v - R_v(\boldsymbol{\theta},\boldsymbol{X})\right] + k_{2,v}^{(p)}\left[(y_v)^2 - \left[(R_v(\boldsymbol{\theta},\boldsymbol{X}))^2 + \mu_2\right]\right]\right\} = 0 \quad (17)$$

where, $p = \overrightarrow{0, Q-1}$, and the optimal coefficients $k_{i,v}^{(p)}$, $i = \overrightarrow{1,2}$ can be written as:

$$k_{1,v}^{(p)} = \frac{\mu_4 - \mu_2^2 + 2\mu_3 R_v(\boldsymbol{\theta},\boldsymbol{X})}{\mu_2\left(\mu_4 - \mu_2^2\right) - \mu_3^2}\frac{\partial}{\partial a_p}R_v(\boldsymbol{\theta},\boldsymbol{X}),\ k_{2,v}^{(p)} = -\frac{\mu_3^2}{\mu_2\left(\mu_4 - \mu_2^2\right) - \mu_3^2}\frac{\partial}{\partial a_p}R_v(\boldsymbol{\theta},\boldsymbol{X}). \quad (18)$$

Substituting the coefficients of (18) into (17), after certain transformations, the system of equations for finding the estimates can be written in the form:

$$\sum\nolimits_{v=1}^{N}\left\{\frac{\partial}{\partial a_p}R_v(\boldsymbol{\theta},\boldsymbol{X})\left[A_2(R_v(\boldsymbol{\theta},\boldsymbol{X}))^2 + B_{2,v}R_v(\boldsymbol{\theta},\boldsymbol{X}) + C_{2,v}\right]\right\} = 0,\ p = \overrightarrow{0, Q-1} \quad (19)$$

where:

$$A_2 = \mu_3,\ B_{2,v} = \mu_4 - \mu_2^2 - y_v\mu_3,\ C_{2,v} = y_v^2\mu_3 - y_v\left(\mu_4 - \mu_2^2\right) - \mu_2\mu_3. \quad (20)$$

The analysis of expressions (20) shows that in the case of symmetry of distribution (or at least identity $\mu_3 = 0$) of the random component of errors of the regression model, the use of the PMM at $S = 2$ is inappropriate, since system (19) degenerates into a system of linear equations of the form (13), equivalent to the system for a conventional OLS. The elements of the matrix $\boldsymbol{Z}_2(\boldsymbol{Y}/\widehat{\boldsymbol{\theta}}^{(k)})$ necessary for the numerical solution of system (19) can be represented as follows:

$$Z_{2N}^{(p,q)} = \sum\nolimits_{v=1}^{N}\sum\nolimits_{v=1}^{N}\left\{\frac{\partial}{\partial a_p}R_v(\boldsymbol{\theta},\boldsymbol{X})\frac{\partial}{\partial a_q}R_v(\boldsymbol{\theta},\boldsymbol{X})\left[2A_2R_v(\boldsymbol{\theta},\boldsymbol{X}) + B_{2,v}\right]\right\} \quad (21)$$

Obviously, when the degree of the polynomial $S = 2$ PMM estimates can only be found numerically, for example, using the Newton-Raphson iterative procedure. As an initial approximation, we can use the ODE estimates.

6 Analyzing the Theoretical Efficiency of the PMM Estimates of Nonlinear Regression Parameters

It is known that one of the main criteria for the effectiveness of estimates of informative parameters is their variances. It has been shown [14] that estimates of vector parameters found using the method of maximizing polynomials generally have the property of correctness and are asymptotically unbiased. To obtain the analytical expressions describing the variances of the PMM estimates, we use the matrix of the amount of information extracted $\boldsymbol{J}_S(\widehat{\boldsymbol{\theta}}^{(k)})$ with elements of the form (9). In the moment description, such a matrix consists of elements:

$$J_{SN}^{(p,q)} = \sum\nolimits_{v=1}^{N}\sum\nolimits_{i=1}^{S}k_{i,v}^{(p)}\frac{\partial}{\partial a_q}\alpha_{iv},\ p,q = \overrightarrow{0, Q-1}. \quad (22)$$

In the statistical sense, the amount of extracted information is conceptually close to the amount of Fisher information. It is shown that the variances $\sigma^2_{(a_p)S}$ of the PMM estimates of the components of the vector parameter $\boldsymbol{\theta}$ in the asymptotic case (at $N \to \infty$) can be obtained as the elements of the main diagonal of the variation matrix, which is the inverse of the matrix composed of the elements:

$$\boldsymbol{V}_{PMMS}(\boldsymbol{\theta}) = \begin{bmatrix} J_{SN}^{(1,q)} & J_{SN}^{(2,q)} & \ldots & J_{SN}^{(1,q)} \\ J_{SN}^{(2,1)} & J_{SN}^{(2,2)} & \ldots & J_{SN}^{(2,q)} \\ \ldots & \ldots & \ldots & \ldots \\ J_{SN}^{(p,1)} & J_{SN}^{(p,2)} & \ldots & J_{SN}^{(p,q)} \end{bmatrix}^{-1} \tag{23}$$

Another important property of the PMM is that when the number of S of the stochastic polynomial, the variance of the estimates decreases, since the amount of extracted information asymptotically (at $S \to \infty$) tends to the Fisher information [14].

Using the obtained relations (13) for the linear variant of the polynomial maximization method (with degree S = 1), we can write the expression for the amount of information extracted in the form:

$$J_{1N}^{(p,q)} = \frac{1}{\mu_2} \sum\nolimits_{v=1}^{N} \left[\frac{\partial}{\partial a_p} R_v(\boldsymbol{\theta}, \boldsymbol{X}) \frac{\partial}{\partial a_q} R_v(\boldsymbol{\theta}, \boldsymbol{X}) \right], p, q = \overrightarrow{0, Q-1}, \tag{24}$$

where μ_2 - is the 2nd order central moment (variance) of random errors ξ_v of the corresponding regression model.

Since the expressions for finding the OLS estimates of the parameters of nonlinear regression models at degree S = 1are equivalent to the OLS estimates, the accuracy of the estimates of both methods is obviously the same. Thus, the expressions (24) defining the accuracy of the OLS estimators at S = 1can be used for a comparative analysis of the efficiency of the OLS estimators obtained using polynomials of higher order. Using expressions (18) for the coefficients $k_{i,v}^{(p)}$, $i = \overrightarrow{1,2}$ which minimize the variance of the estimates of the components of the desired parameter at degree S = 2, we obtain the corresponding expressions for the amount of information extracted:

$$J_{2N}^{(p,q)} = \frac{\mu_2^3 - \mu_2\mu_4}{\mu_2\left(\mu_2^3 - \mu_2\mu_4 + \mu_3^2\right)} \sum\nolimits_{v=1}^{N} \left[\frac{\partial}{\partial a_p} R_v(\boldsymbol{\theta}, \boldsymbol{X}) \frac{\partial}{\partial a_q} R_v(\boldsymbol{\theta}, \boldsymbol{X}) \right], p, q = \overrightarrow{0, Q-1}, \tag{25}$$

where:μ_2, μ_3, μ_4 - are the central moments of random errors ξ_v of the regression model.

A comparative analysis of expressions (24) and (25) allows us to write down the relations for the corresponding variation matrices containing the asymptotic variances of the parameter estimates in the form:

$$\boldsymbol{V}_{(PMM\,2)} = g_{\frac{PMM2}{PMM1}} \boldsymbol{V}_{(PMM\,1)} = g_{\frac{PMM2}{OLS}} \boldsymbol{V}_{(PMM\,1)}. \tag{26}$$

where the coefficient of variance reduction of estimates, which is the same for all components of the vector parameter $\boldsymbol{\theta} = \{a_0, a_1, \ldots a_{Q-1}\}$ can be represented as:

$$g_{\frac{PMM2}{OLS}} = \frac{\mu_2^3 - \mu_2\mu_4}{\mu_2\left(\mu_2^3 - \mu_2\mu_4 + \mu_3^2\right)} = 1 - \frac{\gamma_3^2}{2 + \gamma_4}. \tag{27}$$

The transition in expression (27) from a momentary to a cumulative description is that the deviation of the values of the cumulative coefficients of higher orders γ_r due to the fact from zero shows the degree of difference from the Gaussian distribution. In addition, cumulative coefficients of higher orders cannot take arbitrary values. In particular, for the cumulative skewness and kurtosis coefficients, the following inequality is known $\gamma_4 + 2 \geq \gamma_3^2$. Considering this inequality, based on the analysis of (3.6), we can conclude that the coefficient of variance reduction $g_{\frac{PMM2}{OLS}}$ is a dimensionless value belonging to the range (0; 1]. In general, it can be stated that the relative efficiency of the quadratic version of PMM depends on the degree of difference between the distribution of regression errors and the Gaussian law, which is numerically expressed by the difference in the absolute value (module) of the skewness coefficient γ_3. With its increase, the magnitude of the relative decrease in variance can be quite significant and even asymptotically tend to zero when the absolute value of the skewness coefficient approaches the boundary of the region of acceptable values, i.e. $|\gamma_3| \to \sqrt{2 + \gamma_4}$.

7 Adaptive Estimation of Regression Parameters Using the PMM

It is obvious that the expediency of using the quadratic version (at $S = 2$) of the PMM is only appropriate when the distribution of the random component of regression models differs from the Gaussian law and its asymmetry. We should also note the fact, which is important from a practical point of view, that in order to obtain the PMM estimates, additional a priori information about the error properties of the regression model is required, which is usually absent in real conditions.

It is understood that an adaptive approach can be applied under the assumption that model errors are independent of the regressors [24]. Adaptability is understood in the sense that a number of successive refinement steps are applied. The first step is to estimate the parameters of the deterministic component of the regression model using a conventional OLS that does not consider the specifics of the probability distribution of errors. After removing the deterministic component, the type of random component of the regression model is identified and its parameters are estimated. At the third stage, the refined (adaptive) estimates of the informative parameters of the regression model are found, considering the probabilistic properties of the errors. This approach is based on the property that the inadequacy of the Gaussian model is not a critical factor from the point of view that the OLS estimates of the regression parameters still remain unbiased and valid, although they are no longer optimal. And since the least squares method is inherently linear, the probabilistic properties of the regression residuals are virtually indistinguishable from those of the original random component of the regression model.

We recall the above-mentioned fact that the task of finding the estimated values of higher-order statistics is methodologically much simpler than the task of identifying and estimating the parameters of the distribution density. In this sense, the task of overcoming a priori uncertainty about the probability characteristics of the model is much simpler than the approach based on maximum likelihood. In addition, the degree of uncertainty in the estimates of the moments (depending on their order) may also be less than the uncertainty in the estimates of the parameters of the distribution density. And as will be shown in Sect. 4, this factor can lead to an additional improvement (relative to the

PMM) of the resulting adaptive PMM estimates of the informative regression parameters. It should be noted that the obtained a posteriori estimates of higher-order statistics (in particular, cumulative skewness and kurtosis coefficients) can be used to test the hypothesis of Gaussian distribution and symmetry of regression errors based on Jarque -Bera statistical tests [25].

8 Static Modeling

To verify the theoretical results, we modified a set of functions written in R that implemented a Monte Carlo simulation procedure for finding estimates of linear regression, autoregressive, and moving average models using the polynomial maximization method [19–22]. Now, in addition, it allows for a comparative analysis of the accuracy of three methods (PMM2, SLS, and OLS) when estimating parameters of nonlinear regression models with random component of a non-Gaussian asymmetric distribution.

To find the estimates of the ordinary least squares (OLS) method, the built-in *nls* function of nonlinear regression analysis is used. The implementation of the determination of SLS estimates by minimizing the corresponding functional is based on the use of functions of the *optimx* library containing various quasi-Newton optimization functions, in particular, "BFGS" (Broyden-Fletcher-Goldfarb-Shanno). To solve the systems of nonlinear equations that occur when finding PMM2 estimates, the library of functions *nleqslv* was used, which implements numerical Broyden or Newton methods with a choice of global strategies, such as line search and confidence region [26].

In this section, we present the results of a comparative analysis using two types of nonlinear regression models, which were also studied in [7]. The first regression model has an exponential deterministic function

$$y_v = \theta_1 \exp(\theta_2 X) + \xi_v,\ v = \overrightarrow{1, N} \tag{28}$$

with the true values of the parameters $\theta_1 = 10$ i $\theta_2 = -0.6$;

and the second is a model of species growth:

$$y_v = \frac{\theta_1}{[1 + \exp(\theta_2 + \theta_3 x)]} + \xi_v,\ v = \overrightarrow{1, N} \tag{29}$$

with the true values of the parameters $\theta_1 = 10$, $\theta_2 = 1.5$ and $\theta_3 = -0.8$. The model of the independent variable is used $X \sim \mathit{Uniform}(0{,}20)$.

The centered chi-squared distribution $\xi = (\chi(3) - 3)/\sqrt{3}$ with the scale parameter $\sigma^2 = 2$ was used as the error model. For such a random variable, the degree of non-Gaussianity can be numerically described by the values of the skewness coefficients $\gamma_3 = 1.4$ and kurtosis $\gamma_4 = 2$. According to (27), the expected value of the decrease in the variance of estimates will be $g_{\frac{PMM2}{OLS}} \approx 0.56$. The mathematical formulation of the estimation assumes that the stochastic component has a significantly different asymmetric distribution from the Gaussian, but no a priori information about its parameters. Three criteria are used to compare the effectiveness of different methods: the mean of the estimates (Mean), their variance (Var), and the standard deviation from the true value (MSE). By repeated (for $m = 1000$) Monte Carlo tests with different sample sizes $n =$

30, 50, 100, 200, the results of, the totality of which is presented in Tables 1 and 2. The data presented in Tables 1 and 2 generally correlate with the analogous results of work [7]. They demonstrate that for all the analyzed methods, the resulting estimates are asymptotically unbiased, as their averaged (Mean) values asymptotically (as $n \to \infty$) converge to the true values of the informative parameters. Furthermore, there is a significant advantage in efficiency (reduction in the magnitude of variance and MSE) of the SLS and PMM2 estimates compared to OLS estimates.

This is particularly illustrated by the bar charts (constructed based on the data from Tables 1 and 2) presented in Fig. 1. The charts depict the dynamics of the ratios $\frac{MSE_{SLS}}{MSE_{OLS}}$ and $\frac{MSE_{PMM2}}{MSE_{OLS}}$ and for the estimates of various parameters of the studied regression models, depending on the sample size of statistical data.

Table 1. Results of statistical modeling of the regression model parameters estimation $Y = \theta_1 exp(\theta_2 X) + \xi$

n		OLS			SLS			PMM2		
		Mean	Var	MSE	Mean	Var	MSE	Mean	Var	MSE
200	$\theta_1 =$ 10	9.9099	0.3128	0.3205	9.9835	0.1847	0.1847	9.9845	0.1873	0.1873
	$\theta_2 =$- 0.6	-0.5955	0.0026	0.0026	-0.6001	0.0016	0.0016	-0.5997	0.0016	0.0016
100	$\theta_1 =$ 10	9.8506	0.6514	0.6729	9.9795	0.4136	0.4135	9.9816	0.4142	0.4140
	$\theta_2 =$- 0.6	-0.5959	0.0055	0.0055	-0.6026	0.0035	0.0035	-0.6023	0.0035	0.0035
50	$\theta_1 =$ 10	9.8446	1.3942	1.4167	10.0858	1.1170	0.9998	10.1466	0.9812	1.0015
	$\theta_2 =$- 0.6	-0.5961	0.0127	0.0127	-0.6047	0.0116	0.0084	-0.6050	0.0081	0.0081
30	$\theta_1 =$ 10	9.8111	2.7453	2.7775	10.1339	0.2657	2.2352	10.2456	2.2298	2.2873
	$\theta_2 =$- 0.6	-0.6011	0.0208	0.0208	-0.6132	0.0029	0.0158	-0.6111	0.0154	0.0155

The analysis of the presented results allows us to conclude that, in general, the accuracy of the SLS and PMM2 estimates is practically the same. For most cases, as the sample size n increases, the relative accuracy of PMM2 gradually increases and tends to the theoretically determined (27) almost twofold advantage (in the RMS metric) over OLS. The only exceptions are estimates of the scaling parameter θ_1 of the regression growth model with the relative improvement in accuracy is only 10–15%.

Table 2. Results of statistical modeling of the regression model parameters estimation $Y = \frac{\theta_1}{[1+\exp(\theta_2+\theta_3 X)]} + \xi$

n		OLS			SLS			PMM2		
		Mean	Var	MSE	Mean	Var	MSE	Mean	Var	MSE
200	$\theta_1 =$ 10	10.003	0.0133	0.0133	10.0003	0.0116	0.0116	9.9981	0.0118	0.0118
	$\theta_2 =$ 1.5	1.5145	0.0449	0.0451	1.5038	0.0236	0.0235	1.5039	0.02389	0.0239
	$\theta_3 =$ 0.8	-0.8061	0,0081	0,0082	-0,8031	0,0041	0,0041	-0,804	0,0043	0,0043
100	$\theta_1 =$ 10	9.9979	0.02534	0.0253	9.9912	0.0224	0.0225	9.9866	0.0225	0.0226
	$\theta_2 =$ 1.5	1.5128	0.0723	0.0724	1.4904	0.0415	0.0415	1.4936	0.0426	0.0426
	$\theta_3 =$ 0.8	-0.8032	0.0135	0.0134	-0.7982	0.0075	0.0074	-0.801	0.0079	0.0079
50	$\theta_1 =$ 10	10.019	0.057	0.0573	10.0091	0.0497	0.04968	10.001	0.04869	0.0486
	$\theta_2 =$ 1.5	1.5404	0.1531	0.1545	1.4983	0.0973	0.0972	1.4819	0.1035	0.1037
	$\theta_3 =$ 0.8	-0.8131	0.0281	0.0283	-0.8029	0.0168	0.0168	-0.8	0.0171	0.0171
30	$\theta_1 =$ 10	10.038	0.1003	0.1017	10.0061	0.0826	0.0825	9.9979	0.0837	0.0836
	$\theta_2 =$ 1.5	1.5961	0.2717	0.2806	1.5107	0.1848	0.1847	1.4727	0.2149	0.2154
	$\theta_3 =$ 0.8	-0.8241	0.0474	0.0479	-0.8093	0.0317	0.0318	-0.805	0.0372	0.0372

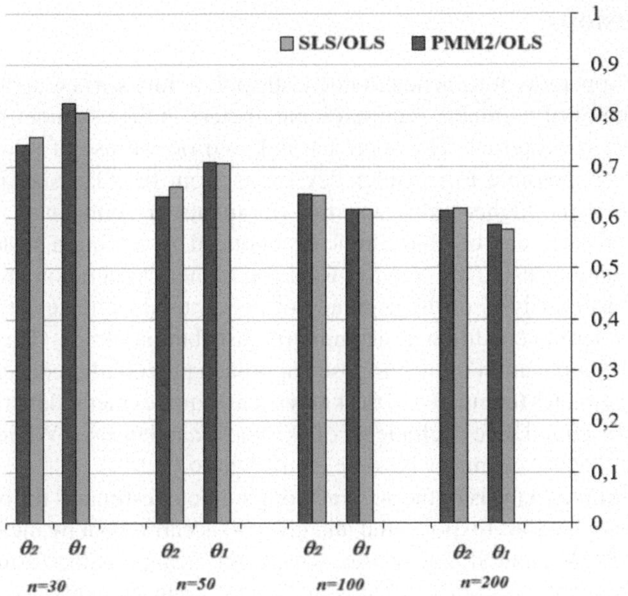

a)

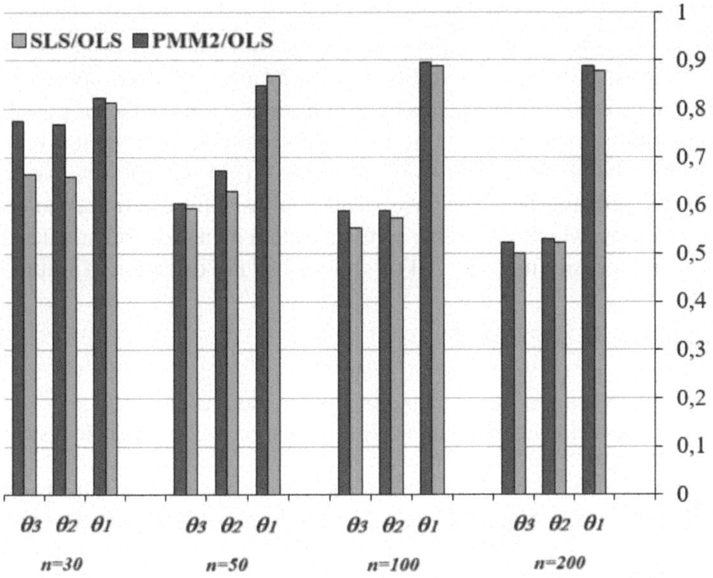

b)

Fig. 1. Experimental values of the MSE ratio of estimates *a) model* $Y = \theta_1 \exp(\theta_2 X) + \xi$, *b) model* $Y = \frac{\theta_1}{[1+\exp(\theta_2+\theta_3 X)]} + \xi$

9 Conclusions

Based on the apparatus of Kunchenko stochastic polynomials, a new approach to adaptive finding estimates of nonlinear regression parameters in the asymmetric distribution of random errors is proposed. The approach is based on the use of power polynomials, which makes it possible to consider deviations from the Gaussian idealization in a realistic way in the higher-order statistics (moments or cumulants). It is shown that the general problem can be algorithmically reduced to solving a system of nonlinear stochastic equations using the Newton-Raphson numerical iteration procedure.

A theoretical analysis of the accuracy of polynomial estimates at degree $S = 2$ is carried out under the condition of asymmetric distribution of non-Gaussian errors. It is shown that the application of the proposed approach potentially reduces the variance of polynomial estimates (compared to the known least squares methods), the value of which depends on the cumulative coefficients of skewness and kurtosis. When the distribution is symmetric, PMM2 estimates become equivalent to OLS.

A comparative analysis of the accuracy of parameter estimates for two types of nonlinear regression models (exponential and growth) is carried out by means of simulation efficiency of PMM2 and SLS estimates, which significantly outperform OLS estimates. The similarity of the accuracy of SLS and PMM2 can be explained by the fact that both estimators use the same additional information about the probabilistic nature of the random component of regression models in the form of moments of regression residuals up to the 4th order.

It should be noted that the polynomial maximization method has an additional potential. It consists in the possibility of scaling the proposed approach by increasing the degree of the forming stochastic polynomials. This makes it possible to use the moment-cumulative description of higher orders algorithmically simply. Another interesting aspect is the possibility of using alternative (not only power) basis functions to form stochastic polynomials. The basis of such a basis, for example, can be trigonometric transformations, which will allow us to operate with a probabilistic description in the form of a characteristic function. All this should be the direction of future research.

References

1. Gallant, A.R.: Nonlinear statistical models. Wiley, New York (1987)
2. Wood, S.N.: Statistical inference for noisy nonlinear ecological dynamic systems. Nature **466**, 1102–1107 (2010)
3. Moon, K.-H., Han, S.W., Lee, T.S., Seok, S.W.: Approximate mpa-based method for performing incremental dynamic analysis. Nonlinear Dyn. **67**, 2865–2888 (2012)
4. Paula, D., Linard, B., Andow, D.A., Sujii, E.R., Pires, C.S.S., Vogler, A.P.: Detection and decay rates of prey and prey symbionts in the gut of a predator through metagenomics. Mol. Ecol. Resour. **15**, 880–892 (2015)
5. Santos, J.D.A., Barreto, G.A.: An outlier-robust kernel RLS algorithm for nonlinear system identification. Nonlinear Dyn. **90**, 1707–1726 (2017)
6. Wang, L.: Estimation of nonlinear Berkson-type measurement error models. Stat. Sin. **13**, 1201–1210 (2003)
7. Wang, L., Leblanc, A.: Second-order nonlinear least squares estimation. Ann. Inst. Stat. Math. **60**, 883–900 (2008). https://doi.org/10.1007/s10463-007-0139-z

8. Kim, M., Ma, Y.: The efficiency of the second-order nonlinear least squares estimator and its extension. Ann. Inst. Stat. Math. **64**, 751–764 (2012). https://doi.org/ https://doi.org/10.1007/s10463-011-0332-y
9. Gao, L.L., Zhou, J.: New optimal design criteria for regression models with asymmetric errors. J. Stat. Plann. Inference **149**, 140–151 (2014). https://doi.org/10.1016/j.jspi.2014.01.005
10. He, L., Yue, R.X.: R-optimality criterion for regression models with asymmetric errors. J. Stat. Plann. Inference **199**, 318–326 (2019). https://doi.org/10.1016/j.jspi.2018.07.008
11. Salamh, M., Wang, L.: Second-order least squares estimation in nonlinear time series models with ARCH errors. Econometrics **9**(4), 41 (2021). https://doi.org/10.3390/econometrics9040041
12. Zhang, J., Cui, L.: Exponential parametric distortion nonlinear measurement errors Models. Commun. Stat. Theor. Methods (2022). https://doi.org/10.1080/03610926.2022.2111526
13. He, L., Yue, R.X., Du, A.: Optimal designs for comparing curves in regression models with asymmetric errors. J. Stat. Plann. Infer. **228**, 46–58 (2024). https://doi.org/10.1016/j.jspi.2023.06.005
14. Kunchenko, Y.: Polynomial Parameter Estimations of Close to Gaussian Random variables. Shaker Verlag, Germany, Aachen (2002)
15. Zabolotnii, S.W., Martynenko, S.S., Salypa, S.V.: Method of verification of hypothesis about mean value on a basis of expansion in a space with generating element. Radio Electron. Comm. Syst. **61**, 222–229 (2018). https://doi.org/10.3103/S0735272718050060
16. Chertov, O., Slipets, T.: Epileptic seizures diagnose using kunchenko polynomials template matching. Progress in Industrial Mathematics at ECMI 2012. Springer International Publishing, Cham, pp. 245–248 (2014). https://doi.org/10.1007/978-3-319-05365-3_33
17. Zabolotnii, S.V., Warsza, Z.L.: Semi-parametric polynomial method for retrospective estimation of the change-point of parameters of non-Gaussian sequences. In: Advanced Mathematical and Computational Tools in Metrology and Testing X, pp. 400–408 (2015). https://doi.org/10.1142/9789814678629_0048
18. Zabolotnii, S.W., Warsza, Z.L.: Semi-parametric polynomial modification of CUSUM algorithms for change-point detection of non-Gaussian sequences. In: Electronic Proceedings of XXI IMEKO World Congress "Measurement in Research and Industry" September 2015 Prague Czech Republic, vol. 30, pp. 2088–2091 (2015)
19. Zabolotnii, S., Warsza, Z.L., Tkachenko, O.: Polynomial estimation of linear regression parameters for the asymmetric PDF of errors. In: Advances in Intelligent Systems and Computing, vol. 743 (2018). https://doi.org/10.1007/978-3-319-77179-3_75
20. Zabolotnii, S., Tkachenko, O., Warsza, Z.L.: Application of the polynomial maximization method for estimation parameters in the polynomial regression with non-gaussian residuals. In: Advances in Intelligent Systems and Computing, vol. 1390. Springer, Cham (2021). https://doi.org/10.1007/978-3-030-74893-7_36
21. Zabolotnii, S., Tkachenko, O., Warsza, Z.L.: Application of the polynomial maximization method for estimation parameters of autoregressive models with asymmetric innovations. In: Advances in Intelligent Systems and Computing, vol. 1427. Springer, Cham (2022). https://doi.org/10.1007/978-3-031-03502-9_37
22. Zabolotnii, S., Tkachenko, O., Warsza, Z.L.: Polynomial maximization method for estimation parameters of asymmetric non-gaussian moving average models. Lecture Notes in Networks and Systems, vol. 630. Springer, Cham (2023). https://doi.org/10.1007/978-3-031-25844-2_21
23. Seber, G.A.F., Wild, C.J.: Nonlinear Regression. Wiley, New York (1989)
24. Cook, R.D., Weisberg, S.: Residuals and influence in regression. Monographs on Statistics and Applied Probability (1982). https://doi.org/10.2307/1269506
25. Jarque, C.M., Bera, A.K.: A tests of observations and regression residuals. Int. Stat. Rev. **55**, 163–172 (1987)

26. Dennis, J.E. Jr., Schnabel, R.B.: Numerical Methods for Unconstrained Optimization and Nonlinear Equations, Siam (1996)
27. Bojarczak, P., Nowakowski, W.: Application of deep learning networks to segmentation of surface of railway tracks. Sensors **21**, 4065 (2021). https://doi.org/10.3390/s21124065, ISSN1424-8220

Hyperspectral Lighting Design for Industrial Applications

Piotr Garbacz(✉), Piotr Czajka, Wojciech Mizak, and Marcin Powroźnik

Łukasiewicz Research Network – Institute for Sustainable Technologies, Pulaskiego Street 6/10, 26600 Radom, Poland
{piotr.garbacz,piotr.czajka,wojciech.mizak, marcin.powroznik}@lukasiewicz.gov.pl

Abstract. The paper raises the issue of lighting in industrial hyperspectral imaging. In the first part of the article, based on the literature review, a general overview of hyperspectral imaging is presented. Examples of applications of this unique technology are briefly listed. Despite significant progress in the area of data analysis and hyperspectral cameras, much remains to be done in terms of reliability and adaptation to industrial conditions. The authors emphasize the role of lighting in machine vision systems. Existing illumination solutions are described, along with their advantages and limitations. Due to specific requirements to be met in the case of hyperspectral imaging the authors propose two types of lighting devices. The prototype of a general usage halogen illuminator characterized by spatial uniformity and broadband spectral range is described in detail. The last part of the article presents the process of designing a dedicated illuminator along with the method of selecting relevant spectral bands for a particular application.

Keywords: hyperspectral imaging · lighting design · machine vision

1 Introduction

The development of innovative imaging methods provides manufacturers with an increasing number of tools that can be used for quality assurance. In addition to classic machine vision systems, in recent years imaging beyond the visible spectrum is getting more attention. Due to low-cost uncooled microbolometer sensors dissemination, thermal imaging cameras have been utilized widely in many industrial branches [1]. The emergence of commercially available compact terahertz radiation sources and sensors has created many new possible applications for this new imaging technology [2]. There is also a noticeable increase of hyperspectral imaging (HSI) applications in various areas of industry, such as agriculture, recycling, pharmaceutical and food manufacturing [3]. HSI also known as chemical imaging (CI) [4] is the combination of spectroscopy and digital imaging. Hyperspectral cameras enable the registration of three-dimensional data structures (hyperspectral cubes) with two spatial dimensions and one spectral dimension (Fig. 1). Spectral signatures are recorded for each pixel in the image of the observed surface, showing the relationship between the reflectance and the wavelength.

R. Szewczyk et al. (Eds.): AUTOMATION 2024, LNNS 1219, pp. 357–365, 2024.
https://doi.org/10.1007/978-3-031-78266-4_31

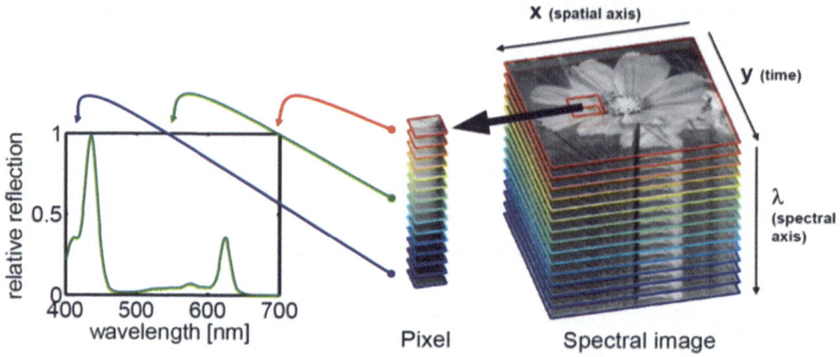

Fig. 1. A hyperspectral data cube and spectral signature at selected pixel [5].

Hyperspectral imaging due to its unique capabilities enables reliable quality inspection of demanding production processes by analyzing the spectral signature of each substance. HSI in comparison to standard RGB imaging provides significantly more information and produces accurate depiction of color images [6]. Furthermore, depending on the type of sensor used, a hyperspectral camera can capture contiguous wavelengths on the visible, near-infrared (NIR), short-wave infrared (SWIR), and mid-wave infrared (MIR) spectra with a spectral resolution approaching a single nanometer (nm). It is worth pointing out that hyperspectral imaging is the only technology capable of identifying different black plastic types [7]. Furthermore, using electromagnetic radiation beyond the visible spectrum enables analysis of the composition of tablets through the blister packaging [8]. Acquiring simultaneously both spatial and spectral information can be also used for industrial process monitoring. For instance, in paper processing, the water content is a critical parameter. A near-infrared spectral imaging enables monitoring multiple spatial positions across the width of the paper web in real-time. This allows the plant control system to automatically adjust its operating conditions for optimum throughput of production and uniform quality of paper [9].

2 Lighting for Industrial Machine Vision Systems

Regardless of the potential benefits of the imaging technology used, industrial vision systems must remain reliable and efficient in long-term running, including harsh environments. Despite the dynamic development of image analysis algorithms and deep learning, image formation is still crucial for the success of any machine vision application, and high-quality images are required to guarantee repeatable outcomes [10]. Acquiring an image characterized by proper contrast and resolution, especially in high-speed mass production, requires a multidisciplinary approach that involves mechanical, electrical, and optical engineering. It is worth noting that in machine vision systems light is an information carrier [11]. A well design imaging optics and illumination are key factors for achieving long-term reliability of machine vision systems.

Over the years, various types of light sources have been used in machine vision systems: fluorescent sources, LED illuminators, xenon lamps, halogen lamps, metal halide lamps, high pressure sodium lamps.

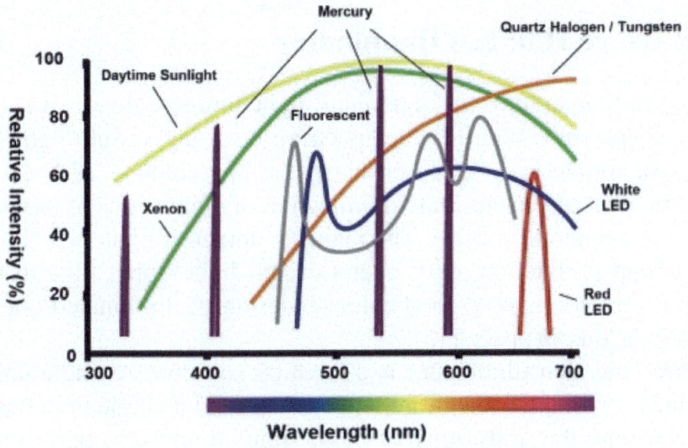

Fig. 2. Light source relative intensity in UV and visible-light spectrum [12].

In addition to significantly different spectral ranges and intensity of the obtained radiation (Fig. 2), the above-mentioned sources are also characterized by unique features that predispose them to specific applications (Fig. 3).

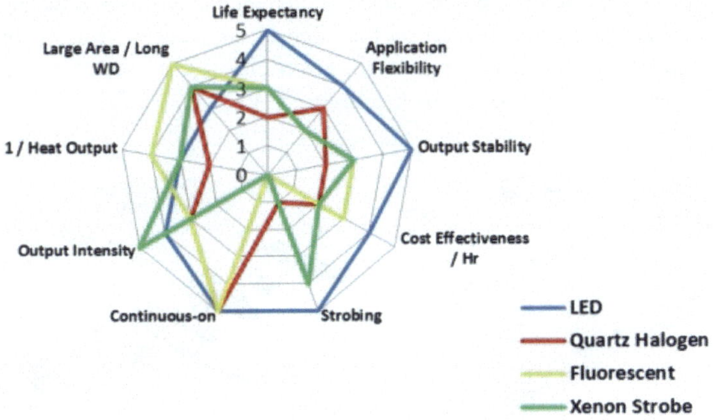

Fig. 3. Comparison of common vision lighting source [12].

The rapid progress of hyperspectral imaging technology places new requirements on illuminator manufacturers. In addition to the above-mentioned factors, ensuring the appropriate spectral range and uniformity also becomes crucial. The halogen illumination technology is characterized by a broad spectrum and high intensity and it is a first choice in a number of HSI applications. The main drawbacks of this technology are low life expectancy and emitted heat which may prevent their use in the inspection of sensitive products. For this reason, industry leaders are working on developing LED-based illuminators to meet customers' requirements.

3 General Usage Halogen Illuminator

Halogen bulbs are temperature (incandescent) light sources. Hence, it is characterized by a continuous spectrum with a wide spectrum covering visible light and infrared. This type of light source is designed primarily for low voltage (12 V, 24 V) which is related to electrical requirements and minimization of dimensions of the bulb to initiate and maintain the regeneration cycle. The main advantages of halogen bulbs are simple construction, cheap production, easy operation, no light ripple (due to high thermal inertia), natural light color, very good color rendering of illuminated objects, uniform distribution of light flux in space [13].

The proposed halogen illuminator is a modular solution. A single unit consists of four halogen bulbs with a narrow beam angle of 10° and a single bulb power of 50 W. In order to determine the uniformity of light, simulations were performed in Dialux software [14]. Since the hyperspectral camera used is linear, the uniformity of light was analyzed on a narrow surface area of 5 × 50 cm. The working distance between the illuminator and the tested surface was 50 cm. Based on the simulation, the appropriate distance between the bulbs was selected [3]. A single halogen illuminator module can be connected to an appropriate number of further units depending on the needs for the width of the analyzed area. The designed solution for connecting modules to expand the illuminated area is shown in Fig. 4. It is important that by placing subsequent illuminator modules next to each other in one line, an equal distance is ensured between individual light sources, including those placed on the edge of adjacent modules (Fig. 4a). Due to the load capacity of the connectors used, electricity can be distributed to up to 12 lighting modules using one phase power source.

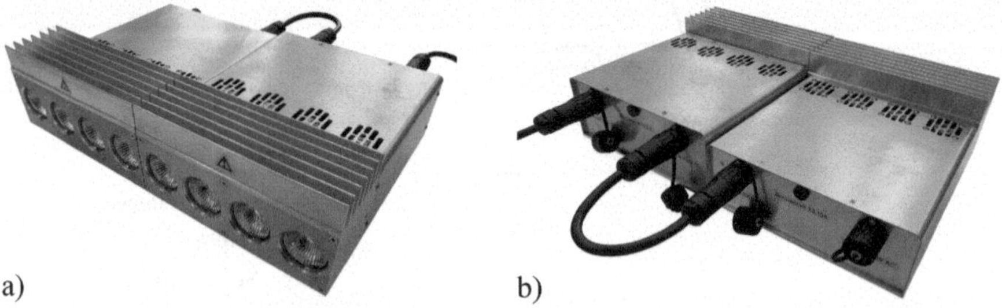

a) b)

Fig. 4. General view of connected halogen modules a) front view, b) rear view.

In addition to the light sources themselves, the housing also includes a switching power supply enabling the illuminator to be powered directly from the 230 VAC, modules detecting damaged bulbs, fans, and overcurrent protection. To ensure effective management of the heat generated by halogen bulbs and protection against external factors, the illuminator housing is made of aluminum. A heat sink is installed on the upper surface of the body. Additionally, a set of four fans dissipate heat from the interior of the illuminator and protect electrical and electronic components against overheating and damage.

4 Design of Tailored MIX-LED Illuminator

The emission of electromagnetic radiation in a light-emitting diode (LED) occurs as a result of radiative recombination of charge carriers [13, 15, 16]. LED emits radiation in a narrow spectral range, hence it is almost a monochromatic light source. Depending on the semiconductor materials used, light-emitting diodes can emit electromagnetic radiation ranging from ultraviolet, through visible light to infrared. The actual recombination conditions cause the observed spectrum to be approximately $30 \div 40$ nm, with a characteristic distribution similar to the Gaussian curve [13, 15]. Although light-emitting diodes emit radiation in a narrow spectrum, it is possible to obtain polychromatic light, e.g. white. The first method involves mixing light from three component diodes (red, green, and blue), which results in white light [13]. The second method involves the use of a blue light-emitting diode, which is placed in a housing covered with a yellow phosphor. The blue light emitted by the diode is partially transmitted and partially absorbed by the phosphor, which in turn converts it into yellow light. The transmitted blue light from the diode is mixed with the yellow light from the phosphor, resulting in white light. This is currently the basic method of producing white light-emitting diodes [16]. The latest research shown that using perovskites, it is possible to produce LEDs with an ultra-wide spectral range, exceeding the visible band [17]. It is likely that nanotechnology will enable replacement of broadband light sources such as halogen lamps. However, it is in the early stages of development and there are no commercially available devices.

Currently in LED lighting devices dedicated for hyperspectral imaging, Mix-LED technology is usually used. This method involves mixing optical radiation from light-emitting diodes characterized by different spectral ranges. For instance, ProPhotonix and Metaphase offer linear LED illuminators with independent control of the intensity of up to twelve different wavelengths of radiation from 365 nm to 1650 nm. A slightly different solution of hyperspectral LED illuminator was proposed by the authors. The number of different types of diodes is reduced to achieve higher radiation intensity on the analyzed surface. According to this assumption, a band selection method is necessary to identify which spectral wavelengths are most responsive for a particular classification task. Identifying a reduced subset of relevant spectral bands allows for a better understanding of the optical properties of the materials and provides information that can be used for designing cheaper task-specific multispectral imagers [18]. The processing time of reduced size hypercube is also significantly lower compared to full-spectrum images. In order to select LEDs for a given application a procedure is used which consists of the following stages:

1. First, preliminary experimental research with the use of a hyperspectral camera and broadband light sources (e.g. using halogen illuminators) is conducted. Spectral signatures are recorded of all classes of objects that will be inspected in the final solution.
2. Recorded data is manually labeled to all classes of objects.

3. Next, a VIF (Variance Inflation Factor) analysis is performed to determine the most relevant wavelengths [19] for labeled objects.
4. From the range of available LEDs, several types of diodes are selected (e.g. 5) characterized by the emission of optical radiation with wavelengths similar to those selected in the previous step.
5. The hypercube is reduced to the wavelength corresponding to the FWHM (Full-Width-Half-Max) range of selected diodes.
6. Based on labeled data classification model is built and evaluated with the use of accuracy and precision metrics.

In the proposed solution, selected diodes are mounted on individual PCB modules. Based on the experimental research carried out, adjusting the light intensity of individual diodes was abandoned in favor of a constant current power supply system using a dedicated LED driver. Two solutions are most often used in LED power supply systems: current sources based on MOS or bipolar transistors and pulse voltage converters with output current stabilization. In the case of illuminators used in machine vision systems, the stability of their parameters is very important. High-speed cameras are often triggered with frequencies from several hundred to several thousand Hertz. Hence, any interference in the form of lighting intensity variations may significantly affect inspection system reliability. For this reason, the pulsed power supply of LEDs was abandoned and a linear current source system was used instead. Individual diodes along with drivers and the required discrete elements were soldered to the modules in the form of printed circuit boards (PCBs). Depending on the required length of the light line, the appropriate number of modules connected in parallel are placed in the housing. Regardless of the number of modules used, the illuminator is always powered by a low-voltage 24 VDC source.

In high-performance light-emitting diodes, all the power is dissipated on a very small junction area (about 1 mm^2), which causes the power density to be very high (at the level of 100 W/cm^2) [16]. To properly dissipate heat, individual illuminator modules are made of MCPBC (Metal Core Printed Circuit Board) plates consisting of a thin layer of laminate with an aluminum core. Laminates with an aluminum core are characterized by a high thermal conductivity of 2 W/mK. For comparison, standard FR4 laminates without a metal core have a thermal conductivity of 0.3 W/mK [20]. Individual PCB modules were screwed to the illuminator housing made of aluminum. Additionally, a thermal paste with a conductivity of 1.5 W/mK was applied. As a result, during the operation of the illuminator, the temperature of the housing of the diodes mounted on the PCB did not exceed 60 °C with an ambient temperature of up to 35 °C. In the front part of the illuminator, above the line of LEDs, a glass rod is mounted. This type of simple focusing lens increases the intensity of the emitted light by concentrating the light flux in a selected area (Fig. 5). Loosening the side screws in the main body enables adjusting the distance of the lens from the diodes. To ensure high chemical resistance, the lens is made of borosilicate glass, which is resistant to acids, alkalis, and other chemicals.

Laboratory tests of the prototype illuminator confirmed that it can be used in high-speed sorting applications that require camera exposure time below 1 ms. A long-term study carried out under industrial conditions is planned for verification of device efficiency and optical performance, especially spectral stability. Further work will be

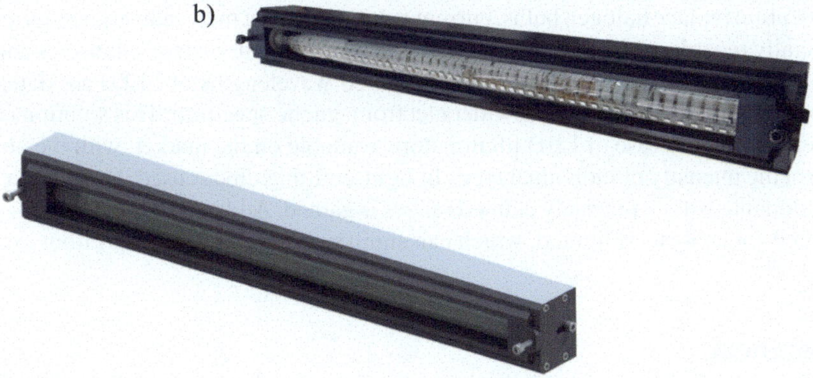

Fig. 5. General view of the developed LED illuminator: a) computer model, b) prototype.

focused on the development of the converging lens. It is planned to use quartz glass, which is characterized by high purity and combines excellent physical and optical properties. Another direction of research is the development of a self-diagnostic system. It is planned to monitor the temperature inside the illuminator using temperature sensors placed next to the LEDs, which will enable signaling of exceeding the established limit parameters and prevent possible thermal damage to the light sources. Depending on the application, selected light-emitting diodes installed in the illuminator may emit only invisible radiation, e.g. in the infrared range. To enable visual verification of the illuminator's mechanical installation an additional module at both ends of the illuminator will be mounted with standard visible-light LED.

5 Conclusions

Machine vision systems require reliable and properly designed lighting solutions. It is particularly important in the case of sorting applications or mass production of goods. Although the hyperspectral camera sensors and new-generation computing units are capable of meeting the requirements of high-speed inspection, there is an issue with the lack of suitable lighting devices. Ensuring the appropriate spectral characteristics of the source is as important as its illumination intensity. An undoubted advantage of halogen light sources is the continuous spectrum of electromagnetic radiation with a wide spectrum covering visible light and infrared. This feature is particularly important in experimental research, but not necessarily in the particular application. Due to limitations of halogen lighting, other solutions are being sought. LED illuminators are an alternative, but in their case, a significant problem is ensuring the continuity of the spectral characteristics in the required band.

The modular halogen illuminator proposed by the authors can be used in a number of laboratory and industrial applications. Depending on the required width of the light line, subsequent units can be connected, maintaining appropriate spacing between light sources. The main drawback is the expected lifetime and generation of undesired heat radiation. Although there is some progress in developing broadband LEDs which could in the future replace halogen bulbs, current solutions do not meet industrial requirements, especially regarding high-speed imaging. The proposed alternative solution is a tailored MIX-LED illuminator. In this case, the required wavelengths of LEDs are determined based on preliminary testing in a wider electromagnetic spectrum. This solution is not as universal as in the case of LED illuminators available on the market, with the ability to control the intensity of individual types of light-emitting diodes used. However, a higher radiation intensity in precisely defined ranges is gained. Additionally, the number of data required for analysis is limited, which can significantly shorten the algorithm execution time.

References

1. Flir: Thermal imaging guidebook. https://www.flir.co.uk
2. Garbacz, P.: Terahertz imaging – principles, techniques, benefits, and limitations. Mainten. Probl. **100**(1), 81–92 (2016)
3. Linowska, B., Garbacz, P.: Hyperspectral imaging system for food safety inspection. automation 2023: key challenges in automation. Robot. Measure. Tech. 204–211 (2023)
4. Inno-Spec: spectroscopic solutions for process analytics. https://www.messe.de/apollo/hannover_messe_2020/obs/Binary/A1014313/1014313_02200733.pdf
5. Polder G., Pekkeriet E.: A spectral imaging system for detection of botrytis in Greenhouses. EFITA-WCCA-CIGR Conference: Sustainable Agriculture through ICT Innovation (2013)
6. Specim: Hyperspectral Technology vs. RGB. https://www.specim.com/hyperspectral-technology-vs-rgb/
7. Specim: Specim FX50 hyperspectral camera – a breakthrough in black plastic sorting. https://www.specim.com/breakthrough-in-black-plastic-sorting/
8. Specim: Hyperspectral imaging in pharmaceutics and medical. https://www.specim.com/hyperspectral-imaging-applications/hyperspectral-imaging-in-pharmaceutics/
9. Photonics spectra: hyperspectral imaging spectroscopy: a look at real-life applications. https://www.photonics.com/Articles/Hyperspectral_Imaging_Spectroscopy_A_Look_at/a25139
10. Dechow, D.: Machine vision systems integration: deep learning. Q. Mag. **60**(9) (2021)
11. Jahr I.: Lightning in machine vision. In: Hornberg A. (ed.) Handbook of Machine and Computer Vision, pp. 63–178. Wiley-VCH Verlag, Weinheim (2017)
12. National instruments: a practical guide to machine vision lighting. https://www.ni.com/en/shop/choosing-the-right-hardware-for-your-vision-applications/a-practical-guide-to-machine-vision-lighting.html
13. Żagan, W.: Basics of Lighting Technology (in Polish). WUT Publishing House, Warsaw (2005)
14. Dialux: Lighting design software. https://www.dialux.com
15. Booth, K., Hill, S.: Optoelectronics (in Polish). WKŁ Publishing House, Warsaw (2001)
16. Czajka, P.: Determination of characteristics of highly-efficient LEDs' impulse work (in Polish). Mainten. Probl. **69**(2), 33–44 (2008)
17. Jin S., et al.: Compact ultrabroadband light-emitting diodes based on lanthanide-doped lead-free double perovskites. Light: Sci. Appl. **11**(52) (2022)
18. Morales, G., Sheppard, J.W., Logan, R.D., Shaw, J.A.: Hyperspectral dimensionality reduction based on inter-band redundancy analysis and greedy spectral selection. Remote Sens. **13**(18) (2021)
19. Morales G., Sheppard J., Logan R., Shaw J.: Hyperspectral band selection for multispectral image classification with convolutional networks. In: 2021 International Joint Conference on Neural Networks (IJCNN), pp. 1–8. IEEE, Shenzhen (2021)

20. Załęski, D.: Circuits with an aluminum core - parameters, applications, design (in Polish). https://tspcb.pl/blog/jakosc/obwody-z-rdzeniem-aluminiowym-parametry-zastosowania-projektowanie

Author Index

R. Szewczyk et al. (Eds.): AUTOMATION 2024, LNNS 1219, pp. 367–368, 2024.
https://doi.org/10.1007/978-3-031-78266-4

The manufacturer's authorised representative in the EU is Springer Nature Customer Service Centre GmbH, Europaplatz 3, 69115 Heidelberg, Germany. If you have any concerns regarding our products, please contact ProductSafety@springernature.com

Printed and bound by CPI Group (UK) Ltd, Croydon, CR0 4YY
01/10/2026
02225911-0009